ABSTRACTIONS AND EMBODIMENTS

STUDIES IN COMPUTING AND CULTURE

Jeffrey R. Yost and Gerardo Con Diaz, Series Editors

ABSTRACTIONS AND EMBODIMENTS

New Histories of Computing and Society

EDITED BY

JANET ABBATE

AND

STEPHANIE DICK

JOHNS HOPKINS UNIVERSITY PRESS | *Baltimore*

9 8 7 6 5 4 3 2 1

Johns Hopkins University Press
2715 North Charles Street
Baltimore, Maryland 21218-4363
www.press.jhu.edu

Library of Congress Cataloging-in-Publication Data

Names: Abbate, Janet, editor. | Dick, Stephanie, 1985– editor.
Title: Abstractions and embodiments : new histories of computing and society /
    edited by Janet Abbate and Stephanie Dick.
Description: Baltimore : Johns Hopkins University Press, 2022. | Series: Studies in
    computing and culture | Includes bibliographical references and index.
Identifiers: LCCN 2021053802 | ISBN 9781421444376 (paperback) |
    ISBN 9781421444383 (ebook)
Subjects: LCSH: Computers—Social aspects. | Computers and civilization.
Classification: LCC QA76.9.C66 A27 2022 | DDC 303.48/34—dc23/eng/20211122
LC record available at https://lccn.loc.gov/2021053802

A catalog record for this book is available from the British Library.

*Special discounts are available for bulk purchases of this book. For more information, please
contact Special Sales at specialsales@jh.edu.*

CONTENTS

ACKNOWLEDGMENTS

Our first thanks go to all of the authors who contributed their insights and hard work to this volume; their scholarship is transforming the history of computing to serve urgent and expanding intellectual goals and community needs. We gratefully acknowledge Valérie Schafer and Nathan Ensmenger, who provided invaluable feedback on our introduction; Matt Priestly and Marc Aidinoff for their thoughtful comments on the afterword; Gerard Alberts for drawing our attention to the potential for focusing on embodiment; and the anonymous reviewers for Johns Hopkins University Press. We also acknowledge the fine work of copyeditor Michael Baker and indexer Enid Zafran. Matt McAdam at Johns Hopkins University Press shepherded this project through changes and disruptions with calm encouragement.

Stephanie would like to thank her history of computing graduate students—especially Sam Schirvar, Zach Loeb, Cheryl Hagan, Sam Franz, Kelcey Gibbons, Aaron Mendon-Plasek, and Katya Babintseva—for their brilliant scholarship and for countless conversations about the history of computing from which I have learned so much. Janet would like to thank Virginia Tech for providing a grant for publication expenses; Trevor Croker for research support in the initial planning stages; and Tom and Wendy for their love and patience.

ABSTRACTIONS AND EMBODIMENTS

Thinking with Computers

Janet Abbate and Stephanie Dick

Computers are good to think with.[1] This is the claim of technologists who created whole industries and academic fields around what it is possible—and not possible—to do with computers. Computers have been centered in discourse and practice everywhere from politics and art to science, law, and activism. They have proven powerful for historical thinking as well: computers concretize social forces and logics that may otherwise be harder to see, to understand, to change. Public attitudes toward computers have revealed deep contradictions at the heart of twentieth- and twenty-first-century life: acclaimed as liberatory, they are also condemned as instruments of control that increasingly dictate the contours of human life; often seen as revolutionary, they also reinforce existing power structures; heralded as engines of innovation, they are also conservative, lingering manifestations of nineteenth-century industrial thinking and Cold War militarism. Computers have been framed both as a mirror for the human mind and as an irreducible "other," so different from people that humanness must be defined (and protected) against them. Historians of computing return again and again to these contradictions: How can a technology be both old and new? Innovative and conservative? Democratic and authoritarian? Historical scholarship has shown that these seeming contradictions often reveal deeper structures. Computers can serve both liberation and control because some

1

people's freedom has historically been predicated on controlling others. Computers can be both alike and unlike human beings when you acknowledge the relative and changing definitions of "humanness" in history. When historians think with computers, we are called to confront and explore foundational structures, binaries, logics, and tensions in modern life.

This volume puts at its center one of the core binaries within computing: between abstractions and embodiments, ideas and materiality, mind and body, software and hardware. On the one hand, computers are inherently abstract: they are informed by the history of Boolean logic, formal languages, and mathematically defined procedures. On the other hand, computing is done by machinery and people—it is entangled with global systems of resource extraction, environmental control, and gendered and racialized labor. Many abstract concepts in computing—"algorithm," "network," "system"—are inherited from the past, and indeed were abstracted *from* particular cultural, economic, and material contexts.[2] Our goal is not to dissolve this tension or to suggest that abstraction is always actually material, but rather to explore how, historically, the material conditions of computing and the abstract ideas and stories that shape its use are co-produced.

Thinking with computers also means thinking about *people*, with whom computers have such intimate and myriad relations. Computing technologies have become constant companions, mediating the interactions people have with themselves, each other, the State, corporations, and the world. Computing has also always been positioned and its capacities defined in relation to the human mind and body, a set of connections we highlight in this volume. Computers have been imagined as a mirror of, or competitor to, the human mind, and have also been promoted as a way to augment the brain or body. While historians have explored in some depth the discourses of "giant brains" and "thinking machines," physical relations between computing machines and embodied users have been relatively neglected.[3] Centering bodies in the history of computing also foregrounds issues of labor, disability, gender, and race that are harder to see in machine-focused histories. Behind the abstractions and the materialities—behind the mathematical and technological dimensions of computing, from algorithms to transistors, that have received so much attention—there are communities and laborers, identities and experiences, that have not, until recently, received the historical attention that they deserve.

Finally, thinking with computers compels historians to think about power. From the earliest digital computers built by military agencies during World War II to the algorithms that increasingly control our lives, computing has been funded by powerful government and private sector actors and harnessed to their political and economic agendas. Yet computing technology, and the skill to wield it, have an equally long history of enabling distributed, countercultural, bottom-up, subaltern, and oppositional power. Historians of computing have a rich corpus of studies that integrate technical and social sources of power and opportunity. A new generation of scholars is deepening those inquiries by bringing computer history into conversation with analyses of power relations through frameworks of colonialism, disability, and race. Throughout the volume, we propose that to think with computers is to think about *social relations*. Each chapter in this volume explores how social relations are materialized in the machines, ideas, concepts, communities, and stories that constitute modern computing.

Old and New Histories of Computing

The field of the history of computing took shape a few decades after the computer itself was thought to have "come of age" in the 1960s.[4] Many of the early works were histories of *computers,* the machines themselves: accounts of what they could do, how they were designed, and by whom. As computers spread through the popular imagination, proliferated into people's homes, and were heralded as "world changing" in newspaper headlines, on magazine covers, and in politicians' speeches, early histories by René Moreau, Paul Ceruzzi, Martin Campbell-Kelly, and William Aspray asked: What *are* these machines that are being hailed as so revolutionary?[5] As seems inevitable in technology history, debates arose about which was the "first computer." These claims quickly became exercises in adjective selection—there was the "first general-purpose electromechanical computer" (the Harvard Mark I), "the first programmable, electronic, digital computer" (Colossus), "the first electronic general-purpose digital computer" (ENIAC), "the first electronic stored program computer" (the Manchester Baby). In crafting these lists of adjectives, historians were forced to confront a core question: What exactly *is* a computer? While early histories tended to define the computer purely in terms of its physical components and technical functionality, they also served to *denaturalize* these machines and the revolution they were purportedly enacting.

Other historians shifted their gaze away from the machines themselves and from the small number of famous institutions and individuals deemed responsible for their creation. Michael Mahoney, in two classic articles, advocated for a decentering of machines and for a focus instead on the people who used them. Mahoney was troubled by the specter of technological determinism that he perceived in early histories of computing, which hinted both that the modern digital computer had emerged naturally and inevitably from the logical and technological paths inaugurated by Leibniz, Turing, Boole, and Babbage, and that computers had a one-sided "impact" on people, as was the claim in so much of the literature about computer "revolutions." Mahoney called on historians to focus on human agency and contingency, proposing that the computer was distinctive in that "its nature is protean; the computer is . . . what we make of it . . . through the tasks we set for it and the programs we write for it."[6] The impact and historical significance, even the character of computing, did not lie in the machine itself but rather in its *use*: "Computers do nothing without programs, and programs do not just happen by themselves. People design them for their own purposes. Hence, the history of computing, especially software, should strive to preserve human agency by structuring its narratives around people facing choices and making decisions."[7] Indeed, some historians were doing just that, such as Martin Campbell-Kelly's history of the software industry or Joseph November's *Biomedical Computing*, which argues that in the process of harnessing computers to a specific application area, both computing and biomedicine were transformed.[8] More recently, Mar Hicks, Nathan Ensmenger, and Jeffrey Yost have emphasized that the successes of computing cannot be explained by famous scientists' contributions but rather depend on the labor of thousands of software professionals and services.[9]

Historians have increasingly sought to put technologies and their uses "in context"—but what "context" do we need to fully understand computing? Recent work insists we have to get beyond technological inputs, beyond biographies of "great men," beyond big-name corporate and industrial actors, beyond Cold War America, beyond "the West," in order to understand the history of computing. Many of these recontextualizations are framed as "decentering the computer," or a progression from what began as insider accounts to critical scholarship, from hardware to software, from inventors to users, from innovators to maintainers, and so forth.[10]

Yet this critique of the historiography *itself* replicates a problematic type of thinking: categorizing a historical contribution by the type of technology it describes. Rather than seeing the historiography of computing as a linear progression from a fixation on hardware to a focus on software and finally people, it may be better described as a spiral, with some historians returning again and again to old topics, actors, and to the machine itself but with new levels of analysis and new questions. For example, during the 2010s, many historians returned to the materiality of hardware in order to reposition it as a part of larger environmental or infrastructural systems. Ensmenger's "The Environmental History of Computing" and Tung-Hui Hu's A *Prehistory of the Cloud,* for example, emphasize that computing did not break from industrial modes of production and resource extraction to achieve some "immaterial" status (though computer manufacturers work hard to create that impression) but rather reconstituted the extraction of rare earth metals, pollution, energy consumption into new forms.[11]

One recurring context for studying invention has been government and military institutions. The Second World War spurred the invention of modern digital computers, and after the war defense funding continued to shape computing research at universities and in industry. Many historians have returned to these original contexts to ask new questions about what problems computing was supposed to solve, and how design and development choices, as well as market share, depended on military and government agenda-setting. Jon Agar's *The Government Machine*, Janet Abbate's *Inventing the Internet*, Paul Edwards's *The Closed World*, Donald MacKenzie's *The Influence of the Los Alamos and Livermore National Laboratories on the Development of Supercomputing*, Rebecca Slayton's *Arguments that Count*, Daniel Volmar's work on computing in the US Air Force, and Jonnie Penn's work on the military and industrial logics at work in artificial intelligence research all explore the influence of government and military planning and problem-solving during the World War II and Cold War contexts.[12]

Other historians have elaborated cultural contextualizations. James Cortada and Alfred Chandler as well as Ronald Kline have situated the computer in longer histories of information management and social organization in the United States.[13] In the 1960s, many who had originally rejected the computer as a dehumanizing, cold, calculating military tool, associated with bureaucracy and the violence of the Vietnam War, began embracing it for its "liberating potential." Fred Turner's work has explored how computing made its way into mainstream American culture by way of countercultural communities such as hackers, new communalists, and hippies, helping to shift the popular meaning of computing to align with democracy, freedom, and individual expression.[14] As Gerard Alberts's and Ruth Oldenziel's collection *Hacking Europe: From Computer Cultures to Demoscenes* explores, computers were also a part of political and ideological transformations in Europe at the end of the Cold War, during which time distinctly European hacker communities took shape.[15] Histories that track such ideological realignments raise the question: Can computing be freed from its origins as a tool of military and government control? Could a technology developed in the belly of the American and British wartime defense establishments, to serve command and control ambitions, capitalist social order, and Cold War centralized planning, be remade to align with other social and political projects? Historical case studies by Eden Medina and Benjamin Peters follow actors who hoped to make computing and cybernetics work for communism and socialism rather than capitalism in Allende's Chile and the Soviet Union.[16] Frank Bösch's edited volume *Wege in die Digitale Gesellschaft* explores how computers factored in a West German vision of democracy and society set in opposition to recent and reemerging fascism.[17] Works by Joy Rankin and by Julien Mailland and Kevin Driscoll explore projects that tried to realize alternative models of resource-sharing and community than the ones that emerged from Silicon Valley, looking to the history of time-sharing in the American Midwest and France for alternative visions of makers, users, and machines.[18] Martin Campbell-Kelly and Daniel Garcia-Swartz similarly show that military-born ARPANET, on which so many histories have focused, was just one of many networking initiatives that shaped the practices, techniques, and communities of the modern Internet.[19]

Computing has also been positioned within the realm of science, in works that examine computing as knowledge and as a way to model natural phenomena. The discipline of "computer science" enshrined many perspectives on computing as a subject of scientific inquiry, but its scientific status was hard won only through negotiations at the borders with mathematics, physics, and engineering. Mahoney, Aspray, Stephanie Dick, and Thomas J. Misa, among others, have explored different facets of that "disciplining" work.[20] A growing body of literature examines the intellectual history of ideas that traveled between computer science, engineering, and the social sciences, as concepts such as "system" or "intelligence" were applied broadly to humans, machines, and social organization. Jennifer Light and Joy Rohde have explored the traffic of defense consultants—and with them, cybernetics, systems analysis, and game theoretic problem-solving tools—into ever expanding domains of urban planning, diplomacy, statecraft, and policy-making, while Hunter Heyck, Jamie Cohen-Cole, and Paul Erickson have explored the development of Cold War social sciences that enshrined and established many core computing ideas and ideals.[21] More recently, historians of science Joanna Radin and Theodora Dryer have explored how data-driven research and algorithmic management have transformed indigenous communities and lands into new sites of extractive epistemology and control.[22] Peter Galison, Stephanie Dick,

Hallam Stevens, Evan Hepler-Smith, and others have also explored how different communities of scientists—physicists, mathematicians, biologists, and chemists, respectively—put computers to work as instruments for knowledge-making in different scientific disciplines.[23] How and what do scientists believe they *know* when they center computers in the work of scientific exploration? Following the "turn to practice" in the history of science, these scholars have worked to recover and reconstruct what different knowledge communities actually *do* with computers, and not just the stories told about them or the final products of their experimentation, simulation, and modeling with computers.

One of the most dramatic trends in recent historiography is a shift of focus from "centers" to "peripheries" of computing activity, while simultaneously questioning these very categories. Whereas older accounts of computing at the periphery tended to emphasize the moment in which a given country got its first computer and therefore "joined the modern era," more recent work explores longer histories, creativity at the periphery, and alternative ideologies at work in computing beyond the West. Thomas S. Mullaney's work on Chinese-language keyboards and Jaroslav Švelch's account of gaming in Eastern Europe demonstrate that computing at the "periphery" was neither derivative nor a product of one-way technology transfer.[24] Rather, as Švelch argues in "Power to the Clones" in this volume, people in these communities actively created computing technologies and practices that drew on local ingenuity and were tailored for local needs. In the process of building a more global history of computing, many of these accounts also explore questions about ideology. Do computers carry their inaugural politics and ideology—namely, American and British Cold War command and control—in their very design, or can they be aligned with alternative aims? The answer turns out to be both: on the one hand, we see elegant demonstrations of how social and political values were designed into computer systems and travel with them, while on the other we admire the ingenuity with which actors on the "periphery" selectively adapt and modify imported technologies.

A similar decentering has occurred with respect to gender, labor, and race in works that challenge the perspective of the white male "innovator" and recover the labor and contributions of communities of computing professionals who were excluded from, or hidden by, the questions earlier histories explored. Hicks, Abbate, Light, and others explore the contributions of women and the role of structural sexism in the fashioning of categories like "technical skill" through which labor is gendered, valued or devalued, and rendered visible or invisible.[25] Increasing and urgent scholarship by Lisa Nakamura, Wendy Chun, Charlton McIlwain, Safiya Noble, Ruha Benjamin, André Brock, Bernard Geoghegan, and others has illuminated the historical intersection of race and technology.[26] These works emphasize that computing, for all its talk of revolution, have often served to reinforce, reproduce, and transmute structural racism, while also centering the agency of communities of color in fashioning computing for their own use. In exploring the technological construction of race, they reveal the default Whiteness of supposedly "universal computing." Related scholarship by Morgan Ames, Daniel Greene, and Virginia Eubanks argues that far from "solving" hard social problems like poverty, homelessness, or underresourced education, utopian visions of computers largely fail and computing systems reproduce inequalities easily and often.[27]

As historians cast their gaze in new places, from new perspectives, at new sources and communities, some have also spiraled back to the question of the place of the computer itself and its constituent processes, technologies, and ideas. A pair of recent publications capture a current dilemma between delving into the black box of computing and pushing it to the side: Eden Medina's 2018 article "Forensic Identification in the Aftermath of Human Rights Crimes in Chile" and Thomas Haigh's 2019 edited collection *Exploring the Early Digital*.[28] Medina's story is a part of the history of truth and reconciliation in the wake of political atrocities and mass death in Pinochet's Chile. Computer automation was introduced to help identify the dead, with the expectation that the technology would radically improve the accuracy, speed, and objectivity of these identifications. But Medina shows that, in fact, computers were often bypassed in favor of older processes, and when computers *were* used, the results were frequently wrong. Her findings emphasize continuity over computer-induced rupture, and she offers a more general exhortation to decenter this machine that so often holds our gaze: to not assume that the computer changes everything or that focusing on it will reveal the important insights about history. She proposes not just decentering the machine, but decentering computing in general, from this history.

In contrast, Haigh and the authors of *Exploring the Early Digital* not only seek to look closer at computing but at the character of "the digital" in particular. Like Medina, the authors recognize that the ubiquity of computing makes it hard to bound its history—computers are everywhere, so what is the history of computing except the history of everything? One contributor, Paul Ceruzzi, "noted that the digital computer was a 'universal solvent,'" referring to the alchemical concept of "an imaginary fluid able to dissolve any solid material." Ceruzzi added that this presented "a challenge because the computer, like any good universal solvent, has dissolved its own container and vanished from sight." But rather than let the field dissolve along with the computer into the histories of everything else, the authors advocate for a focus on the technical specificity of the machines and their uses, so that their significance for other histories might be better understood.

While Medina warns against fixing our gaze too strongly on computing lest we be blinded to other histories, mechanisms of change, and continuities it may obscure, Haigh et al. caution that if we ignore the specificity of the computing technologies involved in different historical episodes, we may not understand them or the historical significance of computing. Taking both insights into account, we argue that no matter where you are looking in computing—directly at the machine, at its users, at the stories or legal concepts that surround it, at inventors and workers, at its theoretical or technical concepts, at the center or the periphery—you can see *social relations*. They are encoded in algorithms, they are embodied in hardware, they are at work in legal proceedings, they operate in the creation of "users" and communities, they are everywhere.[29] The collection of essays we gather here unearths social relations across the spectrum of what might count as the history of computing—across scales and contexts, from labor histories to technical ones, from abstractions to embodiments. Indeed, as we argue in the afterword, even mind–body dualism, the very distinction between abstraction and embodiment, is a social relation. The question of who has a mind and who has a body—who will be remembered for their ideas and who will be remembered for their physical labor—is always at once a historical, a technical, and a social question.

Computing as Social Relations: Volume Themes and Overview

This volume offers fresh approaches to taken-for-granted concepts, actors, and historical milestones of computing history, as well as a more inclusive map of the field that highlights underrepresented groups and geographic areas. The authors engage the history of computing in conversation with other histories such as those of science, gender, race, ethnicity, disability, labor, law, international relations, and media. The volume brings together established scholars whose work has opened up new possibilities for inquiry alongside graduate students and early career scholars whose work signals new directions.

The book is organized around the complementary themes of *abstractions* and *embodiments*, though its chapters reveal that in practice these two ideas are always entangled. The section on abstractions examines foundational concepts—theoretical abstractions such as "algorithm" or "program," technical abstractions like "clone," "network," or "map," and policy abstractions such as "intellectual property" or "risk"—to reveal their underlying assumptions, socioeconomic implications, and unstable meanings. The authors demonstrate that these seemingly technical terms actually reflect power relations, political economies, and cultural identities. The section on embodiments focuses on the physical experiences of human–computer encounters, as well as how employers and computer algorithms interpret bodies as markers of skill and status. Foregrounding bodies as the site where the power and pain of computing technology play out, the authors explore how bodies have become a source of data for algorithmic decision-making and document the ways in which gender, race, and other embodied identities have shaped the opportunities and experiences of computer workers and users. Collectively, the authors mobilize a range of interdisciplinary humanities and social science methodologies to show how both technology and the body are culturally shaped and to argue that there can be no clear distinction between social, technical, and physical aspects of computing.

The first chapter in "Abstractions," Zachary Loeb's "Waiting for Midnight: Risk Perception and the Millennium Bug," revisits a well-known yet underexamined historical milestone to unpack the idea of computing "risk." He uses the lenses of infrastructure, maintenance, and disaster studies to ask how people's varying sense of professional accountability and physical vulnerability mediated their perception of risk. Revisiting the same historical era, Marc Aidinoff recontextualizes the commercialization of the Internet in the 1990s within US political culture, including the Cold War and racial divides, that added nontechnical layers to the meaning of "decentralization." Historians often seek out the "hidden politics" of technologies, but in "Centrists against the Center: The Jeffersonian Politics of a Decentralized Internet," Aidinoff reminds us that we should also attend to explicit politics and "listen to people when they tell us what, and who, they prioritized."

André Brock's reprinted chapter, abridged for this volume, "Beyond the Pale: The Blackbird Web Browser's Critical Reception," challenges the assumption that "browsers" as a category of computing infrastructure are inherently universal and racially unmarked. Brock recounts how the introduction of a browser specifically for African Americans exposed the default Whiteness of standard browsers and the Internet's arrangement of content; the ensuing debate drew on the ideology of technical neutrality to mask a defense of White hegemony. In "Scientology Online: Copyright Infringement and the Legal Construction of the Internet," Gerardo Con Diaz looks to a moment in the 1990s when lawyers and courts scrambled to decide *what kind of thing* the Internet was, so that it could be situated in existing regulatory frameworks for matters of free speech, economic exchange, and liability. Con Diaz explores an infamous case in the history of copyright—*RTC v. Netcom*, in which the Church of Scientology sued an Internet service provider for allowing copyrighted Church materials to be posted online— demonstrating the significance of legal history for determining what computing technologies *are* and how they will and can be used. Tiffany

Nichols similarly turns to the intersection of legal and computing histories in "Patenting Automation of Race and Ethnicity Classifications: Protecting Neutral Technology or Disparate Treatment by Proxy?" Nichols argues that the genre of patents—their language and claims to neutrality—obscure technical decision-making and human agency, and she unpacks how assumptions and intentions regarding racial categories are embedded in these documents.

Several authors problematize common computing terms that work to abstract or hide the messiness and social relations of computing. In "'Difficult Things Are Difficult to Describe': The Role of Formal Semantics in European Computer Science, 1960–1980," Troy Kaighin Astarte denaturalizes formalism in computer science. They identify a historical moment when the meaning of "a program" shifted from a description of a series of hardware operations to a logical abstraction, raising the question of how one could know for sure that a program performed as intended. Examining the rise and fall of formalism as a solution to this crisis, they also expose national differences in the supposedly universal field of computer science. Liesbeth De Mol and Maarten Bullynck similarly work to denaturalize a core triptych of concepts in computing—algorithm, code, and program—by exploring how the meanings of and relationships between these terms have changed over time. They show how these concepts traveled, since the nineteenth century, across technological, industrial, and intellectual contexts, and took on new meanings in the process.

Abstractions in computing often function to hide labor relations. Scott Kushner examines the changing political economy of the concept of "lurking" in online communities, from an early context in which lurking was acceptable to the contemporary context in which, as his title notes, nonengagement is cast as "The Lurking Problem." As social media platforms became profit-driven companies that depended on the visible activity of users to generate data, lurking was redefined to fit changing economic logics. Michael J. Halvorson's "The Help Desk" examines the hidden "customer support" labor that allowed the personal computer to become a viable consumer technology. Support workers were typically women, who were not considered to have the kind of "technical expertise" that merited well-paying jobs, even though they skillfully mediated between developers, machines, and end users on a daily basis. In another study of undervalued labor that blurs the line between social and material aspects of computing, Jaroslav Švelch examines the material and human resources that were required to produce PC clones in Eastern Europe during the Cold War. Challenging the easy dismissal of clones as mere copies, "Power to the Clones" demonstrates that the economic and industrial constraints faced by these countries, which have led many to assume they were slow or derivative in computer development, were in fact *productive*—inspiring creative "bricolage."

The "Embodiments" section begins with a reprint of Lisa Nakamura's foundational article "Indigenous Circuits," which argues that the embodied and racialized labor of women of color in the microelectronics industry formed an integral part of digital infrastructure, a precondition for the very existence and low cost of computer components. She documents how before chip manufacture was outsourced to Asia, there was a lesser-known period from 1965 to 1975 of "insourcing" to Navajo workers in the United States that set the pattern for targeting women of color for this work, arguing that "race and gender are themselves forms of flexible capital." Continuing the theme of racialized labor, Kelcey Gibbons's "Inventing the Black Computer Professional" rethinks the historiography of computing by juxtaposing two revolutions—the rapid spread of computers and the civil rights movement—that rocked the US in the 1960s–1970s but are rarely considered together. She argues that, because Black people were often stereotyped as unskilled, Black computer professionals had to be "invented": both by advocacy groups working with industry to create job opportunities and by Black media, which showcased these professionals as role models of progress. In documenting this history, Gibbons counters the default-White focus of many histories and their narrow view of what "freedom" through computing might mean. In a parallel move, Mar Hicks examines how female programmers in the United Kingdom tried to create professional opportunities and identities in the face of profound structural sexism. "The Baby and the Black Box" historicizes the political economy and materiality of "telecommuting" by focusing on a programming company in the UK that was run by and for home-based women, noting both its successes and the precarity of its female programmers' work–family balance.

Several chapters perform the important work of decentering the West in the history of computing, emphasizing other stories and sites. "Computing Nanyang: Information Technology in a Developing Singapore, 1965–1985," by Jiahui Chan and Hallam Stevens, challenges assumptions based on Western experiences about the role of government in computing and the mechanisms of technology transfer. They show how postcolonial ethnic identity and diasporic networks were significant factors in creating the first educational computing center in Singapore, with government and corporate actors taking a backseat. Ekaterina Babintseva also approaches educational computing from a cross-cultural perspective, contrasting Soviet and US approaches to artificial intelligence in "Engineering the Lay Mind: Lev Landa's Algo-Heuristic Theory and Artificial Intelligence." She explores unique aspects of Soviet scientific culture, particularly cybernetics and psychology, that produced a different version of AI than that pursued by US researchers and an alternative meaning of "algorithm" that focused on how human beings—not computers—think. The relation between human brains and computer algorithms is a theme continued by Xiaochang Li in a US-based case study that charts the history of automated speech recognition. While early approaches sought to mimic "human perceptual faculties," later approaches "reframed speech recognition as a purely computational process" that was divorced from human embodiment or expertise, so that the human body became mere "noise" distorting the "signal" of disembodied speech.

Other authors note how *ideas* about embodied identities can powerfully shape the course of computing history. Cierra Robson situates computer-generated surveillance maps in a longer history of racially inscribed mapping practices and the geographies they produce, contesting the objectivity and neutrality of data-driven maps. Her example, in Oakland, demonstrates that racially distinct communities created by redlining practices in the 1930s are the same communities subjected to new police surveillance technologies in the twenty-first century, revealing how supposedly innovative technologies reinscribe existing social orders. Elyse Graham's "Punk Culture and the Rise of the Hacker Ethic" argues for the influence of punk on the practice and image of 1990s hacker culture, while also revealing contradictions between punk and hacker values. She notes the social privilege underlying cyberpunk culture and goes beyond facile punk symbolism to question how well punk's "power-from-below philosophy" was honored in hacker practice.

The flesh-and-blood body is centered in Elizabeth Petrick's "The Computer as Prosthesis? Embodiment, Augmentation, and Disability," which problematizes how commentors frequently use "prosthesis" as a metaphor for the computer while overlooking the experiences of actual disabled users. Bringing perspectives from disability studies to computing history, they challenge the notion of the "interface" as a pre-defined and bounded piece of technology; rather, what *counts* as an interface emerges in interactions between embodied people and machines. Finally, Laine Nooney's chapter, "'Have Any Remedies for Tired Eyes?': Computer Pain as Computer History," explores how users have experienced "the assemblage of computational life," which includes not only the computer itself but also rooms and furniture, ambient light and noise, and the bodily discipline of work routines. By foregrounding the physical sensation of pain as a significant and enduring part of the computing experience, she challenges us to ask about the costs as well as benefits of computing and to consider the experiences and agency of embodied subjects in making computers "work."

In the afterword, we return to the mind/body dichotomy that is at work throughout computing, and that structures this volume, by revisiting the complicated perspective of Alan Turing. We unpack how embodiment lurks in all of his famously disembodied formulations of "intelligence," "communication," and "computing." Turing's work also reminds us that the binary itself has a history—it has been reconstructed and reconstituted time and again, within different social contexts and political economies, with different abstractions and embodiments at hand and in mind.

Notes

1. The original, "Animals are good to think with," is from Claude Lévi-Strauss, *Totemism* (London: Merlin Press, 1964); the idea was applied to computers by Sherry Turkle in *The Second Self* (New York: Simon & Schuster, 1984).
2. See Gabriele Balbi, Nelson Ribeiro, Valérie Schafer, and Christian Schwarzenegger, eds. *Digital Roots: Historicising Media and Communication Concepts of the Digital Age* (Berlin/Boston: De Gruyter, 2021). See also De Mol and Bullynck in this volume.
3. A recent exception is Jaakko Suominen, Antti Silvast, and Tuomas Harviainen, "Smelling Machine History: Olfactory Experiences of Information Technology," *Technology and Culture* 59, no. 2 (April 2018): 313–37.
4. Early works in the internalist vein include Herman H. Goldstine, *The Computer from Pascal to von Neumann* (Princeton, NJ: Princeton University Press, 1972), and Emerson Pugh's various books on IBM.
5. René Moreau, *The Computer Comes of Age* (Cambridge, MA: MIT Press, 1984); Paul E. Ceruzzi, *A History of Modern Computing* (Cambridge, MA: MIT Press, 1998); Martin Campbell-Kelly and William Aspray, *Computer: A History of the Information Machine* (New York: Basic Books, 1996).
6. Michael Mahoney, "The Histories of Computing(s)," *Interdisciplinary Science Reviews* 30, no. 2 (2005): 119–35, at 122.
7. Michael Mahoney, "What Makes the History of Software Hard," *IEEE Annals of the History of Computing* 30, no. 3 (July–September 2008), 10.
8. Martin Campbell-Kelly, *From Airline Reservations to Sonic the Hedgehog: A History of the Software Industry* (Cambridge, MA: MIT Press, 2003); Joseph A. November, *Biomedical Computing: Digitizing Life in the United States* (Baltimore, MD: Johns Hopkins University Press, 2012).
9. Mar Hicks, *Programmed Inequality: How Britain Discarded Women Technologists and Lost Its Edge in Computing* (Cambridge, MA: MIT Press, 2017); Nathan Ensmenger, *The Computer Boys Take Over: Computers, Programmers, and the Politics of Technical Expertise* (Cambridge, MA: MIT Press, 2010); Jeffrey R. Yost, *Making IT Work: A History of the Computer Services Industry* (Cambridge, MA: MIT Press, 2017).
10. Andrew Russell and Lee Vinsel, *The Innovation Delusion: How Our Obsession with the New Has Disrupted the Work That Matters Most* (New York: Penguin Random House, 2020).
11. Ensmenger, "The Environmental History of Computing," *Technology & Culture*, 59, no. 4S (2018): S7–S33; Tung-Hui Hu, *A Prehistory of the Cloud* (Cambridge: MIT Press, 2016); Kate Crawford, *Atlas of AI* (New Haven, CT: Yale University Press, 2021).
12. Jon Agar, *The Government Machine: A Revolutionary History of the Computer* (Cambridge, MA: MIT Press, 2003); Janet Abbate, *Inventing the Internet* (Cambridge, MA: MIT Press, 1999); Paul N. Edwards, *The Closed World: Computers and the Politics of Discourse in Cold War America* (Cambridge, MA: MIT Press, 1996); Donald MacKenzie, *The Influence of the Los Alamos and Livermore National Laboratories on the Development of Supercomputing* (Annals, 1991); Rebecca Slayton, *Arguments that Count: Physics, Computing, and Missile Defense, 1949–2012*

(Cambridge: MIT Press, 2013); Daniel Volmar, "The Computer in the Garbage Can: Air-Defense Systems in the Organization of US Nuclear Command and Control, 1940–1960" (PhD diss., Harvard University, 2018); Jonnie Penn, "Inventing Intelligence: On the History of Complex Information Processing and Artificial Intelligence in the United States in the Mid-Twentieth Century" (PhD diss., University of Cambridge, 2020).

13. Alfred Chandler and James Cortada, *A Nation Transformed by Information: How Information Has Shaped the United States from Colonial Times to the Present* (New York: Oxford University Press, 2003); Ronald Kline, *The Cybernetics Moment: Or, Why We Call Our Age the Information Age* (Baltimore, MD: Johns Hopkins University Press, 2017).

14. Fred Turner, *From Counterculture to Cyberculture: Stewart Brand, the Whole Earth Network, and the Rise of Digital Utopianism* (Chicago: University of Chicago Press, 2006).

15. Gerard Alberts, Ruth Oldenziel, eds., *Hacking Europe: From Computer Cultures to Demoscenes* (London: Springer, 2014).

16. Eden Medina, *Cybernetic Revolutionaries: Technology and Politics in Allende's Chile* (Cambridge, MA: MIT Press, 2011); Benjamin Peters, *How Not to Network a Nation: The Uneasy History of the Soviet Internet* (Cambridge, MA: MIT Press, 2016).

17. Frank Bösch, ed., *Wege in die Digitale Gesellschaft: Computernutzung in der Bundesrepublik, 1955–1990* [Toward a Digital Society: Computer Use in West Germany, 1955–1990] (Wallstein Verlag, 2018).

18. Joy Rankin, *A People's History of Computing in the United States* (Cambridge, MA: Harvard University Press, 2008); Julien Mailland and Kevin Driscoll, *Minitel: Welcome to the Internet* (Cambridge, MA: MIT Press, 2017).

19. Martin Campbell-Kelly and Daniel Garcia-Swartz, "The History of the Internet: The Missing Narratives" (December 2, 2005), http://dx.doi.org/10.2139/ssrn.867087.

20. Michael Mahoney, *Histories of Computing,* ed. Thomas Haigh, esp. Part Three: *The Structures of Computation* (Cambridge, MA: Harvard University Press, 2011); William Aspray, *John von Neumann and the Origins of Modern Computing* (Cambridge, MA: MIT Press, 1990); Stephanie Dick, "Computer Science," in *A Companion to the History of American Science,* eds. Georgina Montgomery and Mark Largent (Malden, MA: Wiley-Blackwell Publishing, 2015), 55–68; Thomas J. Misa, ed., *Communities of Computing: Computer Science and Society in the ACM* (New York: ACM, 2018).

21. Jennifer Light, *From Warfare to Welfare: Defense Intellectuals and Urban Problems in Cold War America* (Baltimore, MD: Johns Hopkins University Press, 2005); Joy Rohde, *Armed with Expertise: The Militarization of American Social Research during the Cold War* (Ithaca, NY: Cornell University Press, 2013); Paul Erickson, *The World the Game Theorists Made* (Chicago: University of Chicago Press, 2015); Hunter Heyck, *Age of System: Understanding the Development of Modern Social Science* (Baltimore, MD: Johns Hopkins University Press, 2015); Jamie Cohen-Cole, *The Open Mind: Cold War Politics and the Sciences of Human Nature* (Chicago: University of Chicago Press, 2014).

22. Joanna Radin, "Digital Natives: How Medical and Indigenous Histories Matter for Big Data," *Osiris* 32, no. 1 (2017): 43–64; Theodora Dryer, "Designing Certainty:

The Rise of Algorithmic Computing in an Age of Anxiety, 1920–1970" (PhD diss., UC San Diego, 2019).

23. Peter Galison, "Monte Carlo Simulations: Artificial Reality," in *Image and Logic: A Material Culture of Microphysics* (Chicago: University of Chicago Press, 1997), 689–780; Stephanie Dick, "Coded Conduct: Making MACSYMA Users and the Automation of Mathematics," in *BJHS Themes,* vol. 5 (2020); Hallam Stevens, "A Feeling for the Algorithm: Working Knowledge and Big Data in Biology," *Osiris* 23, no. 1 (2017); Evan Hepler-Smith, "'A Way of Thinking Backwards': Computing and Method in Synthetic and Organic Chemistry," *Historical Studies of the Natural Sciences* 48, no. 3 (2018): 300–337.

24. Thomas S. Mullaney, *The Chinese Typewriter: A History* (Cambridge, MA: MIT Press, 2017); Jaroslov Švelch, *Gaming the Iron Curtain* (Cambridge, MA: MIT Press, 2018). See also Ivan da Costa Marques, "Cloning Computers: From Rights of Possession to Rights of Creation," *Science as Culture* 14, no. 2 (2005): 139–60.

25. Janet Abbate, *Recoding Gender* (Cambridge, MA: MIT Press, 2010); Mar Hicks, *Programmed Inequality;* Jennifer S. Light, "When Computers Were Women," *Technology and Culture* 40, no. 3 (1999): 455–83.

26. Lisa Nakamura, *Digitizing Race: Visual Cultures of the Internet* (Minneapolis: University of Minnesota Press, 2007); Wendy Chun, *Control and Freedom: Power and Paranoia in the Age of Fiber Optics* (Cambridge, MA: MIT Press, 2006); Charlton McIlwain, *Black Software: The Internet & Racial Justice, from the AfroNet to Black Lives Matter* (New York: Oxford University Press, 2019); Ruha Benjamin, *Race After Technology: Abolitionist Tools for the New Jim Code* (Medford, MA: Polity, 2019); André Brock, Jr., *Distributed Blackness: African American Cybercultures* (New York: New York University Press, 2020); Bernard Geoghegan, "Orientalism and Informatics: Alterity from the Chess-Playing Turk to Amazon's Mechanical Turk," in *Ex-position* no. 43 (2020): 45–90; Safiya Noble, *Algorithms of Oppression: How Search Engines Reinforce Racism* (New York: New York University Press, 2018).

27. Morgan G. Ames, *The Charisma Machine: The Life, Death, and Legacy of One Laptop per Child* (Cambridge, MA: MIT Press, 2019); Daniel Greene, *The Promise of Access: Technology, Inequality, and the Political Economy of Hope* (Cambridge, MA: MIT Press, 2021); Virginia Eubanks, *Automating Inequality* (New York: St. Martin's Press, 2018).

28. Eden Medina, "Forensic Identification in the Aftermath of Human Rights Crimes in Chile," *Technology & Culture* 59, no. 4S (2018): S100–S133; Thomas Haigh, ed., *Exploring the Early Digital* (New York/Berlin: Springer, 2019).

29. Sam Schirvar's presentation at the 2019 SHOT conference, "The Last Computers: From Computer to Computer-Operator through the ENIAC Project," explored how social hierarchy was embedded in the concept of "automation" and helped our thinking about this framing.

PART I

ABSTRACTIONS

CHAPTER 1

Waiting for Midnight

Risk Perception and the Millennium Bug

Zachary Loeb

Dead computers were raining down upon the streets. Cars, some with suitcases strapped to their roofs, were stuck bumper to bumper, malfunctioning streetlights thwarting their efforts to flee. The scene was chaotic, but no one was staring at the falling machines, they were transfixed by the long-haired figure in a white robe, walking barefoot on the snowy asphalt, holding a cross aloft and wearing a sandwich board which read "THE END OF THE WORLD!?! Y2K insanity! Apocalypse Now! Will Computers melt down? Will society? A guide to MILLENNIUM MADNESS." Luckily, the scene was just playing out on the cover of *Time* magazine.[1]

With just shy of a year until the deadline, *Time* was reporting on and contributing to "millennium madness." Alongside cartoonish images of computer-wrought collapse, the cover story acknowledged the genuine threats but mainly mocked the doom-mongers advising people to head for the hills.[2] The world was not about to end, but *Time*'s own polling showed that 59% of respondents were "somewhat/very concerned" about Y2K.[3] While the cover story treated the subject as amusing, an accompanying timeline featured the less sanguine observation that the "alarmist language . . . may yet be justified."[4]

Beyond being an entertaining artifact of technological eschatology, this issue of *Time* is an example of how the year 2000 computing crisis

(Y2K) was represented then and is remembered now: an overreaction hyped by doomsday prophets, doomsday profiteers, and the media. Yet, within that same issue can be found the largely forgotten story of the people who sounded the alarm, so that a scenario akin to the one on the magazine's cover would be avoided. The dangers surrounding Y2K ranged from the annoying to the catastrophic, but too much emphasis on the catastrophic has often resulted in the basic risks being over-looked. While societies had been growing increasingly dependent on computers throughout the second half of the twentieth century, Y2K forced elected officials and the broader public to confront this reli-ance's hazards. As Congressman Jim Turner (D-TX) observed, by the turn of the century "every facet of our life now depends upon our com-puters working well."[5] Though in the wake of the successful manage-ment of the crisis, what could have been an opportunity to reevaluate societal reliance on computers only led to greater power being vested in those technologies.

At its core, Y2K is the story of an economic problem that became a technical problem, which in turn became a social problem. Y2K was a stumbling block for societies racing into the information age, and schol-ars of computing who were living through the crisis seemed cognizant of the moment's significance; as Paul Edwards pointed out, "Comput-ers have become . . . the infrastructure of our infrastructures."[6] Fur-thermore, Y2K revealed the fragility of this infrastructure, a point driven home by the sociologist Charles Perrow, who framed Y2K as a "normal accident" wherein the increased technical complexity of a sys-tem increases the probability that something will go wrong.[7] And only a few days prior to *Time*'s ominous cover, the historian Paul Ceruzzi argued, in the *Washington Post*, that Y2K showed "we need to pay at-tention not just to those who tell us what our wondrous inventions can do but to the ones who have an inkling of what they can't do."[8]

While the Y2K disaster never occurred—thanks to heroic efforts—the anxiety that surrounded Y2K demonstrates, as Mary Douglas and Aaron Wildavsky have stated that "the perception of risk is a social pro-cess," which "depends on combinations of confidence and fear."[9] The risks of Y2K were abstract, but as various groups sought to make sense of the computers upon which they, and society, had become reliant, they reframed these into concrete threats. For some technical professionals, Y2K was a humiliating crisis of their own making that threatened to mar their field's reputation even as they found themselves the world's best hope of avoiding calamity; for some in the government, Y2K represented a fundamental threat to the systems that kept society running, which in turn required these officials to heed expert recommendations; and to some, who sought to make the crisis legible to the broader public, Y2K was a sign that humans had lost control as computers had taken control. To some, Y2K was a question of professional survival, while for others Y2K was a question of civilizational survival. And though Y2K tested public and political faith in technocrats and technological systems, when January 1, 2000, passed without major incident, that faith was quickly restored.

The process of fixing Y2K involved multiple stages, yet as most of the professionals working on the problem agreed, the first step was generating awareness. Thus, this chapter considers the way that disparate groups framed Y2K's abstract risks as specific dangers, and how that drove them to act. Before the Y2K problem could be fixed, people first had to agree that it was in fact a problem. Fortunately, there were people who saw it as such, and others who were willing to do something about it.

"What we have delivered is a catastrophe"

Before *Time*'s apocalyptic cover story, there was *Computerworld*'s "Doomsday 2000."[10] There, Peter de Jager, one of Y2K's most prominent commentators, described the situation starkly: "The information systems community is heading toward an event more devastating than a car crash."[11] With the deadline seven years in the future, de Jager projected that averting the "crash" would consume tens of billions of dollars and would require an army of technical professionals. To make matters worse, the people who would now have to tell their CEOs that a huge outlay of resources was necessary were the people who had caused the problem. Thus, de Jager warned *Computerworld*'s readers,

"We and our computers were supposed to make life easier, this was our promise. What we have delivered is a catastrophe."[12]

Long before the general public learned of Y2K, computing professionals knew they had a problem. In the 1960s, computer memory was expensive, and memory saved was money saved. Therefore, programmers, especially those working in the Common Business Oriented Language (COBOL), truncated dates by using six characters. This meant, for example, that July 12, 1965, would be represented as 071265. This format worked, but it relied on the assumption, for the computer's calculations, that the century digits were 19. While programmers understood this would eventually cause problems, many programmers working in the 1960s could not imagine their code would still be in operation forty years later. Furthermore, as Troy Kaighin Astarte explores in this volume, these decades were a time of a great deal of development and change surrounding programming languages. Alas, as James Sanders described it in *IEEE Software,* "Old computer hardware goes into museums, old software goes into production every night."[13]

"We knew about the problem long ago. We could have fixed it," Bruce Webster observed, but the crisis demonstrated that the problem had not been addressed ahead of time.[14] And thus there was the possibility of "Date-Related Abend" (abnormal end) failures that might occur due to date-related processing errors, there was the threat of "Garbage" as systems could churn out incorrect data, "Or Nothing" might happen—Leon Kappelman gathered these risks using the fiery acronym "DRAGON."[15] Since the time their eggs had hatched in the mid-twentieth century, "these silicon-based DRAGONs" had "found their way into practically every device and machine around us."[16] And with the year 2000 approaching, these "DRAGONs" were dangerous: abends had "the potential to cause devastating destruction and death," the "Garbage" could cause anything "from minor inconvenience to calamitous chaos and carnage," and if ignored even the "Nothing" could start fires.[17] While the work of correcting the many systems was less exciting than picking up a sword, these alliterative evocations gave notice of the threat.

For those within the computer profession, Y2K posed two sets of risks: foremost were the abend- and garbage-wrought failures, but another risk was reputational. Many in the computer community knew they were being blamed for the crisis, and that should there be a plague of DRAGONs they would be held responsible; thus efforts were made to deflect culpability by highlighting how the problem resulted from clients and executives wanting to save money.[18] Furthermore, the decision to truncate dates truly had saved money, even if companies had a hard time appreciating the millions saved as they prepared to spend millions on repairs.[19] Nevertheless, as Sanders noted, whatever Y2K's consequences, "they will reflect poorly on our profession."[20] A similar sentiment came from Capers Jones, who described Y2K as both "one of the most expensive problems in human history" and "one of the most embarrassing."[21] Computer professionals, Jones lamented, had been discussing the problem "for more than 25 years, and its significance has been hypothesized with increasing alarm for more than 10 years," but the work of fixing the problem was being done with time running out.[22]

Yet, by foregrounding the risks, Y2K also provided an opportunity for redemption. The scale of the work was daunting; it could be, as James Schultz put it, "the challenge of your career," but it allowed computer professionals to "make a critical contribution."[23] Y2K might be, as Howard Rubin explained, "a blessing in disguise," as it was "the first major technology threat we've seen coming"; thus it bolstered the authority of the technicians with the necessary skills.[24] And though Jones lamented that Y2K was an embarrassment, he hoped the crisis would force companies to conduct the computer assessments they had deferred.[25] Instead of being a crisis that would delegitimize computing professionals, fixing Y2K could turn them into the saviors of the information age.

In 1993, de Jager had warned of "doomsday," but by the summer of 1999 his tone had shifted. Y2K still struck him as the "most stupid blunder in the history of technology," but by responding to the threat the computing sector had ensured "society won't fall apart."[26] The DRAGONs had been vanquished. Nevertheless, the victory did not belong only to those working in the computing sector, for they had not been alone in recognizing the seriousness of the problem.

"Disruptions will occur"

"This is clearly one of the most complex management challenges in history." This is how President Bill Clinton described Y2K before the National Academy of Sciences on July 14, 1998.[27] With a touch of humor, Clinton noted, Y2K "is not one of the summer movies where you can close your eyes during the scary parts,"[28] but by the time he made these comments many members of Congress had been pushing him to open his eyes to "the scary parts" for years.

Senator Patrick Moynihan (D-NY) had requested a report on Y2K's dangers from the Congressional Research Service (CRS) in February 1996.[29] While the CRS report lacked apocalyptic fanfare, it stated clearly that "given society's increasing reliance on computers, this problem could have a significant impact on a wide range of activities and interests worldwide," with dangers threatening everything from banking to defense to infrastructure.[30] In Moynihan's estimation, as he described it to Clinton, the CRS study substantiated "the worst fears of the doomsayers."[31] Urging Clinton to act, Moynihan warned, "The computer has been a blessing; if we don't act quickly, however, it could become the curse of the age."[32] And more than a year after Moynihan's warning, Senator Robert Bennett (R-UT) urged the president to focus the Executive Branch on the problem, as he was certain "the year 2000 problem is a national crisis which warrants special government attention."[33] Throughout the twentieth century, as Marc Aidinoff discusses in this volume, there were attempts by Democrats and Republicans to cast the computer as a machine in keeping with their ideological goals, but Y2K presented itself as a bipartisan problem requiring the cooperation of both parties.

While Moynihan's letter to President Clinton had largely been based on the CRS report, Bennett's urgency was informed by the many congressional hearings that had been held after that report was issued. Dozens of hearings were held in the 104th–106th Congresses examining the impacts of Y2K on everything from the postal service to nuclear power plants. These hearings—which brought together experts and bureaucrats culled from government agencies, international bodies, and the private sector—were integral not only for keeping the government informed but for demonstrating that the government was not ignoring the problem. Indeed, Y2K was the perfect apocalypse for the 1990s, with the conclusion of the Cold War, nuclear anxiety was receding; thus the time was ripe for fears to shift from the threatening other to an internal threat.

The pervasiveness of the problem, and the management challenge it represented, were staggering. Representative Constance Morella (R-MD) argued Y2K would "literally affect virtually every human on the planet," for "this problem seems to be endemic to not just computer software or programs but in any product which contains a computer chip."[34] The implications were captured by Harris Miller, president of the Information Technology Association of America, who testified that just as electricity had become an integral part of everyday life, so too had information technology; if computers failed it would be "like losing electricity all of a sudden."[35] The dark irony of Miller's comments was that if the computers went down they would probably knock the electricity out too, as power plants were reliant on computers. Infrastructural concerns were a major topic of congressional hearings; however, Congress saw Y2K's threats everywhere; it posed risks to the financial system, for small businesses, for federal agencies, for the states, for the military, for airlines, for other countries, and there was concern that Y2K would result in a tsunami of litigation. The risks existed wherever there were computers, and as person after person testified in hearing after hearing, computers were everywhere.

Congress's hearings and reports were hardly beacons of light. In its initial, 1996, Y2K report, the House's Committee on Government Reform and Oversight noted that the lack of action meant there was "a high risk of system failure," and though the problem had international dimensions, "as the leading user of computer technology the United States probably has more at risk."[36] Two years later the same committee warned that "the current evidence points to considerable Year 2000 failure unless the rate of progress throughout society improves considerably."[37] This glum assessment was echoed in the Senate's Special Committee on the Year 2000 Problem report, which stated, in February 1999,

"Y2K risk management efforts must be increased to avert serious disruptions."[38] By the time that committee issued its "100 Day Report," the committee believed the United States would not "experience nation-wide social or economic collapse" but still warned "disruptions will occur and in some cases those disruptions would be significant."[39]

In its initial report on Y2K, the Senate committee recognized that "technology has provided the U.S. with many advantages, but it also creates many new vulnerabilities."[40] From Moynihan's request to the CRS, through the arrival of the year 2000, congressional committees focused on raising awareness of, and keeping track of, the "new vulnerabilities." In his comments before the National Academy of Sciences, Clinton had struck an optimistic note, emphasizing that "if we act properly, we won't look back on this as a headache, sort of the last failed challenge of the 20th century. It will be the first challenge of the 21st century successfully met."[41] Though the US government was clearly signaling it was working on the problem, its somber admission that disruptions were likely led many to conclude this headache would not be mild.

"No one knows exactly what is going to happen"

Given Y2K's uncertainty, it was understandable that many people wanted a calm assessment of the problem. Luckily, a figure renowned as unimpeachably logical spoke up, namely: Spock. Or, more accurately, the actor who had played Spock on *Star Trek*. In the foreword to Avian Rogers's *Y2K Family Survival Guide*, Leonard Nimoy pointed to the work being undertaken as a sign that "there is hope," but he couched this in an acknowledgment that "no one knows exactly what is going to happen."[42] Nimoy counseled against panic, highlighting that it was necessary for individuals, families, and communities "to find a reasonable balance between those who call for extreme survival measures and those who advise no action at all."[43] Y2K, in Nimoy's description, was "an opportunity to reflect on where we are headed as a civilization," and though Nimoy closed out his comments with the Vulcan words "Live long and prosper," Y2K posed a threat to both long life and prosperity.[44]

Despite the assurances of computer professionals and governmental leaders that the situation was being handled, plenty of voices urged caution. After all, as the congressional reports made clear, much remained uncertain. In a *New York Times* business bestseller, consumer advocate Michael S. Hyatt informed his readers that as he had researched Y2K he had become "increasingly alarmed," and though he had hoped in the course of his research to be assured "that the problem was well under control, and that I had nothing to worry about" he *"had exactly the opposite experience."*[45] While there were numerous thick technical manuals that provided computer professionals with guides to fixing their systems, books such as Hyatt's sought to inform members of the nontechnical public. And, according to Hyatt, people needed to brace for impact. Hyatt outlined a variety of scenarios on a spectrum from bumps to cataclysm—but he couched these possibilities in references to sober government assessments.[46]

That Hyatt could not tell his readers exactly what would happen was a reflection of the fog surrounding Y2K. And this murky atmosphere provided the space in which prophecies of doom could flourish. There was the chance that everything would be fixed in time, there was a chance that most things would be fixed, and there was the chance that too little would be fixed—but none could say which it would be. Hyatt's *Millennium Bug,* as well as the survival guide and Y2K novel he published, were part of a broader milieu of works responding to Y2K's ambiguity by counseling readers to hope for the best but prepare for the worst. Awareness of the risks had compelled the government to act, but awareness that the government might not get everything finished on time required individuals to prepare themselves. After all, it was better to be stuck with extra beans and batteries, than to find oneself sitting hungrily in the dark. Anxious though Hyatt's book was, some other authors took a much more doom-laden tack, as they predicted the collapse of civilization often in explicitly religious terms. And several members of the computing profession hopped on this publishing bandwagon, penning public-facing guides wherein the author's technological expertise bolstered the credibility of their ominous predictions. In contrast to the punk DIY ethos that infused many hacker communities, as Elyse Graham explores in this volume, the survival guides framed freedom from computing technology, as opposed to mastery of that technology, as the key to the future.

Nimoy had suggested that Y2K provided an opportunity to reflect on the direction in which civilization was heading, and for some Y2K appeared as a chance to change course. This sentiment was captured in the *Utne Reader*'s *Y2K Citizen's Action Guide*, wherein Eric Utne opined that preparing for Y2K could be "quite wonderful . . . we're going to get to know our neighbors . . . we will begin to know what it means to live in real community."[47] By disrupting the march into the information society, Y2K could be "the excuse we've been waiting for . . . to stop our polluting and wasteful ways."[48] Thus, in contrast to the bunker mentality of many survival guides, some saw the dangers of Y2K as an opportunity to rebuild community. This sense, that only mutual support could get people through the crisis, was further detailed by Paloma O'Riley, whose "Cassandra Project" focused on community, not individual, preparedness for Y2K.[49] Eschewing the "predictions of technological doomsday" and the entreaties of "rabid money-hungry consultants," O'Riley framed "preparing for the worst" as being akin to purchasing insurance—one didn't make such an investment because disaster was certain but because one understood the risk was real.[50] And while O'Riley provided tips about what individuals should have on hand in order to be ready, she emphasized that "the best security is a prepared neighbor."[51]

Popular responses to Y2K included some that saw the crisis as the fulfillment of biblical prophecy, a shadowy international conspiracy, or an opportunity to make money; yet others tried to get their fellow citizens ready. Though appeals to stock up on food and ammunition could be seen as having antisocial implications, the efforts to inform undertaken by the likes of Hyatt and O'Riley were not met with universal censure. At the congressional hearing at which the two testified, they were thanked for their efforts in generating public awareness.[52] While Congress had been working to ensure that the business sector, government agencies, and the international community were aware of the risks, people like Hyatt and O'Riley were playing a role in informing the public, and keeping pressure on businesses and politicians who had been dragging their heels. Given the uncertainty, Hyatt testified that "prudent people" needed to engage in contingency planning and "emergency preparedness just like if we knew there was an earthquake coming."[53] While O'Riley urged Congress to share more, not less, information with the public, O'Riley noted that she had not "seen anyone yet panic," and she emphasized uncertainty, not information, would lead to panic.[54]

Computer professionals and government officials publicly stated that they were working on the problem, but they could not guarantee that there would be no disruptions. Thus, as O'Riley stated, regular people needed to take appropriate precautions, in order to "soften whatever Year 2000 problems occur."[55] Taking such precautions was not a sign of alarmism but a prudent response to the siren computer professionals and government officials had been sounding.

Conclusions: "The original sin of technological society"

When the year 2000 began, dead computers did not rain from the sky. Thus, those who had been under pressure to demonstrate that they were doing enough found themselves having to prove that their efforts had been necessary. Responding to the press and the public's skepticism, Representative Morella defended the work of her committee, saying, "In my mind, there is no doubt the problem was real."[56] John Koskinen, chair of Clinton's Y2K Council, justified the compliance efforts in stronger terms, noting, "I don't know of a single person working on Y2K who thinks that they did not confront and avoid a major risk of systemic failure."[57] These sentiments were echoed in the Senate special committee's final report, which expressed the view that "the level of effort was justified, and the expenditures of the public and private sectors were indeed necessary."[58] And given the swift pivot from anxiety to mockery, Frank Hayes, writing in *Computerworld*, captured the frustration of computer professionals: "IT people around the world, and the people who use IT to get the job done, did their jobs so well— nothing went wrong anywhere. And as a result, these clowns will give you credit for exactly . . . nothing."[59]

Y2K was a crisis foreseen. It had been anticipated long before the 1990s, yet a computer-wrought calamity had long been predicted by those who had shuddered at societies' increased reliance on computers. By 1970, the social critic Lewis Mumford was describing the computer as an "authoritarian technic," which would usher in the "computer dominated society" under the all-seeing eye of a high-tech sun god.[60] And such warnings took on extra weight when coming from actual computer scientists, such as Mumford's friend Joseph Weizenbaum, who noted that "if we depend on that machine, we have become servants of a law we cannot know, hence of a capricious law. And that is the source of our distress."[61] While Weizenbaum used his technical expertise to contextualize the dangers of computing, his larger concern was not about computers as such but about the new social order computers were creating.[62]

One need not turn to social critics for such an analysis; the same perspective appeared throughout the Y2K era wherein an appreciation of computerization's benefits became increasingly intertwined with a recognition of downsides. The Senate's Special Report had noted that the core of the problem was a "disconnect between those who use technology and those who create it," with Y2K giving users "a crash course in the fragile mechanics of information technology."[63] In the first installment of his "Diary of a Y2K Consultant" column, Howard Rubin explained how just as the Earth had a physical layer that was threatened by earthquakes, and an atmospheric layer that was threatened by hurricanes, in the "Information Age" the world had acquired an "electronic layer . . . Y2K is just one manifestation of disturbances in this new cyber-geographic layer."[64] *Time* seemed to enjoy mocking the figures fretting over the crisis, but it hit upon something profound by noting that "the Y2K bug is something akin to the original sin of technological society, a mortal flaw bred in the very bones of the modern world."[65]

Y2K was not the first crisis created by computers, nor was it the last. The late 1960s saw the "software crisis," but as Janet Abbate has written, that crisis "was neither a distinct event nor a coherent description of prevailing conditions in the industry."[66] As Abbate, along with Nathan Ensmenger and Donald MacKenzie have noted, the software crisis was a significant moment in terms of the gendered professionalization of the computing field; however, the software crisis never reached the same apocalyptic tenor as Y2K.[67] What makes Y2K so significant is the ways it drew attention to how a society that comes to rely on complex technical systems winds up at the mercy of those systems, and the people who control them. Had Y2K remained an isolated technical problem—quietly and effectively fixed—the event would still warrant scrutiny, but it probably would not have resulted in dozens of congressional hearings, mountains of survival guides, or garnered breathless media coverage. Y2K resonates with Lee Vinsel and Andrew Russell's attention to the history of maintenance within the history of technology,[68] but Y2K is not only a story of technologies being maintained—it is also a story about a crisis that laid bare a shift in societal power. In 1970, Mumford's warning of "computer dominated society" could be scoffed at, but Y2K showed that computers' increasing societal dominance was no laughing matter.

While the memory of Y2K lingers in the popular imagination largely as much ado about nothing, a historical reassessment challenges such ignominy; Y2K is a story about a problem that was fixed, but it is also a story about societies having to reckon with their reliance on computers. One of the key elements that led to that work being completed was the way that various stakeholders understood the crisis's risks. Y2K was a moment where the focus was squarely on envisioning the worst, not merely hoping for the best, and this grim perception helped spur the necessary action. For computer professionals, Y2K was a mundane technical problem made serious by its pervasiveness and the limited resources available—with the threat that technical professionals would be blamed if disaster struck. For government officials Y2K was a wake-up call to the ways in which the "blessings" of computers threatened to become "curses," as societies' essential functions had become dependent on computers. And if computer professionals and government officials were in positions to know about Y2K, for a significant slice of the public the perceived risks of Y2K were magnified by the mixed messages they were receiving as they were instructed not to panic while simultaneously being urged to print out hard copies of essential documents "just in case." The apocalyptic tonality of the reporting on Y2K often shaded into the hyperbolic, yet it still captured an uncomfortable revelation that the old world had ended and a new computer-dependent world had come into being.

Commenting on the challenge posed by Y2K, the Senate special committee noted that Y2K meant "we, as a nation and as individuals, need to consider carefully our reliance on information technology and the consequences of interconnectivity, and work to protect that which we have so long taken for granted."[69] However, when the crisis was averted, this reckoning no longer seemed necessary. In the absence of a calamity requiring postmortems, Y2K's lesson for the government, and much of the public, was that the IT sector would fix the problems of its own creation. Peter de Jager had sounded the alarm in 1993, but in Y2K's aftermath, he commented, "I don't think we've learned anything from this."[70]

Despite the prophecies of doom, when the smoke cleared on Y2K the disaster had been prevented; the computers that had looked like fire-breathing monsters went back to slumbering beneath their mountains of gold, and those who excitedly eyed those riches came to believe that the beasts had been subdued. At its height, the Y2K crisis demonstrated that those pursuing computing's blessings also had to brace for computing's curses; however, the quashing of Y2K made it tempting to believe that all the talk of curses was just ridiculous fear-mongering. Dead computers did not fall from the heavens on January 1, 2000—but in the ensuing years stories of massive data breaches, algorithmic bias, sweeping surveillance, and toxic e-waste have all shown that computers do indeed breathe fire.

Y2K was an opportunity to change directions—to become less reliant on computers, to put in place more stringent oversight, to grapple with computer-exacerbated risks—but instead societies raced onward, tantalized by the glistening promise of computerization. And in this forward rush they ignored the warning that Y2K had chiseled above the slumbering beast's treasure trove: "Here be dragons."

Notes

1. *Time,* January 18, 1999. The cover image is a digital photomontage by Aaron Goodman.
2. Richard Lacayo, "The End of the World as We Know It?" *Time,* January 18, 1999, 60–70.
3. Lacayo, "The End of the World as We Know It?" 64.
4. Chris Taylor, "The History and the Hype," *Time,* January 18, 1999, 72–73. Quote on 73.
5. US Congress, House, Joint Hearing before the Subcommittee on Government Management, Information, and Technology of the Committee on Government Reform and the Subcommittee on Technology of the Committee on Science, *Year 2000 Computer Problem: Did the World Overreact, and What Did We Learn?* 106th Cong., January 27, 2000, 8.
6. Paul N. Edwards, "Y2K: Millennial Reflections on Computers as Infrastructure," *History and Technology* 15, no. 1–2 (1998): 7–29. Quote on 11.
7. Charles Perrow, *Normal Accidents: Living with High-Risk Technologies* (Princeton, NJ: Princeton University Press, 1999).
8. Paul Ceruzzi, "Y2K: Old Hat in New Technology," *Washington Post,* January 11, 1999, A19.
9. Mary Douglas and Aaron Wildavsky, *Risk and Culture: An Essay on the Selection of Technological and Environmental Dangers* (Berkeley: University of California Press, 1982), 6.
10. Peter de Jager, "Doomsday 2000," *Computerworld* 27, no. 36 (September 6, 1993): 105, 108–9.
11. de Jager, "Doomsday 2000," 105.
12. de Jager, 105.
13. James Sanders, "Y2K: Don't Play It Again, Sam," *IEEE Software* 15, no. 3 (May/June 1998): 100–102. Quote on 101.
14. Bruce Webster, *The Y2K Survival Guide: Getting to, Getting Through, and Getting Past the Year 2000 Problem* (Upper Saddle River, NJ: Prentice Hall PTR, 1999), 51.
15. Leon Kappelman, *Year 2000 Problem: Strategies and Solutions from the Fortune 100* (Boston, MA: International Thomson Computer Press, 1997), 1.
16. Kappelman, *Year 2000 Problem,* 1.
17. Kappelman, 2.
18. Capers Jones, *The Year 2000 Software Problem: Quantifying the Costs and Assessing the Consequences* (New York: ACM Press, 1998), 21.
19. Kappelman, *Year 2000 Problem,* 53–54.
20. Sanders, "Y2K," 102.
21. Jones, *The Year 2000 Software Problem,* 207.
22. Jones, 207.
23. James Schultz, "Managing a Y2K Project—Starting Now," *IEEE Software* 15, no. 3 (May/June 1998): 63–71. Quote on 71.
24. Howard Rubin, "Uncovering Weak Links in the Readiness Chain," *Computer* 32, no. 10 (October 1999): 20.
25. Jones, *The Year 2000 Software Problem,* 26.
26. Matt Hamblen, "De Jager: Lighten Up on Y2K," *Computerworld* 33, no. 25 (June 21, 1999): 47.

27. President Clinton's remarks are transcribed on pages 125–30 of US Congress, House, Hearing before the Subcommittee on Government Management, Information, and Technology of the Committee on Government Reform and the Subcommittee on Technology of the Committee on Science, *Year 2000: Biggest Problems and Proposed Solutions,*105th Cong., June 22, 1998. 127.

28. US Congress, *Year 2000*, 128.

29. US Congress, Senate, Special Committee on the Year 2000 Problem, *Investigating the Impact of the Year 2000 Problem: The 100 Day Report,* September 22, 1999, 105th Congress. S. Prt. 106-31, 209–10. Appendix II of this report includes copies of a variety of important letters, several of which are cited in this chapter.

30. US Library of Congress, Congressional Research Service, *The Year 2000 Computer Challenge,* by Richard M. Nunno, 96–533 SPR (1996), CRS 2.

31. US Congress, *Investigating the Impact of the Year 2000 Problem: The 100 Day Report,* 211.

32. US Congress, *Investigating the Impact of the Year 2000 Problem,* 212.

33. US Congress, 214.

34. US Congress, House, Joint hearing before the Subcommittee on Technology of the Committee on Science and the Subcommittee on Government Management, Information, and Technology of the Committee on Government Reform and the Subcommittee on Technology of the Committee on Science, *Year 2000: Risks: What Are the Consequences of Information Technology Failure?*105th Cong., March 20, 1997, 1–2.

35. US Congress, *Year 2000: Risks: What Are the Consequences of Information Technology Failure?* 54.

36. US Congress, House, Committee on Government Reform and Oversight, *Year 2000 Computer Software Conversion: Summary of Oversight Findings and Recommendations,* 104th Cong., 2d sess., 1996, H. Rep 104–857; 11 and 13.

37. US Congress, House, Committee on Government Reform and Oversight, *The Year 2000 Problem,* 105th Cong., 2d sess., 1998, H. Rep 105–827; 12.

38. US Congress, Senate, Special Committee on the Year 2000 Problem, *Investigating the Impact of the Year 2000 Problem,* February 24, 1999, 105th Congress. S. Prt. 106–10, 1.

39. US Congress, *Investigating the Impact of the Year 2000 Problem: The 100 Day Report,* 8.

40. US Congress, *Investigating the Impact of the Year 2000 Problem,* 13.

41. US Congress, *Year 2000 Biggest problems and Proposed Solutions,* 130.

42. Leonard Nimoy, "Foreword," in Avian M. Rogers, *Y2K Family Survival Guide* (Nashville, TN: Rutledge Hill Press, 1999), viii.

43. Nimoy, "Foreword," ix.

44. Nimoy, ix.

45. Michael S. Hyatt, *The Millennium Bug: How to Survive the Coming Chaos* (New York: Broadway Books, 1998), 14. Emphasis in the original.

46. Hyatt, *Millennium Bug,* 159–81.

47. Eric Utne, "I Am Because We Are," in Eric Utne, ed., *Y2K Citizen's Action Guide* (Minneapolis, MN: Utne Reader Books, 1998), 14.

48. Utne, "I Am Because We Are," 14.

49. Paloma O'Riley, "Individual Preparation for Y2K," in Eric Utne, ed., *Y2K Citizen's Action Guide* (Minneapolis, MN: Utne Reader Books, 1998), 70–90.

50. O'Riley, "Individual Preparation for Y2K," 70.

51. O'Riley, 89.

52. US Congress, House, Joint hearing before the Subcommittee on Technology of the Committee on Science and the Subcommittee on Government Management, Information, and Technology of the Committee on Government Reform and the Subcommittee on Technology of the Committee on Science, *Y2K: What Every Consumer Should Know to Prepare for the Year 2000 Problem*, 105th Cong., September 24, 1998.

53. US Congress, *Y2K*, 120.

54. US Congress, 122.

55. US Congress, 122.

56. US Congress, House, Joint Hearing before the Subcommittee on Government Management, Information, and Technology of the Committee on Government Reform and the Subcommittee on Technology of the Committee on Science, *Year 2000 Computer Problem: Did the World Overreact, and What Did We Learn?* 106th Cong., January 27, 2000, 6.

57. US Congress, *Year 2000 Computer Problem: Did the World Overreact, and What Did We Learn?* 14–15.

58. US Congress, Senate, Special Committee on the Year 2000 Problem, *Y2K Aftermath—Crisis Averted: Final Committee Report*, February 29, 2000, 106th Congress. S. Prt. 106–42, 10.

59. Frank Hayes. "Feeling Cheated?" *Computerworld* 34, no. 2 (January 10, 200. "..." in original text.

60. Zachary Loeb, "From Megatechnic Bribe to Megatechnic Blackmail—Mumford's 'Megamachine' after the Digital Turn," *Boundary 2—Online Journal,* special issue on the "Digital Turn," 2018, https://www.boundary2.org/2018/07/loeb/.

61. Joseph Weizenbaum, *Computer Power and Human Reason* (San Francisco, CA: W. H. Freeman and Company, 1976), 41.

62. Zachary Loeb, "The Lamp and the Lighthouse: Joseph Weizenbaum, Contextualizing the Critic," *Interdisciplinary Science Reviews* 46, nos. 1–2 (March–June 2021): 19–35.

63. US Congress, *Investigating the Impact of the Year 2000 Problem*, 7–8.

64. Angela Burgess, Howard Rubin, and James Sanders, "Diary of a Y2K Consultant: Bracing for the Millennium," *Computer* 32, no. 1 (January 1999): 51–56. Quote on 52.

65. Lacayo, "The End of the World as We Know It?" 68.

66. Janet Abbate, *Recoding Gender: Women's Changing Participation in Computing* (Cambridge, MA: MIT Press, 2017), 96.

67. Nathan Ensmenger. *The Computer Boys Take Over: Computers, Programmers, and the Politics of Technical Expertise* (Cambridge, MA: MIT Press, 2010); Donald MacKenzie, *Mechanizing Proof: Computing, Risk, and Trust* (Cambridge, MA: MIT Press, 2001).

68. Lee Vinsel and Andrew L. Russell, *The Innovation Delusion: How Our Obsession with the New Has Disrupted the Work that Matters Most* (New York: Currency, 2020).

69. US Congress, *Investigating the Impact of the Year 2000 Problem*, 13.

70. Anonymous, "Have We Learned Nothing from the Y2K Episode?" *Computerworld* 34, no. 2 (January 10, 2000): 19.

- - - - - - - - -

CHAPTER 2

Centrists against the Center

The Jeffersonian Politics of a Decentralized Internet

Marc Aidinoff

"It's the Jeffersonian idea of democracy," explained a suburban father from Weymouth, Massachusetts, to his local newspaper in 1996.[1] Touting his decision to share information about the local school committee via the World Wide Web, David Higgs knew networked computing was political. His words were not revealing the hidden politics of seemingly benign artifacts, revealed through historical and technical excavation, but rather underscoring the explicit politics that he, and many like him, would eagerly name. Higgs would enthusiastically wrap the World Wide Web in terms that were as overtly political as the language used by duly elected politicians. For $20 per month, the father of two children at the Abigail Adams Intermediate School had Internet access through the Plymouth Commercial Internet Exchange and 10mb for a personal web page. He used that page to post school committee memoranda and meeting minutes. In an article, "Dad Puts School Facts on Internet," the *Patriot Ledger* detailed Higgs's homespun handiwork: "None of the graphic bells and whistles of fancier computer postings on the Web, which frequently boast music and animated characters." Instead, David Higgs of Pleasant Street relied on "bold and italic print and eye-catching capital letters" to solve "what he saw as a lack of communication between the [school] committee and the community."[2] Higgs's claim that the World Wide Web, at least in his hands, could be Jeffersonian borrowed directly from potent existing terminology.

Decentralized networked computing was never simply abstracted. This collection traces the way distributed arrangements of metal pieces and human bodies came to have recognizable meanings—to be legally legible (as Gerardo Con Diaz, Zachary Loeb, and Cierra Robson argue in this volume) and identifiably racialized (as Robson, André Brock, Tiffany Nichols, and Kelcey Gibbons underscore). This work was often political, in the register of US electoral politics. If historians have increasingly drawn our attention to the political economy driving network design, this chapter argues for the role of political narrative shaping those design choices.[3] Just as the politics of the military planning and New Left individualism had given meaning to earlier incarnations of decentralized computing, the private Internet of the 1990s drew political meaning from centrist Democratic politics in the United States. Decentralization, expressed as Jeffersonian decentralization, was rendered legible in both the world of party politics and the world of systems engineering. As such, it is a reminder that the mutual shaping of politics and technology was not always a subterranean project but an explicit one that can be recognized in the loud articulations of those trying to ensure that technology and politics were reinforcing.

Like that of the nascent Internet, the meaning of Thomas Jefferson was hardly self-evident. Upon his death on July 4, 1826, Jefferson was canonized, but, as historian Merrill Peterson notes, the public memory of the prolific Jefferson was "an ill-arranged cluster of meanings, rancorous, mercurial, fertile." Unlike George Washington (deified before death) or Andrew Jackson (whose reputation was formed before his presidency), it would take time before Jefferson would become a more specified symbol. The imagined Jefferson, far more than his peers, was co-produced with contemporaneous politics. As Peterson explains, "Crudely unfinished after his death, his contract untransacted, Jefferson was fulfilled in the procession of the American mind."[4] Jefferson was a protean symbol, but his invocation did have certain potent affordances. In its many forms, the "Jeffersonian" would carry allusions to distinct economic and racialized social orders.

At the core of the narratives about Jefferson was his idealization of the yeoman farmer, the quintessential American icon of democratic decentralization. Contemporary historiography places that farmer, a symbol of anti-statist democratic independence, in the context of robust state interventions.[5] Jefferson saw a strong role for the government to protect the interests of slaveholders in an expanding nation. If Jefferson's image has long benefited from the contradiction of a slaveholder who espoused liberty, Jeffersonian liberty was explicitly enabled by mass enslavement. Jefferson's commitment was to an "empire of liberty." It was to be an empire composed of empires; each decentralized dominion governed by an independent farmer. Narratives about Jefferson bear the imprint of his most consequential action as president, the drastic expansion of that empire with the purchase of the Louisiana Territory. The success of plantation capitalism, as current historiography makes clear, depended on a form of decentralized federal rule that facilitated global commerce but did not intervene in the local dominion of the plantation owner.[6] Jeffersonian decentralization, decades and centuries later, maintained the traces of this earlier referent.

Jefferson the gentleman–scientist, whose leisure was provided for by hundreds of enslaved people, has, since his death, been put to work by a diverse range of political actors. To nineteenth-century antislavery Republicans, Jefferson was a protoabolitionist who could never quite get the words out, but, for secessionists, the founder of the Democratic Party was a defender not just of a state's right to permit slavery but also to secede. Because Jefferson believed forcefully in nullification, the strongest version of rights of states to overrule federal law, secessionists invoked Jefferson to justify leaving the Union and segregationists relied on Jefferson to rally against the concentration of federal power.[7] For much of the twentieth century, Democrats continued to claim Jefferson's mantle. Democrats like Al Smith and Franklin Delano Roosevelt used Jefferson to mean simply populist, unlike the elitist New Yorker Alexander Hamilton. The latter-day elite New Yorker, FDR, was introduced at his nominating convention as the "incarnation" of Jefferson.[8] If FDR was Jefferson-made-flesh, William Jefferson Clinton fashioned himself the third coming of Jeffersonian populism. Jefferson, as a political referent, developed over time but echoed the same historical themes: populist rejection of a strong federal government despite the insistence on government support and, however masked, a nostalgia for the decentralized agrarian life on the plantation.

Politicians hoped invoking Jefferson would help get them elected. For centrist Democrats like Bill Clinton and Al Gore, retrofitting themselves in Jefferson's image was an opportunity to make themselves recognizable to the citizenry as modern heirs of Southern decentralized power. Similarly, those advocating for certain types of government intervention in networked computing, including Al Gore and the Electronic Frontier Foundation, borrowed the same political language to align their technical project with the electoral one. In the 1990s, the term "Jeffersonian" tied the design principle of decentralization to the centrist strain of the Democratic Party, including the racial, economic, and aesthetic commitments of the New Democrats. The shared meanings of "Jeffersonian" helped transform the Internet into something the US political system and public could handle by tying it to democratic (the value) and Democratic (the political party) ideas. Jeffersonian decentralization made a technical arrangement politically recognizable, nonthreatening, and even desirable.

Beyond Cold War Decentralization

The founding myths of networked computing always return to a fundamental design principle: distributed or decentralized communication.[9] The argument for a noncentralized system is encapsulated by perhaps the most famous diagram in the history of computing from a 1964 RAND Corporation report prepared for the US Air Force by Paul Baran (fig.2.1).[10] In the diagram, three models of distributed communication are sketched out: centralized, decentralized, and distributed. Each node in the diagram represents a station, and each line a potential communication link to spread information. Because every node was vulnerable to attack, a successful system would include sufficient redundancy that it could still function even if some nodes were taken down. The RAND report was very clear that these networks should be evaluated on the Cold War "criterion of survivability."[11]

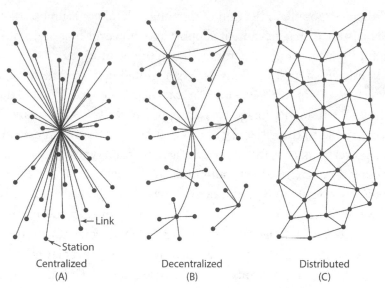

Centralized
(A)

Decentralized
(B)

Distributed
(C)

Figure 2.1. Paul Baran's vision of "Centralized, Decentralized, and Distributed Networks."

Under the conceptual heading of decentralization, Baran was elevating a suite of networking strategies, including packet switching, standardized packets, distributed nodes, headers to identify the packets, and a store-and-forward approach. These techniques became meaningful if they could bolster the capacity of the US military to develop and maintain a hierarchal and efficient military communication and management system known as command-and-control. Since command-and-control required speedy and reliable communication in the event of an attack, a rupture in the communication system itself would leave the hierarchy flailing.[12] Much of command-and-control designation was a post hoc rationalization for technical and organizational choices, and a rationale for funding.[13] In this case, an approach like "store-and-forward," the tried-and-true strategy of the US Postal Service, where letters are temporarily stored at a series of distributed boxes before they are routed to their final destination, would become a potent example of decentralization in the Cold War context. Baran also relied on "the postman analogy," but even this language came only after a framing of survivability. The distribution of store-and-forward was just part of an idealized "network which will allow several hundred major communications stations to talk with another after an enemy attack."[14]

As Peter Galison argues, World War II taught US Americans "to see though a bombsight," and so Cold War planning was marked by this recurring need to eliminate those vulnerable hubs that could be bombed. By this logic, cities should be more dispersed and people should be able to move through the nation without relying on any given route—a new multibillion-dollar interstate highway system was indeed funded. Just like cities and roads, digital communication should not have a vulnerable center.[15] When the government project manager Lawrence Roberts had to craft the US Department of Defense's Advanced Research Projects Agency network, ARPANET, he wholeheartedly embraced Baran's distributed communication as a guiding technical norm.[16] The Internet, therefore, at the core of its conception, was a Cold War network. It was, as Janet Abbate notes, from the world of Dr. Strangelove, a fictional account of an almost fantastical system designed with a driving (even comically absurd) fear of the surprise attack that would cripple the hierarchy of command-and-control.[17]

Historians have carefully documented how, over the following decades, from the early fifties to late eighties, "decentralization" remained a watchword of networked computing. In many of these stories, the Cold War order was not limited to the bombsight fear of the center but mutated to fashion adjacent meanings for decentralization. Devin Kennedy details the close links between a Cold War factory system, which was decentralized in part to prevent union organizing, and computer science.[18] Networked computing again allowed for a form of centralized control, here corporate control of the civilian workforce, through decentralized communications. Outside of the United States, the history of timesharing, packet-switching's parallel origin story, tells of Donald Davies looking for a noncentralized solution for allocating computer time in the Cold War British university.[19] In France, Louis Pouzin named the French Cyclades network for a collection of islands to evoke, as historians Valérie Schafer and Andrew Russell argue, "resisting centralized control by design."[20]

If government control was the dominant narrative for computing in the sixties and seventies, its inverse, decentralization *against* government control, has an equally robust history as the dominant historiographic frame for the history of the computer in the late seventies and eighties.[21] Fred Turner, Thomas Streeter, Steven Levy, and others have laid out how the personal computer (PC) became a symbol of Romantic resistance to the state.[22] Captured in the iconic imagery of the "1984" Apple Superbowl commercial, the PC promised to empower the individual user, in stark contrast to the monolith Big (Blue) Brother mainframe.[23] The PC, as it rapidly entered US homes, did not look like the giant centralized mainframes of the preceding decades, and it promised individualized empowerment. In these heavily marketed narratives, computers offered a story of radical individual-level control and resistance to national systems of state power.

Across meanings of decentralization there was a push and pull between "origin stories," which trapped decentralized computing in the hegemonic military culture from which it came, and narratives that raised the possibility that these same networks could be repurposed and given new meanings and uses.[24] Internet history, the history of the network of networks, has always oscillated between these poles. In the 1990s, as advocates of certain government regulations were looking for new kinds of stories, they turned to the political imagery of the Democratic Party, which had been going through its own search for identity.

The Jeffersonian Rebrand: New Democrats

As the 1980s came to a close, both networked computing and the Democratic Party were getting a rebrand. Democrats kept losing the presidency. With the brief and dispiriting exception of Jimmy Carter's short-lived tour of duty from 1977 to 1981, the White House had been off-limits to members of the Democratic Party since 1968. Centrist Democrats were increasingly raising loud criticism for the party's failures and blamed so-called special-interest groups. For these centrists, the very rainbow coalition that had been highlighted as the Democratic Party's defining positive feature prevented its members access to the presidency. Critiques of a party controlled by these special-interest groups expressed three overlapping fears, often articulated as concerns about electability.[25]

The first concern was racial. "Special-interest groups" meant the American Federation of Labor and teachers' unions, but it more often meant Black voters and civil rights groups. One man, Jesse Jackson, embodied party leadership's fears about special-interest groups better than anyone. Jackson's 1984 and 1988 presidential runs were landmark accomplishments. The first Black candidate of a major party, Jackson won primaries and caucuses in the Deep South, but the prominence of an African American candidate also signaled to some that the South was becoming increasingly unwinnable for Democrats in a general election. Boll Weevil Democrats, the older segregationists like Jim Eastland and John Stennis, critical components of the New Deal Democratic coalition, were on their way out.[26] Southern Democrats were eager to make their party more appealing to white voters by being less explicitly associated with Black civil rights.[27]

The second concern was economic. The Old Liberalism of the Great Society, centrist critics argued, put the quest for equality ahead of economic growth. Established party leaders, epitomized by Ted Kennedy, continued a tradition of government spending and robust federal programs as the primary means to address social ills. At the same time, a new class of liberals, sometimes christened neoliberals, believed that facilitating economic growth was a primary function of government. While that did not mean an embrace of trickle-down Reaganomics, vocal neoliberal Paul Tsongas made it clear that everything should "absolutely" be subordinated to, and a function of, the economy.[28] The new economic order replaced the unionized factory worker with the knowledge worker and the financier as the archetypal worker fueling the American economy.[29]

The final concern was aesthetic. Democrats yearned for symbols of generational change: younger politicians, associated with new ideas, new suits, and fresh smiles. They needed to replace the aged liberal legislators with a fresher model. Many of this new generation saw cutting-edge technologies as the ideal symbol for the path forward. So-called

Atari Democrats recognized that the research corridors of Silicon Valley and Boston's Route 128 offered not only donors but also an inspiring future-oriented vision for the nation.[30] Searching for youth, the party found rejuvenated standardbearers in Bill Clinton and Al Gore, two young Southerners with good heads of hair.

The three core components of Jeffersonianism for 1990s centrist Democrat politics were anti-Black policies, noncentralized economics, and the bright veneer of Americana. This rightward movement of the Democrats is part of the dominant historiographic metanarrative for this period that charts the rise of conservativism.[31] Indeed, much of the party did and would continue to move right. But the rightward pull was always tempered by a counterforce, a conviction that some things about the Democratic Party (and elements of old-fashioned liberalism) were worth saving. For many centrists, concessions to Reaganite conservativism as an economic, racial, and aesthetic regime were conceived of as tactical attempts to resist that very regime. The Democratic Party was driven by a centrist logic of moving rightward enough to draw votes from the Republicans and save the left.

The Clinton campaign was quite conscious about crafting a narrative for the Democrats' potential accession to the White House. For the purposes of this chapter, one nostalgic aspect of that forward-looking messaging stands out: the use of Thomas Jefferson. The candidate—whose middle name, everyone was to hear, was Jefferson—made use of the connection easy. By 1992, as Democrats met to nominate Clinton and Gore, one outlet captured the message with exactly the type of headline centrists were looking for: "Jefferson's Political Heirs Gather in New York City: Democratic Party Convention Will Pin Hopes for Nov. 3 Win on a Son of the Agrarian South."[32]

Republican Ronald Reagan had understood the power of Jefferson imagery to undergird economic policy. Reagan announced his Economic Bill of Rights from the Jefferson Memorial in July 1987. Assured that "if Thomas Jefferson were here, he'd be one of the most articulate and aggressive champions of this cause," Reagan pushed the balanced budget amendment, which would drastically reduce the government spending on social welfare.[33] It was the literal towering silhouette of the third president who would frame Reagan, the fortieth, as red, white, and blue balloons were released. In their campaign, Clinton and Gore were eager to reclaim that iconography as they also embraced balancing the federal budget. The fight over Jefferson was so recognizable that it led to one of former President Reagan's biggest applause lines during speech at the 1992 Republic nominating convention. Addressing Bill Clinton, Reagan riffed off an old debate line: "I knew Thomas Jefferson. He was a friend of mine. Governor, you're no Thomas Jefferson." In response, "the crowd almost blew the roof off the Houston Astrodome with laughter and applause."[34] Despite Reagan's jabs, William Jefferson Clinton won the election and, in January of 1993, it was decided the Clinton and Gore families would hold one final event before their inauguration: a visit to Thomas Jefferson's home plantation in Monticello. From there, they would board a bus to Washington.

The Jeffersonian Rebrand: Networked Computing

Just as centrist Democrats worked to realign their party's relationship with "Big Government," so too did those interested in the future of the TCP/IP Internet. Between 1988 and 1990, the operation of the Internet backbone was transferred from the Advanced Research Projects Agency to the National Science Foundation. Shortly thereafter, Gore sponsored the High-Performance Computing Act of 1991, which sought to fund a federally run National Research and Education Network (NREN) that would give Americans access to the Internet. In the period 1993–1995, amid contentious debate, the network was both turned over to private operation and significantly upgraded in terms of capacity, obviating the need for a separate NREN. In this context, debates on decentralization were at once proposals for network design and political ideologies.

As discussed above, historians of computing have emphasized two opposing strategies for rendering decentralized computing politically legible. The first narrative was one of government control, epitomized by command-and-control computing. The second was one of antigovernment liberation, epitomized by the hacker.[35] Both politicians and activists in the 1990s found these narratives lacking. Neither sufficiently answered the question of what nonmilitary government involvement with network computing could be. Neither offered a compelling hook for New Democratic politics.

While he was still the junior senator from Tennessee, Al Gore Jr., the quintessential New, or Atari, Democrat, hoped technological innovation would form the foundation for his centrist vision of liberalism. Gore did not believe that the private sector would build the appropriate high-speed information network to serve the American public. A fully privatized network, he feared, would end up as a "balkanized system, consisting of dozens of incompatible parts." And that was only if there *were* private investors; Gore and his advisors feared that there would be insufficient private capital investment in a National Information Infrastructure to make such a network feasible. There was, he pointed out, little clear demand for nationwide networked computing until its utility had been demonstrated. The role of the federal government was to provide "federal seed capital" that would activate private-sector investment.[36] Here was a new liberalism that seemed to find a new role for the federal government, as the leader and coordinator of private dollars through federal investment. It rebuked both the old liberalism of a giant government-operated network and the Reaganite preference for a privatized network.[37] Still, Gore needed narratives to sell new policies.

In a 1991 special issue of *Scientific American,* Gore and his staff penned an essay entitled "Infrastructure for the Global Village." To sell voters on the idea of federal seed capital, he drew an analogy with farm policy. US information policy, Gore wrote, resembles "the worst aspects of our old agricultural policy, which left grain rotting in thousands of storage silos while people were starving. We have warehouses of unused information rotting." That information, he believed, would have economic value if it were part of a network. A federally subsidized network would be the infrastructure for commerce. His other rhetorical touchpoint, the decentralized interstate highway system, was not just a convenient metaphor, it was a model of federal construction and development undergirding capitalist growth. It was a vision of the federal government's role that Gore traced to his father, Tennessee senator Al Gore Sr., who had pushed through interstate highway legislation.

The same dispersed highway that had been understood through the bombsight view became a road to take products to market. Gore's position was at odds with both those who wanted a publicly run network and those who thought its construction should be left to the cable and telecom giants. Gore's team was convinced that private industry alone simply would not take the gamble to make more than a private toll road for the elite. He wanted a federal information highway system that leveraged private dollars but ensured broader access.[38]

Gore told a story about the fundamentally similar structures of computing, capitalism, and liberal democracy. Decentralized networks in this framing were quintessentially American.[39] Representative democracy and capitalism relied on citizens functioning as decentralized information processors: "The unique way in which the U.S. deals with information has been the real key to our success. Capitalism and representative democracy rely on the freedom of the individual, so these systems operate in a manner similar to the principle behind massively parallel computers. These computers process data not in one central unit but rather in tiny, less powerful units distributed throughout the computer. Capitalism works on the same principle."[40] For Gore, decentralized computing was All-American not because it was a system to protect the state from attack or to resist the state through individualism. It was All-American because it enabled a certain form of liberal capitalism. Notably, Gore was not making an antigovernment argument nor simply doubling down on laissez-faire capitalism. For Gore in 1991, networked computing would be the infrastructure for capitalism, the roads that got the grain to market, not the object of capitalism, the auctioned-off rights to build the road. Decentralization was not in opposition to the government; a decentralized network would be precisely what the US federal government needed to provide. It was the manner in which the United States, according to Senator Gore, "should lead."[41]

In *Scientific American*, Gore placed himself within striking lineage, not just of his own father, but that of the third president. He writes, "Gutenberg's invention, which so empowered Jefferson and his colleagues in their fight for democracy, seems to pale before the rise of electronic communications." It was by no means obvious that Gutenberg and Thomas Jefferson go together—maybe Thomas Paine—but the shoehorning of Jefferson was rather odd outside of its larger political context. For Gore, following in Jefferson's footsteps meant embracing the notion that "a high-capacity network will not be built without government investment."[42]

A set of technology policy advocates, most notably the Electronic Frontier Foundation (EFF), would follow Gore's lead. In that same issue of *Scientific American*, Mitchell Kapor, co-founder of EFF, wrote a piece on civil liberties in cyberspace. In it, he does not speak in the language of the Founding Fathers when he tells the story of the Secret Service raiding the office of Steve Jackson, who operated an electronic bulletin-board system, but takes a more legalistic frame.[43] Still, what Kapor and others understood so well is that even at its most anti-statist, EFF needed to frame the Internet as something the state could understand, as an entity that fit within the frameworks of both powerful political parties. As two Jeffersonian Democrats, Bill Clinton and Al Gore, won back the White House for the Democratic Party, Kapor moved the office of the Electronic Frontier Foundation from Massachusetts to Washington, DC, and by 1993, Kapor and his colleagues were organizing around the word "Jeffersonian."

In March 1993, Kapor wrote, "Life in cyberspace might be shaping up exactly as Thomas Jefferson would have wanted: founded on the primacy of individual liberty and a commitment to pluralism, diversity, and community."[44] Wrapped in all that patriotic language, Kapor was well aware that "the emerging consensus between business, government, and policy watchdogs" was fragile. Just a few months earlier, at a policy summit in Little Rock, Arkansas, Al Gore and the CEO of AT&T had debated exactly what role the federal government would have in building the information superhighway.[45] Gore and the Clinton administration struggled with this question, and, perhaps, in the winter of 1993 the US Internet could have evolved into a more nationalized system like France's Minitel.[46]

Kapor wanted a clear role for the federal government, involved but slightly less involved than Gore had envisioned. His summary of the emerging consensus worked to draw that line. The network would be private, not public, he argued. The federal government's role would, and should, be "limited to funding research, leading experiments with ultra-high-speed networks, helping promote standards, and protecting the public interest in privacy, freedom of speech, and other areas."[47] That was quite a large set of responsibilities. Historians of computing have paid particularly close attention to the role of standards in creating the contemporary Internet, not to mention legal norms.[48] Still, the cables would be private. Telephone and cable companies, not the federal government, would be "the principal carriers of traffic into the home."[49] This compromise built on Gore's earlier rhetoric and the larger New Democratic plans, but it was still hard-fought. Yes, there would be federal standards, and federal laws, and, though glazed over, many federal tax incentives, but the core infrastructure would be private.

Privatizing the National Science Foundation's Network (NSFNET) which would eventually occur in 1995, was not a rejection of the state but a compromise with it. To make this compromise work, it was aligned with the dominant politics of the time. Kapor made sure to underscore that this was Jeffersonian: "We have a vision that is, we think, true to the spirit of Thomas Jefferson, 250 years old last week ... Instead of a kind of centralized government, where one size fits all, the Jeffersonian view is very decentralized, very non-hierarchical. It says let people do their own thing and let groups of people be free, on a voluntary basis, to do their own thing."[50] This "Jeffersonian" was both capacious and specific. It endorsed Baranian decentralization and the hacker anti-statist instinct, but in a way that laid out bounds for what the government should do with words that would resonate with the ascendant Democratic Party. Talk of Jefferson was beneficial to both the Clinton–Gore administration and the Electronic Frontier Foundation. The narrative structure worked alongside racialized invocations of the digital divide that would guide subsequent policy making.[51] Hardly depoliticized, the Internet was political in exactly the ways that appealed to a centrist understanding of the white electorate.

Later that year when Kapor appeared before the Senate Commerce Committee to testify, his testimony was boiled down to a few sound bites for the evening news, but "Jeffersonian Revolution" made the cut.[52] In Newsday, another EFF leader offered his commentary: "Consider the Internet. No one is in control. You can't go to a central office and say, 'Stop doing that.' This was the start of "a 'Jeffersonian' blossoming of online freedom."[53] By 1994, the Jeffersonian ideal already evoked what had been promised but not achieved by a National Information Infrastructure. The sporadic investment from telecom companies was described as "a far cry from Vice-President Gore's Jeffersonian dream of universal—and affordable—access."[54] A disgruntled University of Colorado computer scientist was concerned about all the Jefferson talk: "The Clinton administration is paying a lot of lip service to the idea that the [information superhighway] will promote a version of Jeffersonian Democracy in which citizens will be able to participate in the political process through taking part in informed on-line debates."[55]

By 1995, the Jeffersonian Internet was so legible, it could articulate a meaningful Third-Way politics for the Internet that made sense in both Britain and the United States. A questioning heading from the *Guardian* read: "What will Tony Blair's promises of new technology and digital socialism actually mean in the future in schools, hospitals, and in your home?" Labour Party leader Tony Blair's "new, evangelical appeal to information technology," the paper explained, "actually owes more to Al Gore" than to Blair's own left-wing predecessors in the Labour Party. "The US deputy president has been preaching 'Jeffersonian' principles of cheap and universal access to the information highway for over a decade . . . It is from these Jeffersonian principles that Blair derives his form of digital socialism."[56]

Still, the political winds were again changing by 1996, and so was the image of Jefferson. Always aware that it wore the costumes of a previous revolution, the Electronic Frontier Foundation was actively pantomiming Thomas Jefferson when John Perry Barlow penned the Declaration of the Independence of Cyberspace. By then the EFF had a more aggressive stance toward the Clinton–Gore administration. It was the new speaker of the house, Newt Gingrich, who was trying on Jefferson's legacy, courting the readers of *Wired*, and proposing a more libertarian, anti-statist, politics for the Internet.[57]

It was Gingrich's conservative take on Jefferson's vision with which David Higgs, the father from Weymouth who opened this chapter, aligned himself in 1996 by leaking information from the school board online. Almost all of the documents on "Friends of the Weymouth Schools" came from two members on the seven-person panel and promoted their view that the textbooks were too critical of the United States. It was their perspective that pushing for an "America First" curriculum would "promote patriotism and a greater appreciation of American culture."[58] "Friends of the Weymouth Schools" advocated for charter schools and school choice: a rather Jeffersonian freedom, echoing the white supremacist roots of the term. These "Friends of the Weymouth Schools" held on to the politics of an empire of liberty with freedom from liberal centralized oversight and hopes for a more nationalist curriculum.

Conclusion

Though the TCP/IP protocol remains core to the conceptual and operational reality of the contemporary Internet, historians of computing have been insistent that technological choices, even powerful standards, were not sufficient to constitute a distributed network. A decentralized network required social rules. These were expectations expressed in manuals, conferences, listservs, and the code itself. They were norms implemented by local institutions and massive corporations. And they were political. It is not simply that the ARPANET was a US government system or that TCP/IP solved the collective action problem of unifying networks only because of government power.[59] The privatized infrastructure of the 1990s was the product of political work: the efforts of politicians and those whose power grew from operating in an explicitly political idiom. Politics, as the mechanism for organizing the network, relied on abstracted appeals to Americana and veiled references to racist economic and social arrangements. Reading the political, as it lies on the surface of technology, as actors directly described it, reveals the white centrist politics that were always right in front of us. As historians of computing become ever more able to parse the hidden meaning of complex systems and hidden biases, Jeffersonian decentralization serves as a reminder, to listen to people when they tell us what and whom they declared self-evidently important. Sometimes, the technopolitics of the Internet turn out to be good old-fashioned politics.

Notes

1. Erin Lee Martin, "Dad Puts School Facts on Internet," *Patriot Ledger,* January 30, 1996, City edition.
2. Martin, "Dad Puts School Facts on Internet."
3. On the political economy of design choices, see Martin Campbell-Kelly and Daniel Garcia-Swartz, "The History of the Internet: The Missing Narratives," *Journal of Information Technology* 28, no. 1 (March 2013): 18–33; Janet Abbate, "Government, Business, and the Making of the Internet," *Business History Review* 75, no. 1 (2001): 147–76. On the homeomorphism between networked computing and political ideology, see especially Eden Medina, *Cybernetic Revolutionaries: Technology and Politics in Allende's Chile* (Cambridge, MA: MIT Press, 2011); Benjamin Peters, *How Not to Network a Nation: The Uneasy History of the Soviet Internet* (Cambridge, MA: MIT Press, 2016); Hallam Stevens, "From RangKoM and JARING to the Internet: Visions and Practices of Electronic Networking in Malaysia, 1983–1996," *Internet Histories,* May 2021, 1–18.
4. Merrill Peterson, *The Jefferson Image in the American Mind* (Charlottesville: University of Virginia Press, 1998), 443.
5. Brian Balogh, *A Government Out of Sight: The Mystery of National Authority in Nineteenth-Century America* (Cambridge: Cambridge University Press, 2009).
6. Sven Beckert and Seth Rockman, eds., *Slavery's Capitalism: A New History of American Economic Development* (Philadelphia: University of Pennsylvania Press, 2016); Walter Johnson, *River of Dark Dreams: Slavery and Empire in the Cotton Kingdom* (Cambridge, MA: Belknap Press, 2017); Ned and Constance Sublette, *The American Slave Coast: A History of the Slave-Breeding Industry* (Chicago: Lawrence Hill Books, 2016); Edward E. Baptist, *The Half Has Never Been Told: Slavery and the Making of American Capitalism* (New York: Basic Books, 2014).
7. Andrew Burstein, *Democracy's Muse: How Thomas Jefferson Became an FDR Liberal, a Reagan Republican, and a Tea Party Fanatic, All the While Being Dead* (Charlottesville: University of Virginia Press, 2015), 65.
8. Burstein, *Democracy's Muse,* 8.
9. As historian of cryptography Gili Vidan has noted, decentralized and distributed are often conflated. Vidan, "Decentralization: The Rise of a Hazardous Spec," *Just Money,* June 2020, https://justmoney.org/g-vidan-decentralization-the-rise-of-a-hazardous-spec/. This chapter tries to use the actors' demarcation between the two, or lack thereof.
10. Paul Baran, "On Distributed Communications" (Santa Monica, CA: The RAND Corporation, August 1964), 2; B. Fidler, "Cryptography, Capitalism, and National Security," *IEEE Annals of the History of Computing* 40, no. 4 (October 2018): 80–84.

11. Baran, "On Distributed Communications," 1.

12. Joy Rohde, "Pax Technologica: Computers, International Affairs, and Human Reason in the Cold War," *Isis* 108, no. 4 (December 2017): 792–813; Paul Edwards, *The Closed World: Computers and the Politics of Discourse in Cold War America* (Cambridge, MA: MIT Press, 1996).

13. Daniel Volmar, "The Computer in the Garbage Can: Air-Defense Systems in the Organization of US Nuclear Command and Control, 1940–1960" (PhD diss., Harvard University, 2018), 6.

14. Baran, "On Distributed Communications," 1.

15. Peter Galison, "War against the Center," *Grey Room* no. 4 (2001): 5–33; Jennifer Light, *From Warfare to Welfare: Defense Intellectuals and Urban Problems in Cold War America* (Baltimore, MD: Johns Hopkins University Press, 2005).

16. Arthur Norberg and Judy O'Neill, *Transforming Computer Technology: Information Processing for the Pentagon, 1962–1986* (Baltimore, MD: Johns Hopkins University Press, 1996).

17. Janet Abbate, *Inventing the Internet* (Cambridge, MA: MIT Press, 1999).

18. Devin Kennedy, "Manufacturing Networks: The Industrial Politics of US Computer Science" (Working Paper, Society for the Social Study of Science, New Orleans, September 2019); Joshua Freeman, *Behemoth: A History of the Factory and the Making of the Modern World* (New York: W. W. Norton, 2018).

19. Abbate, *Inventing the Internet*, chap. 1.

20. Andrew Russell and Valérie Schafer, "In the Shadow of ARPANET and Internet: Louis Pouzin and the Cyclades Network in the 1970s," *Technology and Culture* 55, no. 4 (2014): 886.

21. Bryan Pfaffenberger, "The Social Meaning of the Personal Computer: Or, Why the Personal Computer Revolution Was No Revolution," *Anthropological Quarterly* 61, no. 1 (January 1988).

22. Fred Turner, *From Counterculture to Cyberculture: Stewart Brand, the Whole Earth Network, and the Rise of Digital Utopianism* (Chicago: University of Chicago Press, 2006); Thomas Streeter, *The Net Effect: Romanticism, Capitalism, and the Internet* (New York: New York University Press, 2011); Katie Hafner and Matthew Lyon, *Where Wizards Stay Up Late: The Origins of the Internet* (New York: Touchstone Books, 1998); Steven Levy, *Hackers: Heroes of the Computer Revolution* (Garden City, NY: Anchor Press, 1984); Christopher Kelty, "The Fog of Freedom," ed. Tarleton Gillespie et al. (Cambridge, MA: MIT Press, 2014), 195–220.

23. Ridley Scott, "1984," Apple Computer Advertisement (CBS Sports, January 22, 1984).

24. See Joy Lisi Rankin, *A People's History of Computing in the United States* (Cambridge, MA: Harvard University Press, 2018); Andrew Pickering, *The Cybernetic Brain: Sketches of Another Future* (Chicago: University of Chicago Press, 2011); Donna Haraway, "Cyborg Manifesto," in *Simians, Cyborgs, and Women: The Reinvention of Nature* (New York: Routledge, 1991).

25. The question of electability dominates journalists' and New Democrats' own narration of the period. Steve Kornacki, *The Red and the Blue: The 1990s and the*

Birth of Political Tribalism (New York: Ecco, 2018); Kenneth Baer, *Reinventing Democrats: The Politics of Liberalism from Reagan to Clinton* (Lawrence: University Press of Kansas, 2000); Jon Hale, "The Making of the New Democrats," *Political Science Quarterly* 110, no. 2 (1995): 207–32.

26. Political scientists have been particularly attuned to the relatively rapid shift of the US South from a one-party Democratic system to a one-party Republican system by the start of the twenty-first century. Bruce Schulman, *From Cotton Belt to Sunbelt: Federal Policy, Economic Development, and the Transformation of the South, 1938–1980* (Durham, NC: Duke University Press, 1994); Alexander Lamis, ed., *Southern Politics in the 1990s* (Baton Rouge: Louisiana State University Press, 1999).

27. New Democrats' most aggressive efforts to distance themselves from racial egalitarianism were the "wars" on crime and on "welfare as we know it." Marisa Chappell, *War on Welfare: Family, Poverty, and Politics in Modern America* (Philadelphia: University of Pennsylvania Press, 2011), and Elizabeth Hinton, *From the War on Poverty to the War on Crime: The Making of Mass Incarceration in America* (Cambridge, MA: Harvard University Press, 2016).

28. Randall Rothenberg, "The Neoliberal Club," *Esquire*, February 1982, 40; Paul Tsongas, *The Road from Here: Liberalism and Realities in the 1980s* (New York: Knopf, 1981).

29. Recently historians have disentangled the liberal economics of centrist Democrats from the larger move toward finance. Lily Geismer, "Agents of Change: Microenterprise, Welfare Reform, the Clintons, and Liberal Forms of Neoliberalism," *Journal of American History* 107, no. 1 (June 1, 2020): 107–31; Brent Cebul, "Supply-Side Liberalism: Fiscal Crisis, Post-Industrial Policy, and the Rise of the New Democrats," *Modern American History* 2, no. 2 (July 2019): 139–64.

30. Both Margaret O'Mara and Lily Geismer, anchoring their studies on different coasts, have been particularly effective at documenting the political potency of technological research and development for the Democratic Party. O'Mara, *The Code: Silicon Valley and the Remaking of America* (New York: Penguin Press, 2019); Geismer, *Don't Blame Us: Suburban Liberals and the Transformation of the Democratic Party* (Princeton, NJ: Princeton University Press, 2015).

31. K. Phillips-Fein, "Conservatism: A State of the Field," *Journal of American History* 98, no. 3 (December 1, 2011): 723–43.

32. "Jefferson's Political Heirs Gather in New York City," *Christian Science Monitor*, July 13, 1992.

33. Ronald Reagan, "Remarks Announcing America's Economic Bill of Rights" (Washington, DC, July 3, 1987).

34. Tom Shales, "Reagan, Back in from the Sunset," *Washington Post*, August 18, 1992.

35. Wendy Hui-Kyong Chun argues that we can understand these two poles as bound together, a state of "control-freedom" in cyberspace, in *Control and Freedom: Power and Paranoia in the Age of Fiber Optics* (Cambridge, MA: MIT Press, 2006).

36. For an excellent account of interplay between regulatory bodies and the private sector, see chapter 2: "Honest Policy Wonks," in Shane M. Greenstein, *How the Internet Became Commercial: Innovation, Privatization, and the Birth of a New Network* (Princeton, NJ: Princeton University Press, 2015).

37. On the role of public-private partnerships see contemporaneous accounts, like the popular work of David Osborne, *Laboratories of Democracy* (Boston, MA: Harvard Business School Press, 1988), and Lily Geismer, *Doing Good: The Democrats and Neoliberalism from the War on Poverty to the Clinton Foundation* (forthcoming).

38. Reed Hundt, *You Say You Want a Revolution* (New Haven, CT: Yale University Press, 2000).

39. For the broader history of the informed citizenry through networked information, see the work of Richard John, especially *Spreading the News: The American Postal System from Franklin to Morse* (Cambridge, MA: Harvard University Press, 1995).

40. Al Gore, "Infrastructure for the Global Village," *Scientific American* 265, no. 3 (September 1991): 150.

41. Gore, "Infrastructure for the Global Village," 150.

42. Gore, 150.

43. Mitchell Kapor, "Civil Liberties in Cyberspace," *Scientific American* 265, no. 3 (September 1991): 158–64.

44. Mitchell Kapor, "Where Is the Digital Highway Really Heading? The Case for a Jeffersonian Internet Policy," *Wired*, March 1993, 57.

45. John Markoff, "Building the Electronic Superhighway," *New York Times*, January 24, 1993.

46. Julien Mailland and Kevin Driscoll, *Minitel: Welcome to the Internet* (Cambridge, MA: MIT Press, 2017).

47. Kapor, "Where Is the Digital Highway Really Heading?" 54.

48. Andrew Russell, *Open Standards and the Digital Age: History, Ideology, and Networks* (Cambridge: Cambridge University Press, 2014).

49. Kapor, "Where Is the Digital Highway Really Heading?" 54.

50. Jeff Goodell, "Mitch Kapor: Civilizing Cyberspace," *Rolling Stone*, June 10, 1993.

51. Daniel Greene, *The Promise of Access: Technology, Inequality, and the Political Economy of Hope* (Cambridge, MA: MIT Press, 2021), chap. 1; Alondra Nelson, "Future Texts," *Social Text* 20, no. 2 (2002): 1–15; Rayvon Fouché, "From Black Inventors to One Laptop Per Child: Exporting a Racial Politics of Technology," in *Race after the Internet*, ed. Lisa Nakamura and Peter Chow-White (New York: Routledge, 2012), 61–84; Jennifer Light, "Rethinking the Digital Divide," *Harvard Educational Review* 71, no. 4 (2001): 709–34.

52. *World News Tonight with Peter Jennings* (ABC, September 30, 1993).

53. Jonathan Schell, "The United States Puts Liberty on Call Waiting," *Newsday*, August 11, 1994.

54. Victor Keegan, "All Wired Up for Confusion," *The Guardian*, October 28, 1994.

55. Jennifer Brown, "Big Dollars at the Heart of Information Fray," *Inter Press Service*, June 30, 1994.

56. Victor Keegan, "Labour on Net," *The Guardian*, October 5, 1995.

57. O'Mara, *The Code*, chap. 21.

58. Robert Lee, "Conservative Reformer Won't Run Again," *Patriot Ledger*, April 12, 1997.

59. See Peters, *How Not to Network a Nation*.

CHAPTER 3

Beyond the Pale

The Blackbird Web Browser's Critical Reception

André Brock

> There is no sense to be made of a computer or any other technical
> artefact outside of the social conditions that constitute it as a
> determinate object of experience.
>
> G. KIRKPATRICK, Critical Technology

As 'Xerox' became a generic term for photocopies, the web browser has become the sign for the internet. When people say, 'I was on the internet,' they take for granted that their friends will understand that they used a web browser to access the web. Although early adopters and power users may scoff at this synecdoche, it is not difficult to see why users would understand a complex assemblage of hardware and software through their use of a particular application. Browsers frame the content, media, and protocols we know as the World Wide Web; they have become part of our social life, our work routines, and our leisure activities. As such, the browser is a cultural artifact, defining its users as technologists and as social actors. It is part of our communicative infrastructure, invisible to our information literacy practices until a rupture occurs.

One such rupture found form in the introduction of 40A's Blackbird browser, designed to serve the browsing needs of African Americans. Blackbird's cultural focus engendered a scathing response from Black

Abridged from *new media and society* 13, no. 7 (2011): 1085–1103.

and White internet denizens alike because it apparently contravened popular assumptions of the browser's cultural neutrality. Where most tech products are evaluated in terms of their ease of use or feature set, Blackbird's reception as an ICT artifact was 'colored' by the racial frames of the pundits, bloggers, and commenters who discussed it.

How can these cultural and racialized responses to a technological artifact be understood? . . . Elsewhere, I have argued that technologies should be examined by interweaving a structural analysis of an artifact with a discourse analysis of the cultural means through which users interpellate themselves within relations to the artifact.[1] (By interpellation, I am referring to the Althusserian definition, where ideology (in this case, technology) transforms individuals into subjects through framing and individuals construct themselves through their practices, contexts, or ideologies.)[2] Doing so combines insight into the cultural biases within the culture of the users. Through this approach racial motivations can be apprehended as functional rationales for technology use.

The blogs examined here were critical of Blackbird's feature set for a number of practical reasons, but also because of a number of shared beliefs about what information technology in the age of Web 2.0 should do. In this, they highlight constructions of technocultural identity shaped around ICT practices and technological determinism. Racial frames, however, also shape these technocultural identities. Of particular interest for this article is how, through the racial intentions of the browser, various respondents mediated racial identity through their articulations of information technology. By examining how these web users interpellate identity and technology through their American cultural-framed racial affiliation, we can gain a greater understanding of how belief and ideology shapes information technology use, implementation, and design.

Critical Frameworks: Technocultural and Racial Formation Theory

This article employs an approach called Critical Technocultural Discourse Analysis. CTDA interrogates how ICTs are interpellated by their users by examining an ICT artifact's material aspects, associated practices, and the discourses in online fora of those who use the artifact; these observations are analyzed through a critical race framework. CTDA is a method I have developed to address gaps in humanist analyses of ICT use in information studies. These analyses resort to studying either the ICT artifact or the community of technology users. The first approach leads to instrumental understandings of technology use (e.g. an artifact is only as powerful as its features). The latter approach reduces ICT users to discourse communities that can only be understood based on their technology practices (e.g. how they use the artifact stands as the sole category of identity). While both approaches have their benefits, neither is capable of addressing how ICT artifacts integrate cultural values nor how ICT users employ cultural values while using computational technologies.

Nakamura suggests a different approach for cybercultural studies, where analysts should employ 'critical race theory and theories of cultural difference and close visual analysis of popular Internet objects.'[3] I operationalized her suggestion by drawing heavily on Herring's computer mediated discourse analysis and Van Dijk's critical discourse analysis of racism in media.[4] CTDA has been used to examine the intersection of new media, old media, and race on the *New York Times* opinion blogs; as well as race and public representation in online media.[5]

For Blackbird, I conducted an interface analysis of the browser, briefly reviewed the history and practices of web browsers in general and Blackbird specifically. I also carried out a close reading of blog posts and their associated comments on mainstream and Black-affiliated technology blogs to understand how they constructed the browser in terms of their experience and identities. My close reading focused upon instances of discursive association between culture and technology—positive, neutral, or negative—across the general categories of the blogs they were found upon. These observations and analyses were laid out against a critical race framework that integrates elements of technocultural belief as set forth by James Carey and other scholars of technology.

Carey argued that communication technologies reproduce culture through the transmission of beliefs encoded within information.[6] He added that communication technologies exert ideological (and material) control over time and space while cementing Western imperialism and dominance. Dinerstein extends Carey's argument by adding that American technoculture can be understood as a matrix of six qualities: progress, religion, modernity, Whiteness, masculinity, and the future.[7] These qualities foreground rationales of White technology use and control of the natural and man-made world at the expense of non-Whites.

Pacey conceptualized technology as a tripartite entity comprised of an artifact, its associated practices, and the beliefs encapsulated within those practices and material aspects.[8] He added that popular conceptions of technology neglect the belief aspect. This value-neutral perspective preserves existing hierarchies and structures of domination by obscuring the ideological applications, practices, and beliefs encoded within. . . .

Although the above observations address technology, they can also be viewed as observations about the social construction of racial identity. The rhetorical narrative of 'Whiteness as normality' configures information technologies and software designs. For example, video game protagonists are overwhelmingly White and male[9] in an artistic medium that markets itself as creating 'realistic' characters, graphics, and environments. Omi and Winant contend that race is a matter of social structure and cultural representation, or racial formation.[10] Combining this contention with Pacey's tripartite formulation, we find that racial formation is an undeniable cornerstone of technology: articulated in belief, enacted through certain practices, and embodied in material design.

In the United States, American identity (in particular, Whiteness) is bounded and extended by negative stereotypes of Black identity.[11] Giroux adds that, 'whiteness represents itself as a universal marker for being civilized and in doing so posits the Other within the language of pathology, fear, madness, and degeneration.'[12] Dyer contends that White identity is founded upon a paradox of simultaneously being an individual agent and a representation of the university subject.[13]

I contend that the Western internet, as a social structure, represents and maintains White, masculine, bourgeois, heterosexual and Christian culture through its content. These ideologies are translucently mediated by the browser's design and concomitant information practices. English-speaking internet users, content providers, policy makers, and designers bring their racial frames to their internet experiences, interpreting racial dynamics through this electronic medium while simultaneously redistributing cultural resources along racial lines. These practices neatly recreate social dynamics online that mirror offline patterns of racial interaction by marginalizing women and people of color. While other cultural content is available, the cultural representations available on the internet—where a few corporate-owned websites command the majority of internet visits and page views—obscure alternative perspectives through sheer volume.

Information technologies are particularly susceptible to being influenced and mediated by racial and cultural identity. The browser represents an implicitly unmarked technological space, where each internet surfer configures their browser to conform to their own personal browsing habits through bookmarks, cookies, add-ons, and user scripts. The openness of the platform obscures the reality that most content available through the browser and its technological implements still employs Western notions of race, gender, and class.

The Web Browser

Browsers have not evolved much since the introduction of Mosaic in 1993. Mosaic's graphical user interface (GUI) is still widely copied by today's browsers, while its innovative integration of graphical and textual information transformed the way that people experienced the internet. Although markup languages and media frameworks have evolved since Mosaic's demise, the purpose of the GUI browser remains unchanged: to obscure the technical complexity of the internet's protocols while delivering content mediated by an innocuous, graphically pleasing interface.

Browsers are general-purpose applications designed to display any variety of multimedia resources (print, image, audio, video, code) linked to a specific Uniform Resource Locator. Browsers frame access to the type, amount, and quality of information available online and today's browsers allow near-infinite personalization of interface elements.

. . . The browser's bifurcated universal and individual identity maps closely to Dyer's definition of Whiteness. The universal traits of the browser are evidenced through its electronic and computational command of temporal, geographic, and economic networks. The widespread adoption of the browser as a communication device obscures its utility as an artifact of communication networks aiding in bolstering economic and sociocultural hegemony. Moreover, the browser implies the exercise of democratic values and universal (access to) information through the display of unlimited content. In reality, however, 'democracy' is simply a disguise by volume of the cultural hegemony of content designed by and for mainstream audiences.

From an individual perspective, current generation browsers enact the personalization of web use. Users can change icon sets and themes, block advertising, and even alter style-sheets of existing websites. They are encouraged to store confidential information (password managers) and even maintain a library of notable sites (bookmarking). Thus, while many people use the same browsing software few will experience the web in the same way. The dual experience of universal application and individual preferences, then, prejudices users to assume that the 'universal' web, configured to their liking, is similarly configured for every other user.

The browser should be understood as a software artifact but also as a communication product. Baker argues that communication products offer different benefits depending upon the role of the various stakeholders involved in its design, dissemination, and consumption.[14] . . . Baker adds:

> Even more than its direct value to its audience, the media's greatest value may be for third parties. Even if they do not consume the media content themselves, they can be wonderfully or gravely affected by the media's influence on its audiences' 'construction of reality' and on their resulting behavior.[15]

. . . Although the browser does not directly represent or embody the content displayed within its interface, its role in delivering and framing that content connects it ideologically to the cultural values transmitted within the content.

Baker notes that content commodification migrates traditional information transmission practices from a private sphere with an assumed ethos of caring to a commercial sphere where the ethos is based upon the transactional value of the information (e.g. 'Can we make money from this interaction?'). For minority and underserved groups searching for information online, commodification makes it exponentially difficult to find information that serves their needs, which leads in turn to the introduction of an information application specifically designed for needs of a particular audience.

Blackbird

The Blackbird browser was marketed as addressing the difficulties of finding content oriented towards the information needs and interests of African Americans. Constructed from Firefox's open source codebase, it is aesthetically and structurally similar to the Flock (social networking), and Gloss (women-centric) Firefox-variant browsers. Each variant features custom interface tweaks ('chrome') designed to visually identify the browser, as well as plug-ins, custom searches, and other tweaks designed to enhance the targeted user's experience. Blackbird's creators included social networks such as Facebook and MySpace (but interestingly, not BlackPlanet) to leverage existing accounts and allow African American users to browse their network and the web simultaneously.

Visually, the browser uses a black theme with red accents and white-on-black buttons. By default, the custom features are enabled and visible, adding two toolbars (a ticker and a set of large buttons) to the interface. The layout still resembles a standard Firefox layout, however, with the search bar and the address bar sharing toolbar space and another toolbar offering a selection of bookmarks. A small Blackbird logo—a raven's wing with orange tips—is placed in the upper right hand corner. The interface can get busy, as the ticker streams RSS feed items across the top of the screen and a notification system pops up occasionally in the lower right hand corner.

. . . Because a Blackbird install automatically imports pre-existing Firefox passwords, bookmarks, and plug-ins, but asks whether to import Internet Explorer (IE) settings, it is clear that the designers intended to leverage the growing popularity of Mozilla's browser and take advantage of Firefox's customization features. . . . Blackbird tailors the browsing experience by offering custom features designed around African American content:

- *Blackbird News Ticker*, a pre-loaded (but customizable) RSS ticker toolbar
- *Black Bookmarks*, pre-selected bookmarks featuring African American websites
- *Black Search*, a customized Google Search prioritizing African American content
- *Blackbird TV*, a customized video channel available only to Blackbird users
- *Blackbird Community*, a browser-centered social network allowing users to share content through Grapevine (a Digg clone)
- *'Give Back,'* a feature linking users to designated charities serving African American communities

Blackbird also offers web-service centered features. On the services toolbar, users can configure a button to run Yahoo Mail, Windows Live (Hotmail), or Gmail; the button offers unread email notifications and the ability to switch between accounts without resorting to a bookmark or the address bar. Users can also take advantage of a 'social network' button accessing either Facebook or MySpace with one click. For both buttons, the active service will be represented by the appropriate logo on the button.

Blackbird's design encourages users to integrate their existing social network memberships by offering customizable presence/status notifications for monitoring their social presence while surfing other websites. Facebook can be viewed in the sidebar; the view shows the user's profile picture, status, links to the inbox and invites, as well as a friends list sortable by last updated time, status update time, profile update time, or by name. Logging into Facebook enables the aforementioned browser-oriented notification system that appears to inform users of friend activity. . . .

Blackbird's default search is a customized Google search intended to prioritize results that may be of interest to African American users. Blackbird's home page features a Google search bar and a button for 'Black Search' and 'Google Search'; this option is also available from the toolbar. Using the term 'Barack Obama,' I conducted a comparative search and generated the following results. Table 3.1 lists the first ten URLs returned for each search. The Blackbird search gives greater weight to information coming from sites like BlackAmericaWeb (the internet home of the Tom Joyner Morning Show), BlackVoices (AOL's portal for Black news and lifestyle information), Black Entertainment Television, and Black Enterprise magazine's website. Further investigation revealed that these particular URL results did not show up in the first 50 pages (500 results) of the Google Search, although entries marking Conservapedia's derogatory web page on Obama and a page

Table 3.1 Google search vs. Blackbird search: 'Barack Obama' (conducted May 2009)

Google search	Blackbird search
www.barackobama.com	www.barackobama.com
En.wikipedia.org/wiki/Barack_Obama	www.blackvoices.com/obama-watch
www.whitehouse.gov/administration /President_Obama	Blogs.blackvoices.com/category/barack-obama
www.myspace.com/barackobama	www.samefacts.com/2008/02/race-related-isms /is-barack-obama-black/
www.barackobama.com/issues	En.wikipedia.org/wiki/Barack_Obama
www.chicagotribune.com/topic/politics / . . . /barack-obamaPEPLT007408.topic	Blogs.bet.com/news/pamela/2008/ . . . /the-first -black-president-barack-obama
www.youtube.com/barackobama	www.baystatebanner.com/natl21-2009-02-12
www.reuters.com/news/globalcoverage /barackobama	www.blackvoices.com/news/election/barack -obama-celebrity-endorsers
www.barackobama.com/about	www.blackamericaweb.com/?q=articles/news /movin_america_news
My.barackobama.com	Blackenterprise.com/ . . . /barack-20-barack -obamas-social-media-lessons-for-business/

from Bossip (a Black celebrity gossip blog) were noted. Blackbird's contention that Black content can be difficult to find using regular searches seems to be valid, given the outcome of this comparison.

The Blackbird 'Grapevine,' a feature allowing members to share items to other members and vote on items of interest, is only accessible through the browser. In format, Grapevine resembles Digg.com. Each item is sorted by the date it was submitted to the site, and users can vote on items. Items can be arranged by categories or tagged and sorted by popularity in a tag cloud. When comparing Digg and Grapevine, however, it is easy to see that 40A's aim to encourage cultural content sharing is a viable strategy. Every article on the Grapevine page as of this writing mentions race or racial issues, compared with only 2 of the 20 on the Digg home page. I speculate that the cultural orientation of Blackbird's user base (plus the preloaded content served up by Blackbird's content features) helps to promote content that would otherwise be of no interest to more mainstream audiences.

Another feature of note is the 'Give Back.' Part of Blackbird's promotional strategy touts their intentions to give back to 'charitable and educational organizations that positively impact the African American community.' They do so by noting that they plan to donate 10 percent of 40A's 2009 revenue to their non-profit partners. To encourage a similar charitable spirit among their user base, Blackbird offers a 'Give Back' button in the Services toolbar. This button leads users to a page entitled the 'Do Good Channel,' which allows visitors to enter their location and find charitable organizations in their area. The organizations can be sorted by cause or by ways to participate. The Blackbird 'Do Good Channel' is a branded version of the non-profit endeavor of the same name run by good2gether, a website that offers non-profits a way to advertise their services and content on the web for free plus the ability to generate revenue by adding sponsors.

Blackbird is the first application in recent memory that encourages users through dedicated interface features to engage with philanthropic organizations. Blackbird's version of the 'Do Good' channel, like its other content, pre-selects African American oriented charities and non-profits (when compared to good2gether.com's version) but does not limit its users to selecting those organizations. 'Give Back' is impressive because internet browsers rarely offer users interface features for non-commercial interactions with the outside world, much less offer dedicated channels for charitable giving. While there are social websites and services that work to bring together people with like interests, their emphasis is on leveraging the network effects of the internet to initiate other electronic contacts and those features are not part of the browser's interface.

The features differentiating Blackbird from Firefox speak strongly to 40A's concept of embedded social networking as an electronic definition of a community. The inclusion of content specifically targeting African Americans layers a cultural definition of community on top of the software/internet instantiation, and offers a compelling visualization of the explicit integration of ethnic and technocultural practices. 40A's implementation is a criticism of the structural inequities of 'mainstream' internet content that privileges the information needs of middle-class, male, White internet users. Moreover, Blackbird's inclusion of links to charities and non-profits also speaks to a communal support model that addresses the implicit affluence of web users (those with time to surf and the wherewithal to afford the equipment) and asks them to provide support for their identified cultural communities. This is a paradigm shift, first popularized by moveon.org and other non-profit sites, where the internet's pan-location is used to leverage the power of local connections for civic gain. Blackbird's initiative in tying together non-profit organizations and African American web surfers is a powerful attempt to address the digital divide by asking a community to support its own, using information technology resources.

Online Reactions in Black and White

... As the Web has matured and reached a broader swathe of the population, its interactive nature enables discussions about technology objects that expose technocultural beliefs. ... Blackbird's launch received a fair amount of press from technology blogs as well as from blogs focusing primarily upon racial issues. To understand Blackbird's reception, using purposive sampling I selected six weblogs as examples of how ideological and cultural factors influence users' technology analyses. I found the mainstream sites using the term 'Blackbird browser' across Google, IceRocket, and Yahoo! Search engines; I chose the two blogs that were consistently among the top ten results for that search. For the African American blogs, however, traditional weblog data collection methods (e.g. Google, Technorati, Bloglines) yielded insufficient numbers of African American content blogs, as search engines and blog indexing sites severely under sample them. I turned to the recommendations of established African American bloggers, who recommended the Black interest sites listed here. To operationalize 'online discourse,' the selected blogs have a post specifically addressing Blackbird and there are comments that consistently address the same topic; there were approximately 500 comments in total across all six blogs in this analysis. The URLs of the selected blog posts are included in the endnotes for reference.

The sites I examined for this article include high-profile technology blogs (Tech Crunch and Ars Technica), Black tech blogs (Roney Smith, BlackWeb 2.0), and general interest Black blogs (AroundHarlem, The Angry Black Woman).[16] Roney Smith and BlackWeb 2.0 represent examples of race-oriented, technology-focused blogging emphasizing coverage of technology specifically impacting African Americans. These blogs do not limit themselves to African American oriented tech news, but their intent is to address the perceived lack of coverage of technology by and about African Americans. Meanwhile, AroundHarlem.com has achieved fame for coverage of New York City events, and the Angry Black Woman is a leading online voice among African American blogs addressing racism in various media. . . . For each blog, the original post and on-topic comments addressing Blackbird were analyzed. . . .

The Browser as Racial Apparatus

Responses from the Technology Blogs

Discussions of Blackbird's feature set in the technology blogs were rarely complimentary. They tended to focus on an 'ideal' browser as an information and culturally neutral space for internet consumption, configurable for individual browsing preferences. . . . For example, Tech Crunch's Robin Wauters, a White European male, mentioned Blackbird's content-based add-ons but noted that their addition did not seem like enough of an incentive for African American people to download another browser.

Tech Crunch's commenters also employed racial considerations to articulate their vision of the browser. jdb [*sic*] argued:

> No one is going to convince me that Google is White by default unless you want to argue that being simple, quick and useful is 'white.' LOL. The thing is that from an ideal perspective when a user logs onto the Internet they are starting from a 'unified' and 'unfiltered' position and choose to navigate toward targeted content. The difference here is that someone has developed a 'tool' that controls and filters the 'experience' right from the start. They've found a way to create a segregated experience.

jdb's argument neatly summarizes mainstream perceptions of the internet as a neutral cultural space. It also highlights an aspect of Tech Crunch's discursive position on technology; that information technologies are objective and it is only the intervention of certain social and cultural forces that render them as ideological tools. jdb's use of the word 'segregated' clues us in as to the types of technology considered non-normative (and thus ideological): information or tools exemplifying the interests of Black Internet users. A later comment penned by Max continues this argument:

> It's one thing to build CONTENT targeted at particular target audience . . . It's another thing to build a TOOL that essentially implies that the standard tool (regular Mozilla) is somehow 'too smart' 'too white' or otherwise not good enough for blacks. That's just insulting.

Tech Crunch's comments hosted positive racial interpretations of Blackbird's potential as well, but true to the site's enthusiast ethos, they remained centered on the browser's utility. Que wrote:

> One good thing I can see it has a bookmarks to most Historic African American Colleges everything else looks like this was put together by a

focus group which was asked a bunch of question and they built it from the results and that way you would never get things right.

Bennet A. Joseph added:

> I'm a young Black male who, after using Blackbird, thinks it's not only an interesting idea, but also a decent service. First, a note on some of the comments above about how Blackbird is somehow 'separatist' and/or racist: Blackbird tries to make information more easily accessible that the 40A founders think is useful to the Black community. Like already mentioned, Blackbird's use is not excluded to other peoples, but they'll probably not get as much out of it. This is no different from a Hispanic newspaper or news channel (beyond providing information in a language the target can understand fluently).

. . .

On Ars Technica (AT), David Chartier's Blackbird review begins with 'The Internet may have created a largely color-blind world wide web that connects users with just about any information they could ever want.' He goes on to note that Blackbird's non-standard features are the only significant departures from a typical Firefox installation. Chartier saw the Blackbird custom search as a positive implementation of the developer's intentions to deliver cultural content, but overall argued that Blackbird's feature set was 'nothing new.' Chartier's article is restrained, contextualizing the developer's decision to include culturally specific features within a 'community of practice.'

In the comments following Chartier's review, several Ars Technica audience members offered a less restrained and racialized framework to describe Blackbird's feature set. For example, Murph182 criticized the custom search, asking, 'If Obama starts doing all kinds of nutty stuff, will a standard search return news articles and criticism and the Blackbird search censor such things?' Davidd added, 'so it comes pre-loaded with links to Public Defenders, and tips on how to beat weapons charges . . . Great.' The conflation of technological identity and pejorative Black stereotype is striking; these comments may not be typical of the AT audience's racial attitudes, but that they remain unmoderated indicates that this discourse is acceptable to the community guidelines.

AT commenters also employed axiological formations of technology-based white privilege. rpgspree argued:

> If the browser, as the article states, skews results away from potentially more informative and authoritative sources of information in favor of those that are more culture centric, then it really is doing it's [*sic*] users a disservice.

Although the last comment is less overt than the first two, together they represent the spectrum of color-blind discourse displayed in Ars Technica's comments. rpgspree employs a 'rational' perspective that ignores the cultural perspectives found in mainstream content and privileges mainstream content as being more valid and reliable than 'culture centric' content. Davidd and Murph182 offered examples of deviant Black behavior as cultural touchstones for Blackbird's intended feature set.

Some AT commenters fought back against the tone of these comments. Oluseyi wrote:

> You could argue that the browser is not an 'African American browser,' but rather an 'African American Interest browser.' Nothing precludes non-black Americans from using it, and it's very likely that a large number of its eventual users will be non-blacks.

. . . Anechoic wrote,

> Blackbird isn't about 'walled gardens' or 'separatism'—it doesn't take you to some blacks-only internet, it doesn't wipe your harddrive if a white person tries to use it, it's a product designed to appeal to the needs and wants of blacks. You can disagree with the viability of this model (which I do) but there's nothing wrong with the motivation.

These comments focus on Blackbird's features while eschewing negative stereotypes of Blacks. These comments are closer to Chartier's framing of Blackbird as a community of practice; notice also that these sentiments are critical of the color-blind paradigm of internet use that the earlier commenters deployed. They also serve to highlight another trope of color-blind ICT usage—that Blackbird's users would be 'forced' to segregate themselves from the rest of the internet when making the choice to use Blackbird. The technology blogs' combination of technophilic ethos and color-blind ideology speak to the norming of technology as a White/human discursive enterprise, where efforts by non-whites to stake out space within the realm are unwelcome.

Responses from the Black Blogs

Where the mainstream tech blogs featured comments critical of Blackbird's feature set and Black culture, the Black tech blogs critically assessed Blackbird's features through their potential benefit to the Black community. For example, Roney Smith (roneysmith.blogspot.com) complimented Blackbird's RSS ticker and video channel, but pointed out that restricting social content to only be available in-browser yields 'no newly created value.' His contention was that if a user found a video of interest and sent a link to a non-Blackbird user, that friend would not be able to view the content. Smith adds that many African American users access the internet at work or school where Blackbird cannot be installed, which limits potential use and adoption. This criticism is valid given most corporate/institutional IT policies, which prohibit users from installing unapproved software on company machines for security.

BlackWeb 2.0's initial appraisal of Blackbird, written by Markus Robinson, was very positive. Robinson mentioned Blackbird's customization possibilities enabled through the import of the user's pre-existing Firefox configuration files. A follow-up post by Rahsheen noted Blackbird's Grapevine feature, but also criticized it for being accessible only through the browser. In a related post, Rahsheen also praised Blackbird's video channel and was encouraged by Blackbird's stance on philanthropy. However, he argued that Blackbird is not innovative because its core functions duplicated pre-existing features that power-users could install on their own (e.g. plug-ins).

Segregation Online

Rahsheen raised a 'segregation' argument against Blackbird—one also mentioned on both of the Black general interest blogs. He argues that a browser dedicated to information of interest to Black people limits access to the internet while stifling black innovation and interest in creating content online. To support this argument, the bloggers and commenters pointed to features that constrained their freedom to surf the internet. The Angry Black Woman (TABW) complained that Blackbird hijacked the 'default browser' status. She added that framing content within a specific application segregated Black users from the wider internet:

> If someone wants to de-marginalize news relevant to Black people, videos relevant to Black people, and social networking/bookmarks relevant to Black people, that's great. I am all for it. But I think doing it through a 'Black' browser isn't terribly affective [*sic*].

In TABW's comments thread, some commenters viewed Blackbird's approach as liberatory, rather than segregationist. Jermyn countered TABW's outlook by asking 'When will black innovation avoid criticism and get the respect it so much deserves?' Ben noted:

> Perhaps Mozilla will hire some black developers (these 3 gentlemen?) in the future and bring more culture-based (not necessarily race) ideas into the way we use the internet.

Balabusta added:

> It is true that if one is very interested in African-American perspectives on news and social issues, one has to be savvy in the use of search engines, which do not cough up those results without good Google-fu . . . As a white person with an anti-racist ideology who is interested in reading from Black perspective, I would have downloaded and used the browser just out of curiosity.

Both commenters point out the hit-and-miss nature of using search engines to find cultural content, arguing that Blackbird's approach does a lot of needed groundwork necessary to generate a positive Black internet experience. These comments acknowledge a reality that TABW and Rahsheen may have overlooked: their skill at finding Black cultural content online may not be shared by others.

Over on blog-aroundharlem.com, April remarked that customized searches were counter to the internet's inherent openness. She began her critique by stating 'I don't need anyone helping me find Black content' and noting a perceived lack of innovation . . . Her next statement forms the basis for her critique: 'technology can't be African American. Or, any other ethic/racial group.' . . . April continues by asking:

> How is my web experience enhanced by letting Blackbird filter information through their browser? By visiting African American sites 'they' select? Who are 'they'? What qualifies them to select African American content? Any Black Studies PhD or 'African American experts' affiliated with the site to determine 'the best content'? What is their criteria for acceptable content? Is there any?

April's question regarding the selection of valid and authentic African American content is an important one, given the sale of Black-oriented news sites such as Africana.com and BlackVoices.com to AOL Time Warner. Some Black tech blogger/enthusiasts dropped into the AroundHarlem comments to support April's views. Rahsheen (BlackWeb 2.0) wrote

> How useful would Twitter be if you could only see tweets that have #blck in them? You could only follow people who use the #blck tag. Everyone else disappears. That sound cool? Ok, now do the same thing with the entire Internet. Does that work for you?

Karsh, of blackgayblogger.com, said that Blackbird was 'as inane and untenable a concept to bring to market as any other web product of Saas [sic] which tries to commodify African-Americans.' These comments reveal a similar attitude towards culture-neutral internet use as that expressed on the mainstream technology blogs, but framed in a way that is supportive of Black internet culture and content.

Other AroundHarlem commenters debated the segregationist perspective. Allison wrote, 'Instead of pushing for major browsers or websites to feature AA interest [*sic*], separate browsers and websites are built.' Brick City wrote:

> I see comments about this web browser promoting segregation but
> I do not see a movement of folks talking about closing 'black churches'
> especially if your [*sic*] Christian being that we are worshiping the same
> God. Or perhaps closing all of the 'black universities' since education
> is suppose to [*sic*] neutral I can go on and on with many examples but
> anyone with any commonsense about using a web browser is that your
> choices in terms of search features are limitless, you can use google,
> yahoo, aol, msn.

Brick City's comment highlights the possibility that Black-oriented technology applications can co-exist with other 'historically Black' institutional practices that work to provide intellectual, political, and spiritual support for Black activities without reducing interactions with mainstream America.

For tech pundits, Black bloggers, and interested users, Blackbird's racial interests contextualized its technical features. The racialized perceptions of the feature set reveal a concern over how Blackbird's focus on Black content and users theoretically limited the application's usefulness. This perspective is remarkable because of the linkage between Blackness and limitation, where the internet's value is somehow lessened because users seek Black content.

Analysis: The Internet as Racial Apparatus

The rhetorical positioning of Blackness as a detriment to White technological hegemony of the internet can be seen most clearly on the technology blogs. The technology blog commenters frequently conflated Blackbird's intent to portray Black content as a threat to the internet's perceived cultural neutrality. Some argued for the internet as an artifact that intentionally erased embodied concerns, but others had no problems linking Blackbird to stereotypical imagery of Black deviance. Their arguments draw from a technocultural frame promoting (racial) progress, modernity, and a social status quo that implicitly continues White domination. For example, on Ars Technica commenter JChops wrote:

> Blackbird browser? Next thing you know, they'll have their own computer company. Instead of Apple, it'll be Watermelon. And the CEO will be Steve Jobless. And it'll run OS X BLACK PANTHER. Hell, the browser can send its user agent string as 'Blackbird' and you could tailor your site to shovel KFC ads and overpriced futon furniture at them. Can you see the 404 pages for this thing? Instead of '404,' you'll get 'Nigga, you isn't makin' no sense!'

These arguments were not limited to Whites; Blacks and Latina/os on the technology blogs articulated this position as well. Several commenters identified their racial origins as a warrant for cultural critique: Brandon at Tech Crunch wrote:

> Ok, as a member of the black community, I must say this is not the way to go and I haven't even tried it out. We must stop creating a 'black' this and a 'black' that! Can't we figure out another way for the tech community to recognize our presence?

L. wrote:

> I agree with many people here. To be honest, I think this is the most racist thing I've seen. If this was whitebird, it would be hit with thousands talking about racism, but because it's for african americans it's not racist at all? This isn't a biased opinion considering I'm latin american, just in case you were wondering.

I do not mean to over-emphasize that color-blind discourse is the only discourse of the mainstream technology blogs; there were some excellent anti-racist technology comments posted on both blogs. Sick of Ignorant Racists wrote:

> Equality does not mean that anyone of any race need to leave interests unique to their culture at the door. Ironically, it's only the worst type of racists who try to sell the idea that this is necessary for eliminating racism. Those who truly celebrate equality celebrate the right of every

group to express the uniqueness of their culture—without being so threatened that they have to resort to petty namecalling and thinly (VERY THINLY) veiled racism.

Amber added:

> While I personally think this is a stupid idea (though the news ticker is genius) the comments here have made me sad. I sit here and say wow you know just 40 years ago my grandmother was getting spat on and getting rocks thrown at her for being black but today look how far we have come . . . and then I see really not that far when I see this kind of stuff.

On both tech blogs, counter-discourses against color-blind internet commentary (such as the ones above) were only infrequently encountered. They are remarkable in the amount of thought and detail put into them; some of them were nearly one half to one page in length. . . . These remarks, however, were far outnumbered by comments featuring color-blind ideology and others that used the internet as a racist framework.

Discussion

This analysis of Blackbird melds a cultural analysis of an artifact's user interface with a discourse analysis of online activity around said artifact. I employed close readings of blog-based discourses intent on examining Blackbird's cultural, rather than practical, functionality. Using critical race theory situates these online reactions within wider cultural racial frameworks that highlight offline influences on technology use and adoption. As such, I found that color-blind ideologies and White privilege undergird both White and Black audiences' beliefs about technology use, but the Black blogs uniformly presented an ethos of Black love to contextualize their critical discussions of the browser and Black internet use.

. . .

New technologies often claim to improve some facet of our lives through efficient use of resources. Web browsers brag about speed, polish, and customizability; Blackbird made a claim of personal improvement and empowerment for African American internet surfers. However, Blackbird's reception revealed a level of scrutiny that other browsers (or technology artifacts in general) never undergo. Where browsers are typically measured according to whether they can best render benchmarks, Blackbird was assessed based on its allegiance to an electronically enabled cultural identity.

The responses to Blackbird examined here can be generally divided into three approaches:

A. Appreciation/approval
B. Rejection for racial rationales; and
C. Rejection for lack of perceived utility.

Further examination found that the mainstream technology blogs housed the majority of the comments falling in groups B and C, while the Black general interest blogs housed a greater number of comments falling in groups A and C.

Comments on the technology blogs often revealed anger and despair over the perceived erosion of Whiteness in America; some were outright racist; and many others articulated confusion at Blackbird's temerity for upsetting their vision of a color-blind internet and post-racial future. These comments significantly outnumbered other comments displaying reasoned responses to the browser. When anti-racist appeals were made, they directly addressed the technophilic racist discourses pervading the technology blog comment threads instead of simply lingering on the critiques of the artifact itself. In this, it is heartening to see that the increasing numbers of minority commenters in these enclaves does serve to partially police the color-blind discourses so easily deployed by the audience and often ignored by the blog authors.

The Black bloggers' and commenters' responses to Blackbird evince cultural affiliations as a framework for understanding technology adoption. The Black technology blogs, for example, describe their blogs/websites as interventions against mainstream technology sites that do not cover material of interest to Blacks. April's review of Blackbird promotes positive Black cultural values even while she strongly criticizes the technological and cultural limitations of the browser. Such reactions to the guided nature of Blackbird's interactions with the web conflates the liberatory rhetoric of internet use and adoption with a historical narrative that seeks to delimit American cultural boundaries of Black identity.

Note that the above critics of Blackbird's feature sets—regardless of venue—derided the browser because they assumed that the browser would prejudice access to Black content, which is contrary to both the browser's intent and design. The tech blog commenters conjured up images of Black pathology (e.g. weed locators, 24 inch rims) while arguing that culturally oriented approaches are divisive and racist. The Black cultural bloggers (and their audiences) worried about the consequences of being segregated from the internet through Blackbird's selective features.

As noted in the interface analysis earlier, Blackbird allows users to specify multiple search engine plug-ins and websites. Thus, while the objections were ostensibly directed against the browser's limitations, the limitations discussed are primarily ideological. The objections derived energy from a White racial framework where Blackness signifies a lesser state of being; an all-Black internet is perceived as being less valuable than an internet where Blackness is (at best) a minor presence in a universe of content supporting a White ideological frame. Blackbird's features highlighting African American content were seen as an imposition on the universal appeal of the internet, highlighting the perception of the browser as a social structure limited by Black representation.

. . . The browser served as a site of technological and racial rupture, where longstanding beliefs about racial identity shaped often pejorative reactions to the browser's communicative utility. The deployment of racial beliefs and identities to explicate technology use has much to do with the importance of race to how Western and American culture perceives itself. The internet is not immune to these influences, regardless of how color-blind the online audience wishes the world to become.

Notes

Epigraph. G. Kirkpatrick, *Critical Technology: A Social Theory of Personal Computing* (Aldershot: Ashgate Publishing, 2004), 59.

1. André Brock, "Race Matters: African Americans on the Web Following Hurricane Katrina," *Proceedings of Cultural Attitudes Towards Communication and Technology*, eds. Sudweeks, Hrachovec, Ess (2008): 91–105; Brock, "Life on 'The Wire': Deconstructing Race on the Internet," *Information, Communication, and Society* 12, no. 3 (2009): 344.

2. L. Althusser, *Lenin and Philosophy and Other Essays*, translated from the French by Ben Brewster (New York: Monthly Review Press, 1971), 174.

3. Lisa Nakamura, "Cultural Difference, Theory, and Cybercultural Studies: A Case of Mutual Repulsion," in *Critical Cyberculture Studies*, eds. Silver and Massanari (New York: New York University Press, 2006): 29–36.

4. S. Herring, "Computer-Mediated Discourse Analysis: An Approach to Researching Online Behavior," in *Designing for Virtual Communities in the Service of Learning* (New York: Cambridge University Press, 2004): 338–76; T. A. Van Dijk, "Discourse as Interaction in Society," in *Discourse as Social Interaction*, vol. 2, ed. Van Dijk (Thousand Oaks, CA: Sage, 1997): 1–37.

5. Brock, "Race Matters"; Brock, "Life on 'The Wire.'"

6. J. Carey, *Communication as Culture: Essays on Media and Society* (London: Routledge, 1984).

7. J. Dinerstein, "Technology and Its Discontents: On the Verge of the Posthuman," *American Quarterly* 58, no. 3 (2006): 569–95.

8. A. Pacey, *The Culture of Technology* (Cambridge, MA: MIT Press, 1984).

9. D. Williams, N. Martins, M. Consalvo, and J. D. Ivory, "The Virtual Census: Representations of Gender, Race and Age in Video Games," *New Media & Society* 11, no. 5: 815–34. Available at http://nms.sagepub.com/cgi/content/abstract/11/5/815 (accessed 29 July 2009).

10. M. Omi and H. Winant, *Racial Formations in the United States* (New York: Routledge, 1994).

11. C. Harris, "Whiteness as Property" in *Harvard Law Review* 106 (1993): 1707–91; Toni Morrison, *Playing in the Dark: Whiteness and the Literary Imagination* (New York: Vintage, 1993).

12. H. Giroux, *Fugitive Cultures: Race, Violence, and Youth* (London: Routledge, 1996): 75.

13. Richard Dyer, *White* (London: Routledge, 1999).

14. C. E. Baker, *Media, Markets, and Democracy* (Cambridge: Cambridge University Press, 2002).

15. Baker, *Media*, 47.

16. https://techcrunch.com/2008/12/08/blackbird-is-a-custom-browser-for-african -americans-built-on-top-of-mozilla/; https://arstechnica.com/uncategorized/2008 /12/blackbird-browser-reaches-out-to-african-american-community/; http:// roneysmith.blogspot.com/2008/12/blackbird-browsers-greatest-strength-is .html; https://www.scoopbyte.com/blackbird-the-black-focused-browser-speaks/; https://www.blackweb20.com/2009/05/11/the-blackbird-browser-5-months-later/; http://sheenonline.biz/2008/12/blackbird-is-not-about-tech-its-about-being-blck/; http://blog-aroundharlem.com/2008/12/08/blackbird-the-african-american-web -browser-and-philanthropry-on-the-web/; http://theangryblackwoman.com/2008 /12/09/blackbird-browser-because-the-internet-isnt-black-enough/.

CHAPTER 4

Scientology Online

Copyright Infringement and the Legal Construction of the Internet

Gerardo Con Diaz

Early one morning in February 1995, a former scientologist named Dennis Erlich called 911 because a man in a suit would not stop ringing his doorbell. The emergency operator explained that Glendale Police was outside his door holding a search warrant. Erlich stepped onto his front porch, where a police officer told him that they were there to seize copyrighted materials. Erlich objected, but the officer came into the home along with a private investigator, a camera crew, two armed off-duty officers, and three representatives from the Church of Scientology's Religious Technology Center (RTC): an attorney named Thomas Small, a high-ranking Scientology officer, and a "computer expert."[1]

The visitors spent several hours rummaging through Erlich's small apartment. They confiscated over 300 floppy disks, 29 books, and 100 megabytes of hard drive backups. Despite Erlich's unrelenting protests and his pleas to the Glendale officer, RTC seemed determined to ensure that he would not access his personal archive ever again. They loaded hundreds of files into their own drives and wiped out the copies stored in Erlich's computer. Small gave Erlich an unsigned piece of paper—a partial inventory of seized materials—and the team left as quickly as they had arrived.[2]

A fervent online critic of the Church of Scientology, Erlich knew that it was only a matter of time before the Church acted against him. He had posted the full texts of confidential Church materials on a bulletin board system (BBS, special software that allowed users to access limited online services such as posting messages, uploading and downloading files, and exchanging messages). He thought this raid was the latest example of the Church's well-documented efforts to silence its critics, but the Church's lawyers insisted that they were merely protecting their clients' copyrights (which grant the makers of creative works the right to exclude others from reproducing or distributing their creations).[3] The fiery legal battle that unfolded over the next eleven months, cited in legal circles as *RTC v. Netcom*, was, in fact, both: the enforcement of copyright law as censorship mechanism.

Legal scholars have been documenting the politics of Internet copyright very carefully, but their work has overlooked, and even obscured, a fundamental historical development: how *RTC v. Netcom* became a crucial step in the national effort to carve out a place for the Internet in the US legal system.[4] The infrastructure of the Internet—from the devices and software in users' homes to the networks of servers through which information is transmitted—did not fit neatly into any existing legal categories in copyright law. The California court that heard Erlich's case faced this foundational conceptual problem while, at the same time, handling a very difficult legal question: Who was liable for the copyright infringements in his posts? A legal doctrine called "fair use" allows the limited reproduction of copyrighted material for purposes such as analysis and satire, but this doctrine does not generally grant permission to post entire works or large portions thereof, as Erlich had done. There was little debate over whether Erlich had committed copyright infringement, but it was unclear whether two additional parties—a BBS administrator named Thomas Klemesrud, who hosted a.r.s., and an Internet service provider (ISP) called Netcom, which contracted with Klemesrud—were also liable. After all, their servers hosted Erlich's posts and distributed them to readers around the world.

This chapter argues that the effort to determine who was liable for Erlich's posts enabled the emergence of the Internet as a distinct entity in US copyright law by stabilizing a legal conception of it as a swap meet.

The court could have adopted other ways of conceptualizing the Internet that the different parties' lawyers had proposed, but the swap meet metaphor had the unique advantage of creating a direct metaphorical connection between Erlich's legal problems on the one hand, and cutting-edge opinions in copyright law on the other. In the absence of any clear legislative guidance on the copyright liability of Internet service providers, which would not be passed until 1996, this understanding of the Internet provided the conceptual underpinning necessary for courts to advance a legal reasoning that is all too familiar to legal scholars and industry practitioners today: Internet service providers are not liable for their users' copyright infringements.

This argument shows how the emergence of the Internet as a new technology was tied to broader changes in US legal thought that enabled its introduction into the country's intellectual property infrastructure.[5] This phenomenon—the joint development of conceptions of new technologies on the one hand and legal frameworks on the other—is a longstanding theme in science and technology studies that has recently provided new ways of framing the history of software.[6] However, it has not yet become a central theme in computer network history even though Internet governance is a major subject of study for Internet scholars and legal scholars alike.[7] Correcting this oversight enables the introduction of the history of technology into fields of legal scholarship that traditionally rely exclusively on doctrinal interpretation and the use of legal and regulatory archives as a departing point for the study of Internet history.[8]

To this end, this chapter investigates *RTC v. Netcom* to recount the conceptual emergence of the Internet as a new technology at a crucial junction in its history: the early 1990s, when ISPs grappled with the fact that allowing users to create online content from the comfort and privacy of their own homes would require all the stakeholders involved to articulate and negotiate their rights and responsibilities. The narrative that follows draws on published court opinions, court records, and an assortment of born-digital archival sources, including message board posts, archived websites, and posts on social media platforms such as YouTube. Rather than recounting the story of *RTC v. Netcom* in its entirety, the chapter presents three crucial moments in its development, arranged in chronological order.[9] The first section examines the Church of Scientology's initial complaints against Netcom, Klemesrud, and Erlich. The second recounts Klemesrud's and Netcom's early defenses, which tentatively presented the Internet as a river or a bookstore. The third narrates how Netcom's lawyers deployed the swap meet conception of the Internet to release their client from all liability, leaving Erlich behind as the only defendant against RTC.

The Lawsuits

Erlich had once been an ordained minister, but he had left the Church in the mid-1980s to become one of its most vocal critics. He regularly published his critiques on an online message board called alt.religion.scientology (a.r.s, for short), where he usually signed off as the in-Former. From this digital pulpit, he fostered public discussions on the history, actions, and controversies of Scientology. He also posted the texts of print articles, internal documents that the Church considered private, and works written by L. Ron Hubbard, the founder of the religion. The internal documents disclosed obscure details about the Church's organization and policies, its confidential beliefs, and the training that its high-ranking members received as part of their initiation.

Erlich's online posts raised several red flags at the Church's Religious Technology Center. Founded in 1982, the RTC's main mission is "to espouse, present, propagate, practice, ensure, and maintain the purity and integrity of the religion of Scientology." The articles of incorporation for this religious nonprofit corporation indicate that the RTC is meant to "act as the protector of the religion of Scientology" by enforcing the Church's intellectual properties. This involves both an assortment of trademarks and copyrights for major brands such as Dianetics, and a collection of texts and practices that the articles describe as a "substantial body of confidential advanced religious technology which is a part of a body of truths and methods of application." The RTC treated these textual and ritualistic technologies as trade secrets—though courts would soon conclude that this was an artifice designed to stifle criticism through aggressive litigation.[10]

RTC reacted by filing a complaint at the District Court for the Northern District of California a few days before the raid. The complaint alleged that Erlich was posting copyrighted materials and trade secrets.[11] It listed two more defendants: a man named Thomas Klemesrud, and a Delaware ISP called Netcom Online Communication Services. RTC was therefore targeting not just Erlich himself but also the networks that allowed him to participate in a.r.s. in the first place. Erlich used his telephone connect to support.com, a BBS that Klemesrud operated. Support.com hosted a.r.s., but Klemesrud did not have direct access to the Internet. Instead, he obtained it from Netcom, one of the largest providers at the time. In practice, this meant that Erlich would write a message, connect to the BBS, and upload it to Klemesrud's computers. Klemesrud would periodically connect to the Internet, upload all the messages that users had posted, and download everything that others had uploaded since the last connection.

RTC did not raid Klemesrud's or Netcom's offices, but Church lawyers had been in touch with both for several months before they filed the suits. Back in December 1994, Klemesrud had received an unexpected phone call from Helena Kobrin, one of the Church's main legal counsels. She told him about Erlich's actions and demanded that Klemesrud cut him off from the BBS. Klemesrud told her not to call him again because he considered that harassment, and he hung up on her. Kobrin quickly followed up with an email explaining that any claims Erlich made about fair use were false, accusing Support.com of divulging Church trade secrets, and threatening legal action against both Klemesrud and Netcom if they failed to act.[12] Klemesrud, a frequent contributor to a.r.s. himself, responded that a BBS is "a distributor of information . . . much like a magazine rack, or bookstore." He told her, "Here is what I can do for you. Anytime you see a posting on the support.com BBS, please mail me the original copyrighted work to compare. If I agree, and I will be fair, I will delete the material from the BBS. Public postings average around 3 days life span on the BBS."[13] This did not satisfy Kobrin. The posts included confidential Church documents, so agreeing to this strategy could potentially place her in the difficult situation of having to mail secret texts to a burgeoning community of online critics. Still Klemesrud insisted that this was the only way of complying with take-down requests without having to cut users' access on demand, which amounted in his view to "effectively curtailing their right to free speech."[14]

This response pushed Kobrin and her colleagues to cast a wide net in their complaints. Following Kobrin's lead, RTC accused Erlich of direct copyright infringement, namely, reproducing, distributing, and displaying copyrighted works. The complaint alleged, specifically, that Erlich had, "without authorization, electronically published copies of these works." RTC also accused Klemesrud and Netcom of both direct infringement akin to Erlich's, and of contributory infringement—that is, of enabling others to infringe the copyrights. RTC alleged that Klemesrud and Netcom had helped Erlich by providing the necessary "computer facilities" for his message board and for hosting the documents on their servers. The complaint also alleged that RTC had repeatedly asked Erlich to take these documents down but that he had declined to do so and indicated that "no local government or court" had the power to tell him to do so. Similarly, Klemesrud and Netcom had "refused to cease and desist from receiving, transmitting, and publishing Erlich's unauthorized copies."[15]

A Preliminary Hearing

Erlich arrived at a federal courthouse in San Jose, California, early in the morning of February 21, 1995. The presiding judge was Ronald M. Whyte, who was appointed to the bench by George H. W. Bush and would soon become one of Silicon Valley's most important judges in matters regarding patent law.[16] Erlich, Klemesrud, and Netcom had already received temporary restraining orders that barred them from duplicating Hubbard's works, disclosing any confidential materials, and even transporting any RTC materials outside of their districts. Whyte would now evaluate whether the court would issue preliminary injunctions against all three defendants—that is, orders prohibiting them from certain behaviors until the court decides the case. Specifically, if Whyte issued these injunctions, then Erlich, Klemesrud, and Netcom would be forbidden from things such as storing Hubbard's works in any hardware, databases, network facilities, or archives; displaying any works for which RTC held copyrights, or causing them to be displayed; and destroying or concealing any of Hubbard's works, digitized or otherwise. The hearing would also determine whether the defendants would be required to give RTC all hardware or software connected to the transmission of the copyrighted materials.[17]

Whyte had not yet read all the materials that the court had received. One of the few documents that he did read carefully was a statement by Rick Francis, Netcom's vice president for software engineering. Francis told the court that Netcom was merely an "Internet connectivity provider," and that while it provides its clients full access to the Internet, it "is not involved in any way in determining the content of the information available" online. This distinguished Netcom from ISPs such as CompuServe or America Online, which normally offered limited subscription-based plans that allowed users to access a preset collection of pages and services.[18]

According to Francis, this breadth of access made it impossible for Netcom to ensure the legality of the content that its users posted or downloaded. After all, he estimated, nearly 150 million keystrokes traveled through Netcom and into the Internet each day—150 megabytes of data. Handling this amount of data was especially difficult because it was coming from a vast network of clients. Klemesrud's BBS alone could access 900 newsgroups, and about 2,000 Netcom clients had this level of access. Like all these clients' systems, Klemesrud's BBS allowed individual users to access the Internet without having to deal directly with Netcom. Instead, a BBS user would connect to Klemesrud's computers, which would then post the messages automatically on their behalf to Internet newsgroups and forums. This made it unfeasible for Netcom to establish editorial control over individual BBS users: the only way for Netcom to have shut down Erlich's access to the Internet would be to terminate Klemesrud's access altogether (thereby disconnecting his 500 subscribers).[19]

Francis's statement was enough for Whyte to state, early on, that "it's probably a practical impossibility" for Netcom and Klemesrud to "do any kind of censoring or checking" of the information that their users published or downloaded. This annoyed RTC's lawyers, who insisted that it should not be too difficult to ban users such as Erlich without needing to disconnect anyone else. If messages contained information about their point of origin—that is, about who posted them—wouldn't it be possible, at least, to flag messages that originated from Erlich's account?[20] They told Whyte, "We had a conversation with one of our programmers just the other day who was able in a fairly short period of time to write a program that allowed us to know when Mr. Erlich and other people were actually logging onto the internet. And we believe that it would be possible in very short order to create software that would allow the necessary monitoring to take place, and we're willing to work with defendants on that."[21] RTC's lawyers emphasized that "computer science today would have the ability to flag those messages" and pushed even further to suggest that a widespread surveillance system could potentially be implemented without much technical hassle. Within a decade this would become a commonplace arrangement. However, back in the 1990s, the potential for such a system to exist was not enough to convince Whyte that Netcom and Klemesrud could be compelled to do anything at all.[22]

In contrast to Erlich's case, which would quickly lead to a finding of infringement, defending Netcom and Klemesrud's would require a healthy dose of legal creativity. Their lawyers delivered a broad range of arguments in hopes that Whyte would develop a view of Internet technologies that would favor their clients' interests. Randolf Rice, one of Netcom's lawyers, asked Whyte to think of this lawsuit as a pollution case in which plaintiffs are "pointing to a polluted river and suing the river along with the polluter." In this scenario, Netcom was the river, "simply a means of transmission," not to be blamed for the actions of the polluters who dumped noxious chemicals. Richard Horning, Klemesrud's lawyer, deemed the entire conflict "some form of religious war" in which his client had become an unwilling middleman. RTC's demands were but an attempt "to keep Erlich off the air entirely, whether he's engaged in legitimate criticism or wholesale violations or something in between." Horning's goal was not to defend Erlich but to argue that Klemesrud should not be compelled to take any part in RTC's effort to censor him.[23]

These comparisons mattered because the body of legal precedent for this kind of situation was not well developed enough to provide a clear defense. Previous Internet copyright lawsuits had focused primarily on whether the distribution of photographs through a BBS was copyright infringement and on the trademark and unfair competition issues that arose from the facts of the case.[24] This precedent would likely help RTC win its case against Erlich, but it did not necessarily provide a line of reasoning that would allow Horning and Rice to argue that Netcom and Klemesrud themselves should not be forced to take any action. Earlier legal precedent could work in Klemesrud's favor by establishing that he is not liable for copyright infringement just because the copyrighted materials were stored on his computer, but it did not clarify whether he should explicitly take action against Erlich.[25]

Faced with this dearth of immediately applicable precedent, Rice asked Whyte to think of this suit as a high-tech version of *Smith v. California,* a 1959 Supreme Court case on the publication and sale of obscene material. This case concerned a bookstore owner in Los Angeles named Eleazar Smith, who had been convicted for violating a city ordinance that made it illegal to possess any "obscene or indecent writing," including books. The Court had unanimously sided with Smith, holding, among other things, that the city ordinance was unconstitutional. However, at Whyte's court, what Rice cared about were not the constitutional details of this landmark case but the fact that the Court had essentially treated Smith's bookstore as a middleman. The lawyer insisted that the same was true of Netcom: like Smith's bookstore, Netcom was just a distributor that had just had "strict liability imposed upon it" and whose only recourse (if RTC had its way) was to find a way of monitoring "everything that crosses the internet in a nanosecond, in the blink of an eye."[26]

Rice and Horning's arguments persuaded Whyte to dissolve the temporary restraining order against Netcom and Klemesrud and to deny

RTC's request for a preliminary injunction against them. He ordered RTC to provide a detailed inventory of the items they had taken from Erlich and issued a modified restraining order that allowed him to keep posting messages critiquing the Church while prohibiting him from digitizing, storing, or transmitting any content for which RTC claimed copyright. No further seizures would take place without Whyte's explicit approval, but Erlich would still need to find a way of defending himself more effectively against RTC's accusations of infringement.[27]

At the time this hearing took place, it was becoming very clear that these suits were just the first of many. Several system administrators and users wrote to the Electronic Frontier Foundation (EFF), a civil liberties advocacy group founded half a decade earlier, to report that lawyers for RTC and the Church were reaching out to them with claims of copyright infringement. In the EFF's online periodical, one of its authors (probably Shari Steele, its legal director), reported, "These threats apparently are designed to convince sysadmins to discontinue the carriage of certain newsgroups that involve discussions of the Church of Scientology in its teachings, solely on the ground that some of the messages sent through these newsgroups allegedly involve infringement of [Church] copyrights or other intellectual property rights."[28] Even EFF itself had received a letter; the Church had written to say that it would not be threatening system administrators with lawsuits if there were any other way to deal with the infringing messages. The EFF deemed it unacceptable for RTC to handle the matter with "the threat of litigation to shut down entire newsgroups," but it also recognized that there was no culture of conflict resolution in place to handle these Internet-born disputes effectively.[29]

The EFF noted that these new and unprecedented kinds of conflicts showed how important it was "to search for new paradigms for handling disputes." After all, electronic communications at this scale were still "in their infancy," and service providers such as Klemesrud were "not big corporations with substantial funds to spend on expensive litigation." For instance, the EFF suggested that Church members could respond to their online critics by hosting discussions of their own; this would allow them to counter wrongful allegations about them "with more speech." Another option would be to engage in mediation or arbitration, perhaps even through an online message board discussion that EFF could help arrange.[30]

These proposals were not calibrated to the social and cultural stakes of the situation, as the Church's efforts to maintain the secrecy of its sacred texts were not compatible with the goal of fostering more discussions about them. Still, this misalignment signaled the coming of a major battle for civil liberties advocates: regardless of what happened at Whyte's court, it was only a matter of time before copyright law became a central consideration in the quickly intensifying national debate over online speech.

The Limits of Liability

In partial alignment with the EFF's views, Netcom's lawyers filed a motion to dismiss the suit against them. Their new argument abandoned comparisons with bookstores and rivers in favor of claiming that ISPs and their clients have a relationship similar to that between a leaser and a lessee. Specifically, Netcom "leases access to the Internet" by giving users the software necessary to use "the NETCOM gateway to access to the Internet." Users could choose one of two monthly pricing options depending on the kind of access they wanted: if they knew some UNIX, they could pay $17.50 for a shell account; otherwise, they could lease a more user-friendly graphical interface for $19.95. Emphasizing this arrangement, the brief stated, "NETCOM does not post or control the contents of the information posted or its destination. NETCOM does not control where, when or whether the information may be 'downloaded.' NETCOM does not provide the means for downloading the information. NETCOM, through a fixed-rental arrangement, leases the equipment necessary to enable its subscribers to communicate on the Internet."[31] This characterization of Netcom as a lessee was especially effective because of the metaphor that the lawyers were able to build on top of it: the company was leasing equipment (software) that allowed users to enter a space (the Internet) and complete a specific task (send and receive messages). In other words, the company "merely leases its premises to subscribers who wish to communicate on the Internet and does not in any way select, edit or benefit from what they post."[32]

This framework made Netcom's situation a digital counterpart to *Fonovisa v. Cherry Auction*, a recent opinion from the Eastern District of California.[33] This unusual case concerned a physical swap meet called Cherry Auction. The meet's operators leased space to vendors who traded all kinds of goods, including counterfeit records and movies. Cherry Auction's staff knew that the vendors were committing copyright infringement, but the staff did not promote, advertise, or encourage them. After a record label called Fonovisa sued Cherry Auction for infringement, the court found that "merely renting booth space is not 'substantial participation' in the vendors' infringing activities."[34] This meant that Cherry Auction was not liable for contributory infringement, and Netcom's lawyers hoped that the same reasoning would apply to their case. This was an unprecedented move: legally conceptualizing the Internet itself as an intangible swap meet akin to a flea market, thereby bringing into courts of law longstanding metaphors for online message boards.[35] In the process, Netcom's lawyers bypassed RTC's arguments about the possibility to monitor individual communications by arguing that Netcom bore no responsibility at all.

Whyte took several months to issue his opinions. In September 1995 he ordered that RTC return Erlich's property, but he issued the strict preliminary injunction that Kobrin had requested and noted that a full trial on the merits was still pending. He also allowed Erlich to continue posting online commentary, but he warned him not to reproduce any of the materials covered by the injunction and not to post anything beyond "the extent necessary to carry out his critical purpose." Three months later, he dismissed RTC's suit against Netcom and Klemesrud, adopting a very substantial enrichment and expansion of the arguments that Netcom's lawyers had proposed. This opinion effectively established that BBS operators and ISPs are not liable for their users' copyright infringement if they do not actively copy the infringing works or receive financial benefits from the infringement.[36]

Over the next few months, while Erlich negotiated a settlement to avoid a full trial, the Internet advocacy landscape expanded so quickly that it even became difficult for civil liberty activists and corporate lawyers to keep track of all the ongoing conflicts.[37] RTC continued to file copyright infringement lawsuits against message board users, generating a collection of complaints so large that some of its critics started creating websites to archive all the briefs that the courts were receiving. One of the greatest changes, however, was that advocacy organizations like the EFF would soon bring together two interrelated communities of advocates under the same roof: those concerned with a recent surge of legislative proposals to create censorship regimes for the Internet, and copyright activists who continued to think of RTC's copyright suits as a harbinger of legal regimes to come. The copyright battles over online speech were just getting started.

Conclusion

The problem of determining whether ISPs are liable for their users' copyright infringement reached the US Congress soon after *RTC v. Netcom*. In 1996, lawmakers passed the Communications Decency Act (CDA). An overt attempt to regulate the spread of pornography through the Internet, much of the CDA was struck down at the Supreme Court less than a year later as a violation of the First Amendment right to freedom of speech. However, Section 230 of the Act establishes that "no provider or user of an interactive computer service shall be treated as the publisher or speaker of any information provided by another information content provider."[38] In practice, Section 230 enshrined the reasoning in *RTC v. Netcom* at the statutory level, transforming Whyte's conclusions of law into the first of major legislative pillars on which Internet copyright stands today—the other one being the Digital Millennium Copyright Act.[39]

The swap meet metaphor for the Internet entered the canon of US law at the end of a long process born from the extraordinary expansion of stakeholders that accompanied the Internet's transformation into a home technology. The conflict that became *RTC v. Netcom* was not initiated by ISPs, universities, or government agencies. It was one of many Internet-related legal battles initiated by a church, albeit one with a long and deep history of litigiousness and controversy. Scholars of religion have recently intensified their reliance on media studies to inquire how digital technologies have transformed what it means to attend a religious service, engage with religious texts, or even feel spiritually and socially connected with others.[40] In contrast, *RTC v. Netcom* invites inquiry into the converse problem, namely, assessing how the history of religion has shaped information technologies and their place in the world.

The Church of Scientology's efforts to stifle criticism through copyright lawsuits prompted Whyte's use of metaphors to grapple with the Internet without having to open the technical black boxes found in and around it. This suggests that metaphors for the Internet matter not just because they have shaped users' experiences and expectations but also because they have codified the Internet's place in the broader legal and regulatory environments that frame its operation, limitations, and development.[41] Investigating how this occurs requires infusing major themes in the history of computing—the politics of user experience and engagement, the discursive emergence of new media and networks as distinct entities, and the incorporation of new technologies into established communities—with the archival spaces and methodological priorities of legal history.[42]

RTC v. Netcom is both a textbook opinion in US copyright law and a key step in the legal construction of the Internet as a new technology. Its story unveils the serious conceptual work that took place in the effort to transform the Internet into a legally legible entity (that is, one that could easily be compared and contrasted with objects and spaces for which there was clear legal precedent). The metaphors considered at Whyte's court (the river, bookstore, and swap meet) were all aimed at characterizing what happened on the Internet: Is information exchanged, or does it flow? Are Internet users more like visitors to a landscape like a riverbank, or are they participants in a highly regulated commercial space like a swap meet?

These questions and metaphors are extraordinary because they offer a new way of thinking about the discursive emergence of information and telecommunication technologies as new and distinct intellectual properties. Previous scholarly work on the matter has developed, in part, as a response to the trend in the legal academy to debate which conception of a given technology should reign supreme at the courts.[43] Unlike the lawyers and legal scholars leading those debates, historians have refocused the question away from *which* conception should win, to *how* each of these conceptions grants commercial and sociocultural advantages to some communities of developers and users over others. In the case of *RTC v. Netcom*, though, all metaphors for the Internet were designed to serve the same goal: to free ISPs from legal liability regardless of how the battle between copyright holders and alleged infringers would turn out. In other words, this is not a case wherein each conception of a technology is a stand-in for a different cluster of commercial and technical interests. It is, instead, a case in which all the metaphors at hand were designed to advance the same goal of releasing ISPs from infringement liability—a goal that would yield entirely new statutory law in the form of Section 230, which the administration of President Donald Trump sought to weaken as part of his disinformation tactics.

The swap meet's victory at Whyte's court offered a glimpse of the arguments that ISPs would be delivering at Congress on the eve of the CDA's passage, but it also illustrates the profound historical contingency of the legal construction of technology. Perhaps rivers or bookstores could have attracted Whyte's attention, but Netcom's strategy of relying on *Fonovisa v. Cherry Auction*—an unusual case being considered by the same higher court that would handle an appeal to Whyte's opinion—was very effective. It positioned Netcom's arguments at the cutting edge of legal thought and signaled that the federal appeals court overseeing Whyte's work was already leaning toward a vision of copyright infringement that aligned with Netcom and Klemesrud's claims. In other words, the Internet legally became a swap meet not because this conception was somehow superior or more technically sophisticated but instead because it allowed Netcom's lawyers to connect their case to trends in legal reasoning that happened to be developing at the right place and the right time.

Notes

This work was made possible by the Alfred P. Sloan Foundation, the Hoover Institution at Stanford University, and the UC Davis Department of Science and Technology Studies.

1. Dennis Erlich, ca. February 1997, https://web.archive.org/web/19970302163351 /http://www.eff.org/pub/Censorship/Scientology_cases/erlich_experience.article.
2. Dennis Erlich, ca. February 1997.
3. Dennis Erlich, ca. February 1997; Dennis Erlich to Ronald Whyte, February 21, 1995, https://web.archive.org/web/19970302163706/http://www.eff.org/pub /Censorship/Scientology_cases/erlich_whyte_022195.letter; "Exhibit A: List of Published Literary Works," Religious Technology Center v. Netcom (907 F. Supp. 1361, herein after RTC v. Netcom).
4. See, for instance, Eva Wirtén, *No Trespassing: Authorship, Intellectual Property Rights, and the Boundaries of Globalization* (Toronto: University of Toronto Press, 2015); Bill Herman, *The Fight over Digital Rights: The Politics of Copyright and Technology* (New York: Cambridge University Press, 2013); Laura DeNardis, *The Global War for Internet Governance* (New Haven, CT: Yale University Press, 2014); Patricia Aufderheide and Peter Jaszi, *Reclaiming Fair Use: How to Put Balance Back in Copyright* (Chicago: University of Chicago Press, 2018); Matthew Sag, "Internet Safe Harbors and the Transformation of Copyright Law," *Notre Dame Law Review* 93, no. 2 (December 2017): 499–564. There is, of course, a thriving literature in East Asian studies dealing with online speech. See Ya-Wen Lei, *The Contentious Public Sphere: Law, Media and Authoritarian Rule in China* (Princeton, NJ: Princeton University Press, 2017); Rongbin Han, *Contesting Cyberspace in China: Online Expression and Authoritarian Resilience* (New York: Columbia University Press, 2018).
5. This complements the focus on sociotechnical, cultural, commercial, and political phenomena found in works such as Janet Abbate, *Inventing the Internet* (Cambridge, MA: MIT Press, 1999), and Shane Greenstein, *How the Internet Became Commercial: Innovation, Privatization, and the Birth of a New Network* (Princeton, NJ: Princeton University Press, 2015); Julien Mailland and Kevin Driscoll, *Minitel: Welcome to the Internet* (Cambridge, MA: MIT Press, 2017). See also Bradley Fidler and Andrew Russell, "Financial and Administrative Infrastructure for the Early Internet: Network Maintenance at the Defense Information Systems Agency," *Technology & Culture* 59, no. 4 (2018): 899–924; Maximilian Hosl, "Semantics of the Internet: A Political History," *Internet Histories* 3, no. 3-4 (2019): 275–92; Benjamin Burroughs, "A Cultural Lineage of Streaming," *Internet Histories* 3, no. 2 (2019): 147–61; Valérie Schafer and Benjamin Thierry, *The 90s as a Turning Point Decade for Internet and the Web*, Special issue of *Internet Histories* 2, no. 3-4 (2018).
6. Gerardo Con Diaz, *Software Rights: How Patent Law Transformed Software Development in America* (New Haven, CT: Yale University Press, 2019).
7. See, for example, DeNardis, *The Global War for Internet Governance*; Laura DeNardis, *The Internet in Everything: Freedom and Security in a World with No Off Switch* (New Haven, CT: Yale University Press, 2019); Julie Cohen, *Configuring the Networked Self: Law, Code, and the Play of Everyday Practice* (New Haven, CT: Yale University Press,

2012); Julie Cohen, *Between Truth and Power: The Legal Construction of Informational Capitalism* (New York: Oxford University Press, 2019); Benjamin Peters, *How Not to Network a Nation: The Uneasy History of the Soviet Internet* (Cambridge, MA: MIT Press, 2016).

8. This is the main motivation for my upcoming book, *Uploaded: How Copyright Battles Shape the Online World* (Yale University Press, expected 2023).

9. Journalists and popular commentators have been writing about Erlich's story for years, focused on how the Church handles its critics. See Wendy Grossman, "alt.scientology.war," *Wired,* December 1, 1995, https://www.wired.com/1995/12/alt-scientology-war/.

10. Elizabeth Rowe, "Trade Secret Litigation and Free Speech: Is It Time to Restrain the Plaintiffs?" *Boston College Law Review* 50 (2009): 1425–55.

11. Complaint, RTC v. Netcom.

12. Thomas Klemesrud, "HELENA KOBRIN CONTACT," December 30, 1994, https://groups.google.com/forum/#!searchin/alt.religion.scientology/klemesrud$20%22helena$20kobrin$20contact%22%7Csort:date/alt.religion.scientology/vgtOssuCinA/4xFN442FzGkJ4; Helena Kobrin, "Infringements through Support.com," December 30, 1994, https://groups.google.com/forum/#!msg/alt.religion.scientology/ElgUPnNNrBg/l1DcJKsA3eAJ;context-place=topic/alt.religion.scientology/9fBFJ4fMrlM.

13. Klemesrud, "HELENA KOBRIN CONTACT."

14. Thomas Klemesrud, "Infringements through 1/2," December 30, 1994, https://groups.google.com/forum/#!msg/alt.religion.scientology/yObA-VXE468/jx8KwgItp88J;context-place=msg/alt.religion.scientology/ElgUPnNNrBg/l1DcJKsA3eAJ.

15. "Verified Complaint for Injunctive Relief and Damages," RTC v. Netcom.

16. "Reporter's Transcript of Proceedings," RTC v. Netcom.

17. "Order for Temporary Restraining Order, Order to Show Cause Re: Preliminary Injunction, and for Order of Impoundment," RTC v. Netcom.

18. "Declaration of Rick Francis in Support of Defendant Netcom's Opposition to Plaintiff's Request for Injunctive Relief," RTC v. Netcom.

19. "Declaration of Rick Francis."

20. "Reporter's Transcript of Proceedings."

21. "Reporter's Transcript of Proceedings."

22. "Reporter's Transcript of Proceedings."

23. "Reporter's Transcript of Proceedings."

24. Con Diaz, *Uploaded.*

25. Con Diaz, *Uploaded.*

26. Con Diaz, *Uploaded.*

27. "Amended Temporary Restraining Order," RTC v. Netcom.

28. Electronic Frontier Foundation, "EFF Opposes Scientology Censorship and Attacks on System Operators," *EFFector Online* 8, no. 2 (February 23, 1995).

29. Electronic Frontier Foundation, "EFF Opposes Scientology Censorship."

30. Electronic Frontier Foundation, "EFF Opposes Scientology Censorship."

31. "Memorandum of Points and Authorities of Defendant Netcom On-Line Communication Services," RTC v. Netcom.

32. "Memorandum of Points and Authorities."

33. Fonovisa v. Cherry Auction, 847 F. Supp. 1492 (E.D. Cal. 1994). See also Fonovisa v. Cherry Auction, 76 F.3d 259 (9th Circuit, 1996).

34. Fonovisa v. Cherry Auction, 847 F. Supp. 1492 (E.D. Cal. 1994).

35. See Kevin Driscoll, *Hobbyists Inter-Networking and the Popular Internet Imaginary: Forgotten Histories of Networked Personal Computing, 1978–1998* (PhD diss., University of Southern California, July 16, 2014).

36. RTC v. Netcom.

37. This expansion is recounted in Con Diaz, *Uploaded.*

38. See Jeff Kosseff, *The Twenty-Six Words That Created the Internet* (Ithaca, NY: Cornell University Press, 2019).

39. See Jessica Littman, *Digital Copyright* (New York: Prometheus, 2001).

40. See Andrew Ventimiglia, *Copyrighting God: Ownership of the Sacred in American Religion* (New York: Cambridge University Press, 2019); Heidi Campbell, "Contextualizing Current Digital Religion Research on Emerging Technologies," *Human Behavior and Emerging Technologies* 2, no. 1 (2020): 5–17.

41. See the chapters by Marc Aidinoff and Scott Kushner in this volume, and Annette Markham and Katrin Tiidenberg, eds., *Metaphors of Internet: Ways of Being in the Age of Ubiquity* (New York: Peter Lang, 2020).

42. See the chapter by Tiffany Nichols in this volume.

43. An overview of this literature can be found in Con Diaz, *Software Rights.*

CHAPTER 5

Patenting Automation of Race and Ethnicity Classifications

Protecting Neutral Technology or Disparate Treatment by Proxy?

Tiffany Nichols

"Technology is neither good, nor bad, nor is it neutral," but what of the patents that define and protect it?[1] Patents are designed to provide neutral—value free—descriptions of technologies for use by inventors, companies, courts, attorneys, legal scholars, and the general public.[2] This apparent neutrality masks and obscures human decisions and assumptions that are incorporated in the disclosed technologies. Consideration of this phenomenon is particularly important when patents protect practices that entrench racial inequality and technological redlining, a concept developed by Safiya Umoja Noble to conceptualize the ways in which digital practices and technology "reinforce oppressive social relationships and enact new modes of racial profiling."[3] This article focuses on patents—and their seemingly neutral language[4]—for algorithmic methods that classify individuals into racial and ethnic categories. Specifically, I explore patents—owned by Google, NetSuite, Facebook, and Verizon—claiming proprietary protection for the predictive assignment of race and ethnicity categories to images, identification documents, and last names, and for the purposes of microtargeted advertisement and online content.[5] This chapter reveals how the neutral language of these patents obscures problematic decisions and assumptions, linking these practices to a longer history of race science and technology that inform how these automated racial and ethnic classifications are implemented and structured.

Noble has called for the application of a Black feminist perspective to contemporary technologies by "contextualizing information as a form of representation, or cultural production," rather than as seemingly benign and neutral data, software, or algorithms.[6] Patents similarly lend credibility, objectivity, and authority to automated race and ethnicity classifiers, when in fact, as I will show, these methods in fact both reproduce and hide-from-view racial injustice. I also draw from Anjali Vats and Deidré Keller's framework of Critical Race Intellectual Property (IP), which focuses on the intersection of IP, race, and social justice, and "how intellectual property law protects whiteness as property."[7] Prompted by these frameworks, I look beyond the façade of neutrality in patents, especially those applying racial and ethnic classifications, and suggest that patents are products of inventors, patent attorneys, and patent examiners, as well as contingent historical context, and not objective documents protecting standalone and neutral "inventions."

Noble has also highlighted the difficulties in determining code processes and code implementation because such information is proprietary, and thus hidden from public view.[8] I show that patents offer a powerful opportunity to open the "black box" of proprietary computing by providing epistemic insights into how those involved in the authorship of patents conceptualize race and ethnicity to uncovering their activities, agency, and assumptions.[9] I do this by turning to the invention disclosures made in patent documents. In order to receive the twenty-year monopoly afforded with a granted patent, the patentee must engage in a trade-off. They must disclose their proprietary invention in a patent application, which will be published eighteen months from the filing date allowing for public review. Such public disclosures are meant to fuel innovation and allow for incremental improvements on the patented technology while allowing the patent holder to use the monopoly period to generate a return on their research and development investments. These disclosures provide detailed insights into how inventors and patent attorneys, agents, and examiners conceptualize race and ethnicity. They also provide insights into how algorithms and software code function, and the decision processes contained therein.

My approach builds on Kara Swanson's innovative historical work with legal treatises, court decisions, and the documented labor practices of patent practitioners, which uncovered gender dynamics among patent office clerks, and friction between scientific and technical expertise among US patent examiners.[10] Vats has also applied historical techniques to constitutional and statutory law and court decisions to show that "creatorship" in the US IP context serves whiteness, excluding creators of color.[11, 12] I build on these historical approaches by focusing on patent documents themselves, which I show can be read for epistemic insights into the inventor's concepts and assumptions through the illustrative case of patents covering automation of race and ethnicity classifications.

Through utilizing a Black feminist and Critical Race IP perspective in combination with history of science and legal analysis techniques, I show that while the technologies and subject matter within these patents are new, the practice of ignoring and excluding the social context of the automated racial and ethnic coding techniques mirrors the fraught practices of the past and profits on the further marginalization of certain groups legitimized and allowed by the limited state-granted monopoly that comes with the issuance of a patent. Just as nineteenth-century race scientists collected and categorized data, combined with an "accumulation of inferences," to produce what was seen as an objective and "a scientific truth" about race and ethnicity classifications,[13] the patents explored here show similar assemblages of data and assumptions in applying a race or ethnic label to a user's information, justified and naturalized by the genre of patents themselves, and their neutral and authorizing language. These patented classification practices are particularly troubling in the context of big data and microtargeted advertisements, especially when combined with law enforcement data and systems in ways I document here.

Techniques of Neutralizing Patents and Obscuring Their Human Assumptions and Decisions

Patents are seen as neutral documents consisting of unadorned descriptions of allegedly novel technologies and methods. Their dry style, language, and uniform figure aesthetics give the impression of a uniform source. These conventions, among others, are both legally required and serve to afford the broadest scope possible for patent protection. The resulting semblance of objectivity and neutrality obscures the very human decisions, assumptions, and labor involved in the production of patent documents and the technologies they describe. It is particularly troubling when algorithms incorporating assumptions of race and ethnicity are present in issued patents. This section interrogates the façade of neutrality by providing a framework for opening the black box of patent documents.

One way that patents masquerade as neutral and objective legal instruments is through their language, which is often jokingly referred to as patent-lese. This is an overly legal and abstract prose—partially mandated by law and partially reflecting conventions of the field—that uses, for example, *comprising* instead of *has* and *embodiment* instead of *version*.[14] Patent-lese foregrounds technology while obscuring the human activities and assumptions that constitute the disclosed invention.[15] Patents are written in the passive voice, reinforcing the conceptual disassociation of hardware and algorithmic processes from those who control and design these devices and components. Hidden actors design the algorithms, supply the data, decide what data to use, and determine the objectives of the processes described in the patent. The labor of patent attorneys and agents, who work with inventors to draft the patent application and negotiate the scope of the patent with the patent examiner, and also the patent examiners, who determine whether the statutory requirements are met, is also hidden from view. Patents themselves and the technologies and methods they describe are not static documents but instead take shape through often prolonged negotiations between these hidden actors.

Software is patented in ways that particularly hide human agency from view. This is because standalone algorithms are patent ineligible subject matter, stemming from legal mandates that require such algorithms to interact with tangible elements (e.g., computer/server hardware) to make them eligible for patenting.[16] As such, software routines are patented as generic "engines"—understood to be algorithms *operating on machinery* rather than abstract standalone algorithms, emphasizing machine agency while obscuring the agency and design decisions of programmers and software engineers.[17]

The way in which patents are *granted* also contributes to the apparent neutrality and credibility of the patented techniques because the examination process parallels the peer-reviewed system used by refereed academic journals.[18] In the United States, patents are only granted to those inventions deemed new, useful, and nonobvious as determined through a rigorous review by examiners at the United States Patent and Trademark Office (USPTO) that can often take several years.[19] When it comes to software patents, applications often receive multiple rounds of review due to the vast number of existing software patents and pending applications. These repeated rounds of review are often perceived as rigorous but actually indicate a field overcrowded with applications.

Three patent features that are central—to protect inventions and grant a limited monopoly to make, use, or sell the invention protected therein—also contribute to their authoritative neutrality. First, the claims section, is required by administrative law governing patent content, defines the legal metes and bounds of an invention through reliance on incredibly abstract language negotiated during the examination process.[20]

Second, the figures are often reviewed with reference to the claims to interpret the scope of the patent. Based in practice and law, diagrams are seen as an expedited way to understand the written description section of the patent and the corresponding claims.[21] Therefore, it is common practice to review the drawings and to skip over the written description to the claims at the end of the document when discerning whether the patent is relevant to one's needs. Thus, the very format gives more weight to the figures at the beginning of patents and the claims at the end.

That third component, the written description, is my focus in this article. Although the written description is given less attention, it nonetheless is required by law.[22] Although the written description also frames the patented invention in neutral terms, being written in patentlese and the passive voice, it offers the most opportunity to read against the grain, between the lines, and recover the human agency and social context that informed the described technology. Here, the inventor is permitted to be their own lexicographer and describe their invention, revealing motivations, underlying assumptions, and epistemic commitments.[23]

In the remainder of this chapter I reveal the human decisions, assumptions, and labor masked and obscured by the construction of neutrality and objectivity in the patent genre, by focusing on patents for automated racial and ethnic classification. These patents provide timestamped, epistemic insights into how inventors—often engineers and programmers—understand and define race and ethnicity, and they reveal how automated race and ethnicity classifications are entangled with longer histories of race science and marginalizing activities, resulting in the re-creation and reinforcement of racial disparities. Patents are a rich and largely unexplored source base for scholars seeking to understand the mechanisms through which racial structures are replicated and reinforced by technology.

A Patent Claiming Last Names as a Proxy for Race and Ethnicity

The techniques of developing the content of training data sets are under increasing scrutiny for inequalities present in the data and appearing in products. Meredith Broussard has stated, "If machine-learning models simply replicate the world as it is now, we won't move toward a more just society."[24] This is because data reflects the inequalities of the society from which it was collected. The invention of US Pat. No. 10,636,048 ("'048 Patent"), assigned to Verizon, describes a technique for assigning ethnic classifications to names collected from contact lists extracted from users' phones, to improve targeted advertisements. The patent states that "automatic rather than manual labeling of training data set is desired in the context of constructing a large scale training data set with ethnicity labels,"[25] because—according to the text—parsing through and labeling a set of 1 million names with the "appropriate ethnicity label" will take "8333 working hours, and such manual labeling places personal information contained within the email at risk."[26]

Throughout the patent, the sentences are structured such that the actor is the machine learning algorithm masking the human contributions to the system. For example, the patent states, "The machine learning algorithm regressively learns correlations between the features and the ethnicity label in the training data and establishes a model that may be used to classify an unlabeled name into a most probable ethnicity classification."[27] Because the sentence was written to make the machine learning algorithm the actor, it is unclear who is supplying the ethnicity label, the training data, and the parameters for the model. The patent further states:

> Last names within each country are sorted by their popularity (such as number of occurrences). The popularity of a last name may be represented by a ranked name ratio of the last name in terms of popularity, between 0 and 1, with 0 representing the highest popularity and 1 representing the lowest popularity. . . . [E]thnic compositions or ethnicity ratios of last names may be obtained by looking up the U.S. census data. Ranked name ratio–ethnicity ratio curves for each country may be constructed.[28]

The patent offers no critical interrogation of whether ethnicity can be discerned from last names or whether such a classification system and training set should be developed at all. These questions of concern are preempted by the neutral authority of the patent. The patent justifies the automatic application of race and ethnicity labels through the *benefits to the user*—protecting their personal information—and *to the company*—an 8333-hour manpower reduction. These moves frame the invention as both helpful and benign, while also deterring inquiries into how such categories are applied and whether they should even be applied at all.

The classifications do not stop there. The patent discloses that a country of origin is predicted based on the location where a user's email account was created.[29] US census data is then used in combination with the collected names to determine the "ethnic compositions or ethnicity ratios of last names" and constructs "ratio–ethnicity ratio curves for each country."[30] The only categories provided are Black, White, Hispanic, API (although not defined in the patent, this category most likely corresponds to Asian Pacific Islander), and Unknown—all selected without justification or definition for user data originating from the United Kingdom, India, Mexico, and the United States. At first these labels appear to mirror census labels, but upon further review, there are important differences. For example, the census labels at the time of the filing of this patent were White, Black or African American, Asian, and Hawaiian/Pacific Islander as racial categories and Hispanic as an ethnic category.[31] Catherine Bliss explains that several of these categories were originally created by the Office of Management and Budget (OMB) with the aim to "monitor and redress social inequality" with the warning that the categories should not be viewed as scientific and anthropological.[32] There are fewer and different ethnic categories listed in the patent than used on the census, yet the invention still uses census records to predict the ethnicity category through name data. Moreover individuals usually select their own racial and ethnic categories when responding to the census, a flaw that has been raised time and time again by social scientists.[33] As Dorothy Roberts explains, when the underlying criteria for classifying data or people into categories are not consistent, researchers make "subjective decisions about where to place some subjects."[34] Roberts also warns that racial and ethnic categories vary by country, stating that it is "hazardous" to compare and link global data labels based on US perspectives, and variable census labels.[35] It is clear that these census labels used in the patent are not compatible with how racial groups are defined in the United Kingdom, India, or Mexico, which is left unacknowledged in the patent.

What is most telling about the inventors' understanding of race and ethnicity classification is their concept of "homo-ethnicity" to describe those countries where the inventors have assumed that "the vast majority of its population had the same or similar ethnicity."[36] Using this assumption presented as fact, the inventors claim that labeling last names as the single identified ethnicity category will produce a "high accuracy" of classification.[37] Countries are designated as homoethnic when one group makes up 80% of the population. Based on this threshold, the most frequently occurring last names are labeled with that "dominant" ethnicity.[38] The inventors offer the United Kingdom, India, and Mexico as examples of homoethnic countries.[39] However, even these countries use different ethnic and racial categories than the United States. For example, in the United Kingdom, "White" is not an omnibus category and instead Welsh, English, Scottish, Northern Irish, and British are used. For the category of "Black" the UK census employs African and Caribbean.[40] Such distinctions are not made in the United States or the patent. These discrepancies reveal that racial and ethnic identities are mutable and are influenced by social context, all of which is obscured in the patent. The patented method also ignores those individuals who are multiethnic and/or multiracial, as well as the extensive histories of migration (forced and unforced), colonization, and social norms that dictate how last names are assigned, lost, given, acquired, or used at all.

Similar to the nineteenth century, race science attempts to correlate behavior and race, the '048 Patent claims that communication activity indicates a user's race and ethnicity through using machine learning algorithms. The patent states, "A user is more likely to communicate frequently with a contact of his/her ethnicity (e.g., to send email to his/her contacts)" and thus "an ethnicity classifier may be developed by producing a learning document containing account holders' names and their frequent contact names in close proximity."[41] There is no supportive evidence provided for the claim that those frequent contacts in close proximity are of the same ethnic group. Further, the patent is silent on the history of redlining, suburban formation, and migration that may amplify racialization of the data without empirical support that a user frequently contacts those with whom they are in close proximity.

With these potential fallacies in play, the inventors also suggest crawling Wikipedia for names and ethnic identifiers. Once found, the categories are organized into Mexican, Colombian, Spanish, Cuban, and Argentinian to Hispanic and Taiwan, French, German, Indonesian, and Indian to non-Hispanic recategorization, which may not mirror the source data.[42] Assigning ethnicity as a stand-in for race in this way raises further red flags, since national boundaries do not correspond to racial categories, and they provide the misconception that race and ethnicity are static and based on nation-state boundaries.[43] Without critical reflection, the inventors claim that once the last names are associated with the chosen ethnic labels, "cultural, social, economic, commercial, and other preferences of the person" can be linked to the user data.[44]

When used for targeted advertisements—including for jobs, real estate, news, and public health content—the patented technique furthers technological redlining, reinforcing and expanding the marginalization of some groups. These methods mirror the scope of the claims, which means that Verizon, as the current holder of the patent, has a limited federal monopoly granted by the USPTO on making, using, and selling products and methods falling within the scope of the claims for at least twenty years from the earliest filing date. Verizon has already been fined by the Federal Communications Commission (FCC) for selling user location data, cell usage, and cellular activity data to a wide array of entities including those engaged in pretexting, i.e., entities not permitted to access such data but who nonetheless gain access by posing as those who do have such permission.[45] This patent also engages in a move to privatize methods that use census data—a public resource collected with taxpayer dollars—for private profit, running counter to and beyond the primary purpose of the census and even limits what others can do with the data because of the threat of litigation.[46]

Patent Assigning Racial and Ethnic Categories to Users for Microtargeted Advertisements

NetSuite's US Patent No. 10,430,859 ('859 Patent) also describes a method for identifying race for the purposes of advertising, and it too engages in the privatization of publicly available source data. NetSuite offers data-driven analysis of customers and their purchases that lends itself to targeted advertising, especially on platforms such as Facebook. The inventors of this patent claim a method of "inferring" a user's ethnicity or nationality based on their first and last names, zip code, address, and purchase history—albeit only a single purchase is necessary.[47] The first few figures in the '859 Patent are of generic hardware and do not readily reveal these aims.[48] Nowhere do these figures depict how race and ethnicity are the focus of the patent, reinforcing its neutral appearance. The claim scope is "a computer-implemented method for generating a recommendation for a product or service to a customer" that uses census data to gain "information regarding an ethnicity or ethnic group of a person based on one or more of their first name, last name or zip code."[49] Claiming the use of census data, just as the Verizon patent did, potentially bars other similar uses of public domain data while providing monopoly access to this method to the patent owner, NetSuite. Issuing a patent with such claims legitimizes the proprietary use of public census data for for-profit targeted advertisements.

The written description of the '859 Patent focuses on hardware and uses the passive voice, obscuring the inventor's agency and underlying assumptions. The patented method begins with customer address information obtained from an online purchase or creation of a loyalty account—often users provide this information without so much as a second thought.[50] When the user does not provide their address, the patented method uses their IP address instead.[51] This information is then compared with publicly available resources such as census data[52] and a "probability" of the user's "specific demographic characteristic of interest" is calculated—referred to as "educated guess" in the patent—based on correlations between demography, locations, and

Last name	prob. of being male	prob. of being female	ethnic background/prob.
Smith	not available	not available	not available
Ivanov	100%	0%	Eastern European / White–100%
Thompson	not available	not available	not available
Ivanova	0%	100%	Eastern European / White–100%
Lee	not available	not available	Chinese / Asian–100%
Gupta	not available	not available	Indian / Asian–100%

Figure 5.1. Table from the '859 Patent Showing Assignment of Ethnicity to Last Names.

purchases.[53] Based on this "guess" of the user's ethnicity, the automated system provides a "product or service recommendation," a neutral term for targeted advertisements.[54] The patent would lead one to believe that the algorithm acts alone, obscuring how human decisions are black-boxed in this method. The patented technique performs technological redlining and can only be understood in light of the history of segregation in the United States and its many instruments.

The patent provides examples of predicting ethnicity based on last name, that reveal just how arbitrary these "educated guesses" can be. For example, figure 5.1 shows that the algorithm assigns "Lee" to users labeled as "Chinese/Asian" with a 100% probability.[55] This racial assignment is troubling, given that individuals with the last name Lee are members of a diverse range of groups.[56] This discrepancy also brings to mind Ruha Benjamin's observation that the last names of African Americans and Filipinos are not predictive of their racial or ethnic group due to histories of slavery and colonialization.[57]

Techniques for assigning race and ethnicity based on IP address, geographic location, zip code, and products purchased are contained in the patent. This aggregation is merely a form of technological redlining that results from extensive guesswork and assumptions based in careless categorization and disregard of social context and history. In a society built on racial hierarchies, this patented invention is revealing: the method prioritizes race and ethnicity classifications to determine what advertisements and online users see.

Patenting Racializing Images

US Pat. No. 8,989,451 ('451 Patent), assigned to Google, provides insights into the very technologies and the inventors' understanding and biases associated with race that likely contribute to the biased image search results that mispresent Black women featured in Noble's *Algorithms of Oppression*.[58] The invention covers analyzing photographs of people to find similar images based on "visual characteristics" and "information that is known about the person."[59] The invention takes advantage of the popularity and ubiquity of digital images and their metadata including location, filename, etc. The described algorithm parses an image for "visual characteristics," which include eye color, facial type, skin tone, hair color, and hair length, using vector mathematic methods that appear neutral but in actuality are far from being free of biases.[60]

The patent first describes a system for "performing similarity searching of persons using various kinds of information."[61] As with the previous patent, the figure simply denotes a computing system and offers biased assumptions and their resulting implications. The claims merely state that the scope of the patent is directed to "a computer-implemented method for determining face similarity."[62] The written description, however, depicts a system that will find similar images to any given image, based on information stored in the "collection 120." The method for determining facial similarity searches for similarities between images such as eye color and hair color and length through the use of color profiles and histograms analyzed over a subset of pixels. It then moves to face shape, which considers the width and height across pixels that are "assigned as skin during segmentation" of the image.[63] These techniques are modernized versions of nineteenth-century physiognomy. More troubling is the provision that "other sources of information, rather than the query image," such as biographic information, are used, as they "may yield clues to classifications about a person's appearance."[64] These clues "enable similarity searching to be performed in a manner that a viewer deems to be more reliable."[65] Such clues include matching biographical information collected when an account is registered, information from a user's profile, or analysis of the environment surrounding the face or figure in an image. In addition to believing race and ethnicity appear from biographic information and photographs associated with an individual, the inventors also use names as a proxy for race, ethnicity, and gender by claiming—without support—that "two people with similar names have a better chance" of "being part of the same ethnic group (some Asian countries have little variability in their last name)."[66] As explained with respect to the '048 and '859 Patents, assumptions about race and ethnicity based on last names do not consider the historical context showing that last names are unreliable predictors of a person's race and ethnicity. To additionally inform classifications, the method uses information provided by the user's account and discerned from their platform-based activity—all without their awareness—to assign a racial and ethnic classification.

These racial classification methods present as "scientific" and "empirical" what are in fact colloquial and contingent constructions and expectations of identity.[67] This patent further provides a glimpse into the lengths to which Google's engineers will go to racially and ethnically classify its users through its image assets. It also shows how Google engineers fail to avoid replication and exacerbation of racial inequalities already present in the data.

Patent Linking Racial and Ethnic Classifications to Surveillance Technologies

Of the patents analyzed in this article, this Facebook patent—US Patent No. 10,706,277 ('277 Patent)—is the most troubling. It describes a surveillance method for automatically predicting the race and ethnicity of a person in an image in a government-issued document, such as a driver's license, based on all available aggregated data, including arrest records, demonstrating that Facebook directly profits from public carceral data. Such aggregation of data is what Vaidhyanathan warns of when stating, "Facebook's playbook has seemed to be slowly and steadily acclimating users to a system of surveillance and distribution that if introduced all at once might seem appalling."[68]

As with the patents described above, the figures of the '277 Patent provide no indication of the racial and ethnic surveillance technologies at play in the patent. They summarize the invention as extracting characteristics from a captured image and comparing those characteristics with "priori knowledge," failing to include the predictions of race and ethnicity, health insurance information, and arrest records used.[69] However, in the written descriptions, the inventors acknowledge that such information is confidential—"The nature of personally identifiable information can be confidential or otherwise sensitive," it can "include one or more of a name, an address, a social security number, an identification number, banking information, a date of birth, a driver's license number, an account number, financial information, transcript information, an ethnicity, arrest record, health information, medical information, email addresses, phone numbers, web addresses, IP numbers, or photographic data associated with the person." Depending on this sensitive data, the patent explains how to racially label users and link to arrest records.[70] The patent even claims that such information would be *safer* when placed in the care of companies such as Facebook, even though these companies have compromised user data on numerous occasions, and each day we learn more about how Facebook pushes the boundaries of privacy for profit.[71]

Once a personal identification document has been captured by the patented system—often with additional cooperation of the user who provides a self-portrait on their smart phone—all possible personal identifying information (shortened to PII) is extracted.[72] Noble calls such collective activities between a user and platform "labortainment."[73] If the identification document is a driver's license, then the system also extracts the sex, height, and full-face photograph.[74] Based on this information, the user is linked with a race or ethnicity and also with known arrest records. Just as correlating zip codes and census data to predict race and ethnicity amplifies historic social injustices, linking a race and ethnicity to the exhaustive list of personal identification information extracted and aggregated with carceral data by companies like Facebook magnifies such injustices as well.

Facebook, a for-profit social media platform, holds patent rights (until at least twenty years from the earliest filing date) for the extrapolation of personal information from government documents such as driver's licenses and arrest records as aggregated with user-generated data using its platform. The patent law system allows Facebook to profit from public data in this way, without consideration of the unethical and potentially discriminatory nature of its technologies when issuing patents. This injustice is further exacerbated by Facebook's own Terms of Service, which include provisions allowing the sharing of user profile information with other Facebook companies[75] and with law enforcement, which currently was using social media to identify and arrest protesters calling for justice in the killings of Breonna Taylor and George Floyd and the shooting of Jacob Blake. This is yet another troubling example of patenting techniques of technological redlining, and evidence that these automated techniques participate in, benefit from, and reinforce inequity and injustice. Michelle Alexander has shown that police arrests and incarceration of Blacks in the United States are disproportionately higher than for any other group.[76] Each encounter can lead to an arrest or entry of personal information into police records such as field interrogations, for example, having negative downstream results, especially if there are repeated stops of individuals who have already been recorded in the law enforcement in the system.[77] Cierra Robson also explores this phenomenon in this volume.

Facebook and the police benefit from one another. For example, the Department of Justice has released a report exploring and promoting the use of social media, specifically Facebook, by law enforcement to "mine" its data "to identify victims, witnesses, and perpetrators. Witnesses to crime—and even perpetrators—often post photographs, videos, and other information about an incident that can be used as investigative leads or evidence." In turn, Facebook even provides an informational pamphlet for law enforcement on how to create a successful Facebook presence, including directing law enforcement to make posts with images of wanted individuals. Based on the '277 Patent, it is clear that these images will likely be analyzed and coded by

Facebook's algorithms and combined with other data assets it may have.[78] Facebook not only seeks to determine each users' characteristics and eccentricities to craft and profit from microtargeted advertisements, but it also increases its profits and the accuracy of its predictions by incorporating and generating the data for a racist policing system.

Where Do We Go from Here?

On September 12, 2020, a manhunt was underway for the shooting of two deputies of the Los Angeles Sheriff's Department in Compton. The LA Sheriff's Department announced that the suspect was "a dark skinned male."[79] Posts immediately appeared across social media platforms naming an alleged culprit. They were photographs of Compton resident Darnell Hicks, his driver's license photo, address, a photo of his car, and license plate number along with claims that he was armed and dangerous. Hicks's phone received nonstop notifications with messages like "be on the lookout" along with images of the social media posts. As a father, football coach, and son, who lives with his immediate family and 93-year-old grandmother, Hicks thought this was a prank, but he quickly realized it was not. The post even found its way to (or possibly originated in) the Lakewood, California Regional Crime Awareness, Prevention & Safety Facebook Group. Although Hicks was cleared by a tweet from the Sheriff's Department, he received death threats, including social media posts by private citizens calling to "shoot on sight," for over two weeks. He has also had to retain an attorney.[80]

The news articles, attorneys, sheriffs, newscasters, the victim, and those disseminating this false information all refer to posts, their content, and those who posted and propagated the false information. Facebook and Twitter were never implicated in the harm to Hicks and his family. Once Hicks's photo and personal data were posted to Facebook and shared without his permission, this information was likely incorporated into Facebook's big data and algorithmic assets.[81] Although the Sheriff's Department announced that the information was erroneous, the damage had been done. Hicks had been falsely accused, and then through the all-pervasive Facebook platform, his personal information was made available, leading to death threats and inducing a fear that may never be resolved.

This post will have a life well beyond this situation. Now Darnell Hicks's image and location information are part of social media data assets, where they are tagged with a race, a geographic location, and keywords including "armed," "dangerous," "murder," "wanted," "black male," "gang member," and "crime." Even though the Sheriff's Office exonerated him, Facebook likely will not. Reviewing the numerous crime announcements on publicly facing crime prevention pages of Lakewood, a predominantly white area in Los Angeles with close ties to the Sheriff's Office and next to Compton, a predominantly Black area, provides a glimpse into what the 18,000+ members of the Facebook group post.[82] The posts mostly announce crime suspects, who are Black and Brown according to those members of the group. This confluence of a crime prevention group of a predominantly white area of Los Angeles, utilizing and posting an unconfirmed Sheriff's suspect's information on Facebook, and Facebook's information holdings and data analysis—along with the insights gained from analysis of the patents—reveals how these systems amplify racial injustices.

Using patents as historic sources and focusing on patents covering the automation of racial and ethnic classifications, I have shown that patents focused on microtargeted advertisements—along with their described methods—are not neutral. Rather they enshrine, reproduce, naturalize, legitimate, and authorize colloquial and problematic assumptions about racial and ethnic classification, reminiscent of nineteenth-century race sciences. This is evident both in the racial categories and identifying features listed in the patents, and in the application of these classification techniques to targeted advertising. In the place of critical definitions of race or ethnicity beyond the envisioned sorting categories, the patents are populated with inventors' (e.g., engineers' and programmers') colloquial understandings of race and ethnicity and proxy categories that further enforce everyday understandings of race and ethnicity and increase the effects of racial inequity. In conflating categories of race, ethnicity, and nation identification, these patents lay proprietary claim to public information such as census data, in which the outcomes of segregation and racial injustices as an indicator of race and ethnicity are embedded. When such data is used to determine racial and ethnic classifications, the effects of segregation and redlining are amplified. These patents use false proxies for race, including last names, phone contacts, and text associated with posted images, among other factors. What used to be racial categorization techniques using pen and paper are now adjusted for new big data technologies and placed into the private domain of the largest data-holding companies in the world through their incorporation in issued patents. The US patent system legitimizes tools that reproduce racial inequity and allows proprietary gain from public data like census and arrest data, enabling the silent proliferation of these techniques. This exemplifies the imperative need for critical and active intervention by technology companies and their employees to implement antiracist policies that avoid seeking proprietary gain and monopoly profits from race and ethnicity. I hope that the approaches in this chapter will be considered for any patented automated classification scheme to avoid such pitfalls and address such concerns in other areas, including but not limited to gender, socioeconomic class, and health status.

Notes

1. Melvin Kranzberg, "Technology and History: 'Kranzberg's Laws,'" *Technology and Culture* 27, no. 3 (July 1986): 545.
2. I thank Janet Abbate, Alex Csiszar, Gerardo Con Diaz, Stephanie Dick, Avriel Epps-Darling, Jovonna Jones, Erica Sterling, and Cori Tucker-Price for their insightful comments and feedback.
3. Safiya Umoja Noble, *Algorithms of Oppression: How Search Engines Reinforce Racism* (New York: New York University Press, 2018), 1.
4. My exploration of patent "neutrality" is inspired by Noble's demonstration that search engine results "are deeply contextual and easily manipulated," in spite of being presumed neutral and transparent (Noble, *Algorithms of Oppression*, 45) and Siva Vaidhyanathan's exploration of how social media platforms and targeted advertisement companies also lay claim to neutrality but actively influence how people think. *Antisocial Media: How Facebook Disconnects Us and Undermines Democracy* (Oxford: Oxford University Press, 2018).
5. Patents analyzed include US Pat. No. 10,636,048, Ye et al., "Name-Based Classification of Electronic Account Users," filed Jan. 27, 2017, and issued Apr. 28, 2020;

US Pat. No. 10,430,859, Olesksiy Ignatyev, "System and Method of Generating a Recommendation of a Product or Service Based on Inferring a Demographic Characteristic of a Customer," filed Mar. 24, 2016, and issued Oct. 1, 2019; US Pat. No. 8,989,451, Vincent Vanhoucke et al., "Computer-Implemented Methods for Performing Similarity Searches," filed Sept. 14, 2012, and issued Mar. 24, 2015; US Pat. No. 10,706,277, Raphael Rodriguez, "Storing Anonymized Identifiers Instead of Personally Identifiable Information," filed Feb. 19, 2019 and issued July 7, 2020. These patents include race, ethnicity, and automations in the claims section, which defines the legal scope of a patented invention.

6. Noble, *Algorithms of Oppression*, 30–34, 106; see also Yarden Katz, *Artificial Whiteness: Politics and Ideology in Artificial Intelligence* (New York: Columbia University Press, 2020), 6–7, 9, 94–97.

7. Anjali Vats and Deidré A. Keller, "Critical Race IP," *Cardozo Arts & Entertainment Law Journal* 36, no. 3 (2018): 764; Anjali Vats, *The Color of Creatorship: Intellectual Property, Race, and the Making of Americans* (Stanford, CA: Stanford University Press, 2020), 13.

8. Noble, *Algorithms of Oppression*, 4.

9. For other uses of patents as historical sources, see Peter Galison, *Einstein's Clocks, Poincare's Maps: Empires of Time* (New York: W.W. Norton & Co., 2003); William Rankin, "The 'Person Skilled in the Art' is Really Quite Conventional: U.S. Patent Drawings and the Persona of the Inventor, 1870–2005," in *Making and Unmaking Intellectual Property: Creative Production in Legal and Cultural Perspective*, ed. Mario Biagioli, Peter Jaszi, and Martha Woodmansee (Chicago: University of Chicago Press, 2011): 55–75. Alex Wellerstein, "Patenting the Bomb: Nuclear Weapons, Intellectual Property, and Technological Control," *Isis* 99, no. 1 (2008): 57–87; Mario Biagioli and Marius Buning, "Technologies of the Law/Law as a Technology," *History of Science* 57, no. 1 (2019): 3–17. Calling for the historization of intellectual property, see Kali Murray, "A Welcome Conversation: Toward a New Historiography of Intellectual Property," *Law & Social Inquiry* 43, no. 3 (2018): 1113–29.

10. See Kara W. Swanson, "The Emergence of the Professional Patent Practitioner," *Technology and Culture* 50, no. 3 (July 2009): 519–48; Swanson, "Rubbing Elbows and Blowing Smoke: Gender, Class, and Science in the Nineteenth-Century Patent Office," *Isis* 108, no. 1 (2017): 40–61.

11. Kelcey Gibbons, Mar Hicks, and Michael J. Halvorson show, in this volume, that notions of "technical skill" are similarly constructed along racial and gendered lines, to include what educated white men are doing and exclude others.

12. Vats, *The Color of Creatorship*. On the patenting of racialized technologies, see, e.g., Shubha Ghosh, "Race-Specific Patents, Commercialization, and Intellectual Property Policy," *Buffalo Law Review* 56, no. 2 (2008): 409–94; Jonathan Kahn, "Race-ing Patents/Patenting Race: An Emerging Political Geography of Intellectual Property in Biotechnology," *Iowa Law Review* 92, no. 2 (2007): 353–416; Kahn, "Revisiting Racial Patents in an Era of Precision Medicine," *Case Western Reserve Law Review* 67, no. 4 (2017): 1153–70; Kahn, "Inventing Race as a Genetic Commodity in Biotechnology Patents," in *Making and Unmaking Intellectual Property: Creative Production in Legal and Cultural Perspective*, eds. Mario Biagioli, Peter Jaszi, and

Martha Woodmansee (Chicago: University of Chicago Press, 2011): 305–20; Kahn, "Mandating Race: How the PTO Is Forcing Race into Biotechnology Patents," *Nature Biotechnology* 29, no. 5 (2011): 401–3.

13. See John S. Haller, Jr., *Outcasts from Evolution: Scientific Attitudes of Racial Inferiority, 1859–1900* (Carbondale: Southern Illinois University Press, 1971), 16–17, 128–29, 140; Ann Fabian, *The Skull Collectors: Race, Science, and America's Unburied Dead* (Chicago: University of Chicago Press, 2020); Evelynn Maxine Hammonds, Rebecca M. Herzig, eds., *The Nature of Difference: Sciences of Race in the United States from Jefferson to Genomics* (Cambridge, MA: MIT Press, 2008), 107.

14. Dan L. Burk, Jessica Reyman, "Patents as Genre: A Prospectus," *Law & Literature* 26, no. 2 (2014): 163–90. Gerardo Con Diaz also argues in this volume that legal history, including copyright decisions and litigation, can establish and stabilize the definition and scope of a technology.

15. Mary Louise Pratt, "Scratches on the Face of the Country; or, What Mr. Barrow Saw in the Land of the Bushmen," *Critical Inquiry* 12, no. 1 (1985): 119–43; Nancy Lays Stephan and Sander L. Gilman, "Appropriating the Idioms of Science: The Rejection of Scientific Racism," in *The "Racial" Economy of Science: Toward a Democratic Future*, ed. Sandra Harding (Bloomington: Indiana University Press, 1993), 174.

16. 35 U.S.C § 101; *Alice Corp. Pty. Ltd. v. CLS Bank Int'l*, 573 U.S. 208 (2014) (citing *Ass'n for Molecular Pathology v. Myriad Genetics, Inc.*, 569 U.S. 576, 589 (2013); US Patent and Trademark Office, "2106 Patent Subject Matter Eligibility" in *Manual on Patent Examination and Procedure*, 9th ed. (Alexandria: USPTO, 2019 [2020]).

17. Con Diaz's research shows that software patents focus on the "programmed machine" rather than the human practice of programming, obscuring and devaluing the human contributions. Con Diaz, "Intangible Inventions," 787–88. In contrast, De Mol and Bullynck provide in this volume an earlier computing patent from 1937 describing computer operators and their programming before changes in technology that shifted programming from a physical activity to one of intangible algorithmic activity.

18. This is even codified in statutory law. The statute states, "A patent shall be presumed valid." 35 U.S.C. 282.

19. US Patent and Trademark Office, "2103 Patent Examination Process," in *Manual on Patent Examination and Procedure*, 9th ed. (Alexandria: USPTO, 2019 [2020]).

20. 37 C.F.R. 1.75.

21. 35 U.S.C. 113.

22. 37 C.F.R. 1.71.

23. US Patent and Trademark Office, "IV. Applicant May Be Own Lexicographer and/ or May Disavow Claim Scope," in *Manual on Patent Examination and Procedure*, Section 2111.

24. Meredith Broussard, *Artificial Unintelligence: How Computers Misunderstand the World* (Cambridge, MA: MIT Press, 2018), 115.

25. Ye et al., "Name-Based Classification," col. 5, 13–16.

26. Ye et al., col. 5, lns. 16–27.

27. Ye et al., col. 5, lns. 5–9.

28. Ye et al., col. 5, lns. 37–46.

29. Ye et al., col. 5, lns. 28–31.

30. Ye et al., col. 5, lns. 37–46.

31. Pew Research Center, "What Census Calls Us" (February 6, 2020), https://www.pewresearch.org/interactives/what-census-calls-us/.

32. Catherine Bliss, *Race Decoded: The Genomic Fight for Social Justice* (Palo Alto, CA: Stanford University Press, 2012), 20–21.

33. Dorothy Roberts, *Fatal Invention: How Science, Politics, and Big Business Re-Create Race in the Twenty-First Century* (New York: The New Press, 2011), 71; Dan Bouk, "Error, Uncertainty, and the Shifting Ground of Census Data" in *Harvard Data Science Review* 2, no. 2 (2020): https://doi.org/10.1162/99608f92.962cb309; Abram Gabriel, "A Biologist's Perspective on DNA and Race in the Genomics Era," in *Genetics and the Unsettled Past*, 52.

34. Roberts, *Fatal Invention*, 71.

35. Roberts, 20, 71.

36. Ye et al., "Name-Based Classification," col. 5, lns. 56–59.

37. Ye et al., col. 5, lns. 56–63.

38. Ye et al., col. 5, ln. 60-col. 6, ln. 11.

39. Ye et al., col. 6, lns. 3–8.

40. Office for National Statistics, "Ethnic Group, National Identity and Religion" (January 18, 2016), https://www.ons.gov.uk/methodology/classificationsandstandards/measuringequality/ethnicgroupnationalidentityandreligion.

41. Ye et al., "Name-Based Classification," col. 6, lns. 59–65.

42. Ye et al., col. 12, lns. 1–12.

43. Roberts, *Fatal Invention*, 75.

44. Ye et al., "Name-Based Classification," col. 12, lns. 15–20.

45. Federal Communications Commission, "Notice of Apparent Liability for Forfeiture and Admonishment," *In re Matter of Verizon Communications*, FCC 20-25 (February 28, 2020): 4, https://docs.fcc.gov/public/attachments/FCC-20-25A1.pdf.

46. Ye et al., "Name-Based Classification," col. 13, lns. 34–40.

47. Ignatyev, "System and Method of Generating a Recommendation of a Product," title page, including abstract, col. 16, lns. 8–19.

48. Ignatyev, "System and Method," Figure 3.

49. Ignatyev, col. 28, lns. 47–48.

50. Ignatyev, col. 16, lns. 33–45.

51. Ignatyev, col. 16, lns. 46–55.

52. Ignatyev, col.14, lns. 31–37; col. 16, lns. 8–19.

53. Ignatyev, col. 16, lns. 48–57.

54. Ignatyev, col. 16, lns. 58–64.

55. Ignatyev, col. 20, lns. 25–33.

56. For example, the Oscar-winning director and screenwriter Shelton Jackson "Spike" Lee and Tommy Lee of the band Mötley Crüe.

57. Ruha Benjamin, *Race after Technology: Abolitionist Tools for the New Jim Code* (Medford, MA: Polity Books, 2019), 146.

58. Noble, *Algorithms of Oppression*, 107–9.

59. Vanhoucke et al., "Computer-Implemented Methods for Performing Similarity Searches," Abstract.

60. Vanhoucke et al., "Computer-Implemented Methods for Performing Similarity Searches," col 4., lns. 25–28; lns. 49–56. Benjamin, *Race after Technology*, 112–13, 124.

61. Vanhoucke et al., "Computer-Implemented Methods for Performing Similarity Searches," Figure 1.

62. Vanhoucke et al., col. 12, lns. 41–42.

63. Vanhoucke et al., col. 5., lns. 10–12.

64. Vanhoucke et al., col. 5., lns. 53–58.

65. Vanhoucke et al., col. 5., lns. 58–60.

66. Vanhoucke et al., col. 9., lns. 30–33, col. 8., lns. 28–36.

67. Vanhoucke et al., col. 6., lns. 30–47. For results testing Google's deep-learning image captioning system, see Katz, *Artificial Whiteness*, 113–14.

68. Vaidhyanathan, *Antisocial Media*, 73–75.

69. Rodriguez, "Storing Anonymized Identifiers," Figure 7.

70. Rodriguez, col. 1, lns. 20–27, col. 2, lns. 30–35.

71. See Vaidhyanathan, *Antisocial Media*, 73–76.

72. Rodriguez, "Storing Anonymized Identifiers," col. 19, ln. 61, col. 20, ln. 1.

73. ". . . when users consent to freely give away their labor and personal data for use of Google and its products, resulting in incredible profit for the company." Noble, *Algorithms of Oppression*, 36.

74. Rodriguez, "Storing Anonymized Identifiers," col. 1, lns. 1–5.

75. Facebook, "Terms of Service," https://www.facebook.com/terms.php.

76. Michelle Alexander, *The New Jim Crow: Mass Incarceration in the Age of Colorblindness* (New York: The New Press, 2010 [2012]), 6–7, 113, 192.

77. Isaiah Thompson, "Black People Made Up 70 Percent of Boston Police Stops, Department Data Show," *WGBH*, June 12, 2020, https://www.wgbh.org/news/local-news/2020/06/12/black-people-made-up-70-percent-of-boston-police-stops-department-data-show; Zak Cheney-Rice, "California Police Are Falsely Labeling People as Gang Members. It's Part of a Bigger Crisis," *New York Magazine*, January 7, 2020, https://nymag.com/intelligencer/2020/01/lapd-falsely-labeling-gang-members.html.

78. Community Oriented Policing Services (COPS), US Department of Justice, "Social Media and Tactical Considerations for Law Enforcement" (July 2013), https://cops.usdoj.gov/RIC/Publications/cops-p261-pub.pdf; Facebook, "Building Your Presence with Facebook Pages: A Guide for Police Departments," https://developers.facebook.com/attachment/PagesGuide_Police.pdf.

79. Alex Villanueva, "Advisory: Manhunt Underway for Suspect in Ambush Shooting of 2 LA Sheriff's Deputies in Compton," *Los Angeles County Sheriffs Dept Information Bureau (SIB)*, September 12, 2020, https://local.nixle.com/alert/8240660/?sub_id=0.

80. Richard Winton, "Social Media Accused Him of Ambushing Two Deputies. It Was Fake News, but He's Paid a Steep Price," *Los Angeles Times*, September 16, 2020,

https://www.latimes.com/california/story/2020-09-16/falsely-accused-on-social
-media-of-being-a-shooter-he-feared-for-his-family; Erika Martin and Kimberly
Cheng and Nexstar Media Wire, "California Father Falsely Accused in Ambush of
Sheriff's Deputies Speaks Out," *Fox 8 Local News, Los Angeles*, September 15, 2020,
available at https://myfox8.com/news/california-father-falsely-accused-in-ambush
-of-sheriffs-deputies-speaks-out/; Street TV, "Who Is Trying to Frame Darnell
Hicks as the Shooter of 2 Deputies in Compton?" September 16, 2020, https://www
.youtube.com/watch?v=eWSHaDUqmuA (an interview of Darnell Hicks); Snejana
Farberov, "Compton Father-of-Two, 33, Says He Is Receiving Death Threats after
Being Falsely Accused on Social Media of Shooting of Two LA County Sheriff's
Deputies," *Daily Mail*, September 15, 2020, https://www.dailymail.co.uk/news
/article-8735021/Compton-father-falsely-accused-social-media-ambush-two-LA
-County-sheriffs-deputies.html; Los Angeles County Sheriff's Office Twitter Page,
available at https://twitter.com/ LASDHQ/status/1305366021310570496/. I have
opted not to reproduce images of the posts to avoid causing further harm to
Mr. Hicks.

81. Vaidhyanathan, *Antisocial Media*.

82. Facebook Group, LAKEWOOD CA. REGIONAL CRIME AWARENESS, PREVEN-
TION & SAFETY GROUP, Facebook, https://www.facebook.com/groups
/167687367111156/.

"Difficult Things Are Difficult to Describe"

The Role of Formal Semantics in European
Computer Science, 1960–1980

Troy Kaighin Astarte

In 1970, Friedrich Bauer sketched a history of computation from abacuses to ALGOL 68, which he presented as a pinnacle of achievement in computing.[1] ALGOL 68 was the culmination of the ten-year international ALGOL project, "the seed around which computer science began to crystalize as an academic discipline."[2] A key aspect of the project was the use of formalism in the language description; formalism meant abstractions, fixed structures, and rule-based reasoning borrowed from formal logic. ALGOL 68's predecessor ALGOL 60 had a formalized syntax, and the ALGOL Committee had unanimously resolved in 1964 that the new language must be "defined formally and as strictly as possible in a meta-language."[3]

Since the late 1950s, a certain community of influential computing researchers began to view formal specification of programming languages as a crucial part of research into computation, and central to the practice of language design and implementation. Many in this community worked on formal semantics during the 1960s, but by the end of the 1970s most had moved off the subject. This mirrored a general trend in computing research away from languages and toward programs or programming techniques.[4] Part of this trend was the rising interest in program verification, a topic which is better known to historians than semantics.[5] Indeed, the two most-cited works in verification both show a concern for semantics.[6] The study of formal semantics established an intellectual agenda which carried through into work on verification.

The history of computing shows the programming community in the late 1950s grappling with their work, which was turning increasingly professional, scientific, and impactful through its embedding in society. Programming, a nebulous practice somewhere between material and abstract, cried out for understanding: an ontological gap yawned. The opportunity to bridge the gap with tools and agendas from mathematics and especially formal logic also provided legitimization for the nascent "computer science," allowing it to be rooted in, but distinct from, existing intellectual traditions.[7]

This chapter explores the context of formal semantics and situates the work within the history of computer science, discussing how and why the community involved changed its focus to program correctness. The inability of formal language semanticists to embody their work in broader programming practices gives us clues to why formal semantics declined. An immediate reason was expressed in Hoare's remark "difficult things are difficult to describe."[8] Programming languages of the sort being developed in the 1960s contained many complex, interacting features that challenged easy abstractions; the same reasons made formal description embodiments in implementations difficult. More subtly, the community's strongly held belief that formal specification was a prerequisite for producing correct and reliable programs was challenged by industry's de facto policy that "good enough" programs could be produced without the lengthy and costly processes many saw as preliminary at best.[9]

A Crisis in Programming

At a NATO-sponsored conference in 1968 a crisis was declared: software had become too important to society for the sloppiness of modern programming.[10] One solution was "software engineering," interpreted by some as "the reduction of programming to a field of applied mathematics."[11] This tension between "technological apocalypticism" and a genuine fear in the "fragility of infrastructure" mirrors and perhaps foreshadowed the later Y2K crisis.[12] Though the threat was not as immediate in the 1960s, both crises served as motivators to action. Unlike the very real danger posed by Y2K, however, the software crisis may have been a retroactive construction by academics, especially Dijkstra,[13] to justify their own research agenda: mathematical abstractions for specifying and verifying programs.

One cause for concern was programming languages, called by Dijkstra the "basic tool" of computing.[14] Early programming codes had a direct correspondence to machine operations; programs were written to fit the peculiarities of specific machines.[15] The high-level languages of the 1950s, more abstract in their notation and automatically translated to run on machines, departed from this direct meaning. Dissociating program from machine was a consequence of machine-independent languages, but it opened up the possibility of new ontologies of programs. If a program was written in terms of abstract entities and its execution was understood in terms of hardware function, could a truly believable correspondence between the two be established? One critical step was the creation of the ALGOL languages, intended from their genesis to be machine-independent. This made the definition documents the ultimate reference, but fears remained that documents would prove insufficient to solve language ambiguities and determine implementation correctness.[16] Decoupling languages from machines reified them into objects of study in their own right, and ALGOL became the keystone of European programming: a paradigm shift in thinking about programming.[17]

Formalism addressed the ontological problem of programming by creating an abstraction of machine operations, thereby subjecting it to the (then prevalent) logico-mathematical view of human thought.[18] The use of computers by, for example, Newell and Simon in the early 1950s to generate mathematical knowledge represented a new kind of epistemology; Babintseva problematizes the way in which this "mathematical reasoning [was accepted] as the standard of human thinking."[19] This American view of computation was moulded and reshaped in the European scientific environment to form a different path for theoretical computer science: program correctness and "formal methods" rather than complexity theory.[20]

Exploring Semantics

In the 1950s, John McCarthy developed LISP as part of the same mathematical reasoning effort as Newell and Simon.[21] McCarthy then realized LISP could also be used to form abstractions for generic language features: recursive functions manipulating a program state.[22] McCarthy wrote these functions using the notation (but not the conversion rules) of lambda calculus; he later admitted he did not properly understand the system.[23] Lambda calculus comprises a notation for mathematical functions, and a series of conversion rules for its terms. Subsequent research demonstrated its expressive power to be equivalent to Turing machines; this "Church-Turing" thesis was a significant result in logic and mathematics. By borrowing lambda notation—if not its substance—McCarthy placed his work in that lineage.

This was deliberate: McCarthy wanted to create a "mathematical science of computation." While logicians and mathematicians such as Gödel and Turing had worked on computability problems in the 1930s, the embodiment of computation in the electronic machines of the 1940s and 1950s created a new problem: What exactly could these computers *do*? McCarthy's goal was to find the basic notions of computation and then explore what deductions could be made from there. One way to gain insight into the way practitioners think about their work is to examine the historical models they choose to frame their own work.[24] McCarthy demonstrates this presented pedigree by referring to Kepler's laws of planetary motion being derivable from Newton's laws of general motion, and placing his own work in the same tradition.[25] This desire to "scientize" computing lent a legitimacy to the nascent field and was embodied by the growing use of the term "algorithm," a word associated with the scientific method, Newton and Leibniz, and procedural practice.[26]

McCarthy's work was discovered by Peter Landin in 1958; Landin was impressed by LISP, particularly its abstract interpreting machine.

However, he spotted the problems with McCarthy's lambdas and wanted to see if he could "[tie] up some loose ends."[27] At this time, Landin was working for a consulting business run by Christopher Strachey—ostensibly developing an Autocode compiler for a Ferranti computer.[28] Drawing on his background as a mathematics graduate, Landin decided to try using lambda calculus as an "interlingua" and hoped a compiler would emerge from the semantics of this language.[29] By 1963, he published his lambda calculus approach to semantics: translate parts of a language into a "syntactically sugared" form of lambda notation, and then interpret this with an abstract computing machine.[30]

Concurrently, Strachey was also becoming interested in programming languages. Having co-authored a paper with Maurice Wilkes criticizing some aspects of ALGOL 60, he was invited by Wilkes to take part in developing a new language, CPL.[31] Strachey threw himself into the task but became more interested in formalizing the language's semantics than writing a compiler.[32] Landin's approach appealed to Strachey, who wrote with pride that he was sponsoring "the only work of its sort [language theory] being carried out anywhere (certainly anywhere in England)."[33]

In September 1964, McCarthy, Landin, and Strachey attended an International Federation for Information Processing (IFIP) working conference near Vienna, entitled Formal Language Description Languages (FLDL).[34] This conference brought together programmers, mathematicians, and linguists, allowing sharing of ideas and setting of a research agenda: namely, that programming languages were worthy objects of study and that formalism provided the toolkit for that study.[35] It is no coincidence the ALGOL Committee (IFIP Working Group 2.1) decided in 1964 to fully formalize ALGOL 60's successor: its autumn meeting coincided with FLDL to draw from the conference's findings.[36]

One important outcome of FLDL was its influence on the new IBM Research Laboratory in Vienna (internally abbreviated VAB—not an acronym). Headed by Heinz Zemanek, the group originated in a research team at the Technical University of Vienna. Prior to the IBM move, the group had developed an ALGOL 60 compiler for its Mailüfterl computer. Two members of the group recorded their experience in a technical report; they felt that without a "complete and unambiguous" formal definition, the task was long and difficult.[37] Zemanek, chair of the FLDL organizing committee, staffed the conference with members of the VAB, seeing an opportunity for his team to learn the state of the art in language formalism.[38] They needed this, because in late 1964 the VAB had agreed to write a formal definition of IBM's new programming language, PL/I. This was an ambitiously large and complex language intended to "meld and displace FORTRAN and COBOL," freeing IBM from the need to maintain compilers for multiple languages.[39] Vienna was instructed to develop a formal definition that would aid the compiler development.[40]

The first version, finished in 1966, formalized nearly every aspect of the syntax and semantics of PL/I.[41] It was a huge document—over 300 pages—known internally as the "Vienna telephone directory."[42] Drawing influence from McCarthy and Landin, PL/I was defined in terms of its actions on a large and complex abstract machine.[43] Successive versions refined the approach, adding in abstractions of storage and the new features constantly being added by language control.

By 1968 the Vienna group had defined the entirety of a complicated language, managing its storage, tasking (concurrency), scoping, and rich type system. The VAB had produced a definition of ALGOL 60 and a notational guide,[44] and, critically, the PL/I definition had been used to prove properties about an implementation.[45] Later work by Henhapl used the definition to find bugs in the compiler; he and Jones, temporarily visiting the Lab from compiler development in Hursley, proposed an algorithm to fix them.[46] These were first steps toward a use for formal semantics: while the VAB had demonstrated "practical feasibility" for proofs about a compiler, they became "unsurmountably [sic] lengthy and tedious" when attempted on a full-size language like PL/I.[47] The Viennese work was not received positively and mostly served as a counterexample in many technical responses: a nightmarish behemoth to scare readers into preferring shorter formalisms.[48] Both Strachey and Hoare cited Viennese work but only in the context of an argument that a better way to define semantics must exist.[49]

Tony Hoare began his career in computing working for Elliot Brothers, where he wrote an ALGOL 60 compiler.[50] He had learnt about ALGOL at a workshop in Brighton taught by Landin, Dijkstra, and Naur. Hoare was impressed by the ALGOL report's formal syntax and wanted to solve semantics similarly formally.[51] Hoare joined the ALGOL committee and was also present at FLDL, where he made a comment that presaged his later work. Hoare noted that any complete implementation of a language (such as a compiler) could be viewed as a precise definition, and vice versa. However, for real utility, a definition should not be concerned with machine-specific details and furthermore should have "a mechanism for failing to define things."[52] In other words, Hoare explicitly argued for the utility of abstraction over embodiment when considering a programming language.

Hoare began sketching an alternative to the Vienna approach while attending a 1965 workshop on the method. Citing inspiration from Peano's axioms of arithmetic, Hoare directly opposed the "constructive," state-based style used by Landin and the VAB, arguing instead for choosing appropriate basic concepts—axioms—that represented key properties of programming languages. Hoare's approach was presented in contrast to a state-based one because of the difficulty in specifying every single aspect of a state. Rather, Hoare advocated specifying only properties of interest for a given program construct and defining the feature by its effect on these properties.[53]

After working through a series of unpublished drafts for a semantic method,[54] Hoare first presented his work at a 1969 meeting of IFIP's Working Group 2.2, which had been formed after FLDL as a continuation of that conference's formal language description agenda. Hoare's "axiomatic method" met with much interest, positive and negative.[55] Some praised the simplicity and elegance of the approach, while others were skeptical about its ability to scale to more complex features.

At the same meeting, Christopher Strachey met Dana Scott. Strachey had continued working on language description after his consultancy closed and presented some ideas at FLDL.[56] Influenced by his experience with CPL and Landin's lambda calculus work, Strachey started by considering properties of machine storage, deciding that functions were an appropriate abstraction for the core of computation. Computer stores could be seen as functions mapping addresses to values, and statements were functions mapping stores to stores. Like McCarthy, Strachey used lambda calculus to represent these functions, and he called his method "mathematical semantics." Strachey had not worked out an appropriate basis for these functions, however, and he later acknowledged his approach was at that time "deliberately informal, and, as subsequent events proved, gravely lacking in rigour."[57] Scott, a logician who had studied under Alonso Church, was intrigued by Strachey's propositions at Working Group 2.2, and joined his Programming Research Group (PRG) in Oxford for the autumn term of 1969.

Thus began a period described by Scott as "feverish activity" and "one of the best experiences in [his] professional life."[58] Scott developed a theory of domains with the properties required to model programming languages, and this gave a foundation to Strachey's semantics.[59] Further extensions to the approach were made by students and researchers at the PRG, and by 1974 it was used to model ALGOL 60[60] and the large pedagogical language Sal.[61] Scott/Strachey semantics was more concise than the Vienna work, but it involved complex mathematics—and inclusion of features like jumps required extensive convolutions of the functions. Although others carried forward research in mathematical semantics, Strachey's contributions came to an abrupt end in 1975 with his sudden death.

Community Commitments

Strachey's decision to label his semantics "mathematical" (despite his concession that he was "not a mathematician"),[62] exemplifies the desire amongst semanticists to provide a "mathematical theory of computation" to legitimize the emerging field of computing.[63] As McCarthy had referenced Newton and Kepler, so Hoare evoked mathematical "greats" to justify his work: "[this approach] may represent the same magnitude of advance in Computer Science as the axiomatic geometry of Euclid compared with the crude land-measurement of the ancient Egyptians."[64] Zemanek, too, gave his group's work a prestigious Viennese intellectual lineage, referring to the *Wiener Kreis* and Wittgenstein.[65]

As computers became involved in the practice of writing,[66] practices were borrowed further from linguistics. Formal linguists such as Chomsky were invited to the FLDL conference; VAB's Lucas later wrote that the view was that "by viewing computers as language interpreting machines it [became] quite apparent that the analysis of programming (and human) languages [was] bound to be a central theme of Computer Science."[67]

Bringing tools and techniques from one field to another also carries the source field's agenda.[68] Tarski and Carnap's semantics of formal logic influenced formalization in computing; the bridging of these practices and the formalist agenda generated what some saw as a new science of computing.[69] However, as Babintseva observes, mathematicians do not rely solely on demonstrative reasoning, and indeed computing formalists show a greater obsession with formality than is present in mathematics.[70] DeMillo and Lipton, critics of formalism, explained the pervasiveness of this formalistic attitude: "The amount of influence that [formalism] had in computing at the time was enormous. It was *the* only way to understand programs; it was *the* only way to understand the foundations of programming."[71]

We might observe a kind of performative mathematization in the works discussed here, where references to mathematics are used to legitimize the authors' own beliefs in the "theory of programming language semantics" or "axiomatic basis for computer programming." This is evidenced by Strachey's initial unconcern about the lack of foundation for his function-based approach and in McCarthy's incorrect application of lambda binding. Hoare, too, initially presented his system of axioms without consideration of consistency or completeness, which made it gravely lacking by mathematical standards.

There was tension within the formalist community about precisely which values embodied its mathematical ideals, as the following exchange in response to Hoare's WG 2.2 presentation shows:[72]

Caracciolo: A reduction to simpler questions would mean to omit the proper problem.

Scott: Only the most primitive, non-problematic things have been dealt with using this approach.

Laski: A language definition should specify as little as possible.

Hoare was asked whether his method had been applied (so far) only to simple languages:

Hoare: Yes. But, of course, difficult things are difficult to describe.

Strachey: What is "difficult" very much depends on the frame-work of thinking. For example, assignment is difficult in the λ-calculus approach, recursion is difficult in other systems. But both occur in programming languages and are simple to use.

In other words, while there was agreement on the importance of abstraction, the choice of abstraction was not universally agreed. The intellectual values held by computer scientists came into conflict: elegance and parsimony ("specify as little as possible") contrasted with expressiveness and power ("cover as much ground as possible"). A subtler value is the cachet of working on problems that were recognized as difficult. Strachey's later description of his earlier work as "deliberately informal" and "lacking rigour" shows him framing his current work as more substantial and worthy—even though at the time he had derided the excess of formality in other works.

These values of intellectualism and abstraction were also in tension with a desire to present formal semantics work as practically important: embodying the formalisms would demonstrate their worth. Influenced by the product-directed atmosphere of IBM, the Viennese had PL/I compilers as ultimate end goals, and their work on proof was intended to improve the reliability of these. Hoare wanted his work to be useful for the creation of machine-independent standards for programming languages;[73] and Landin saw his methods as a way to improve language design.[74] Strachey, meanwhile, called practical work performed without knowledge of fundamental principles "unsound and clumsy" and theory work with no connection to practical programming "sterile."[75] This goal of dual relevance was typical of computer science in the 1960s and '70s: many computing academics were theoreticians hoping to build an embodied science of computing by establishing a theoretical, abstract base and working upward. This "reflective closure" was intended as a rationalization of existing practice through theory-forming.[76]

Description vs. Design

From the first presentations, formal semantics faced heavy criticism, most commonly that descriptions were too large and complex. This is illustrated by an exchange at an ALGOL 68 meeting:[77]

> *Turski:* In Grenoble we decided that the proposed description method is a milestone in the development of the language.
> *Randell:* A milestone or a millstone?
> General laughter follows.

IBM's Language Control reacted similarly poorly to the 1966 PL/I definition, one member commenting that using the document to handle PL/I was akin to reading *Principia Mathematica* to perform addition.[78]

Frequently, this negativity was in response to models of whole programming languages. Toy examples showed it was relatively easy to give semantics to trivial chunks of programs, but at scale semantic descriptions became unwieldy. The distinction between complexity in definition method and language under definition was often overlooked: much of the bulk in the ALGOL 68 and PL/I definitions was due to richness of features in those languages.[79] Strachey and Milne addressed this: "A superficial glance will show that our essay is long and our notation elaborate. The basic reason for this unwelcome fact is that programming languages are themselves large and complex objects which introduce many subtle and rather unfamiliar concepts."[80] In other words: "difficult things are difficult to describe."

The embodiment of formalism in language design was rare (ALGOL 68 is a notable example, but the language was hamstrung by its arcane presentation). Jones suggested that a reason for this was that semantics workers had not created a literature that was useable by and useful to language designers.[81] Plotkin agreed, explaining that engaging with formal language description was unnecessarily difficult for most language designers.[82]

Work on formal semantics did not cease as a result of this criticism,[83] and the ANSI PL/I standard, while not fully formal, shows clear inspiration from the Vienna work.[84] But many of the semanticists discussed here were disheartened by the reception to their work. McCarthy's interest in semantics dwindled at the tail end of the 1960s, partly due to the cool response his ideas had received, and he instead became interested in Manna's program proving techniques.[85] This was typical: rather than giving up on abstraction, many reinvented themselves as program verificationists. Later, McCarthy reframed his work on formal semantics as "ultimately aiming at the ability to prove that an executable program will work exactly as specified."[86] He added that he had believed that a situation would emerge where "no-one would pay money for a computer program until it had been proved to meet its specifications."[87]

The Vienna group's interests changed similarly throughout the 1970s. Having experienced difficulty proving theorems using its 1960s approach,[88] the group was given the chance to try again in late 1972, when IBM wanted a PL/I compiler for its new "Future Systems."[89] (The need for a commercial embodiment of its formalism was a constant source of friction.) Jones rejoined the Vienna Lab from Hursley, and the group adapted Strachey's method. Using this "denotational" approach, the group produced a definition of PL/I which was shorter and more amenable to formal reasoning.[90] The project ended abruptly when IBM cancelled Future Systems, but Jones and colleagues collected the work into an edited volume.[91] The collection includes some discussion of programming language definition but is more concerned with "correct by construction" program development. This took ideas from the semantics work, but working with proofs at a program level was easier due to the smaller scale of the problem.[92]

Hoare's early drafts of his axiomatic method showed an emphasis on semantics, but when the ideas saw publication in 1969 his focus had shifted. The paper highlighted a "logical basis for proofs of the properties of a program," with language definition relegated to the final section. Hoare became one of the loudest advocates of the "strong verificationist" position: that all computing was reducible to mathematics and could be subject to formal proof.[93] Hoare's position was shared by Edsger Dijkstra, who viewed programming languages as mathematical objects with important aesthetic properties[94] and had worked for a time on formal semantics.[95] Like Hoare, however, Dijkstra's agenda shifted from languages to programs; while his work on "predicate transformers" included mention of semantics, the aim was "the goal-directed activity of program composition."[96]

This change in "technocratic paradigm" from language to program was typical of the 1970s.[97] An example is the formation of IFIP's Working Group 2.3, which followed the publication of ALGOL 68 as an IFIP-sponsored language. A significant minority of the ALGOL Committee was concerned that the ALGOL 68 team had become too focused on its language and its definition method, losing sight of what the Minority Group felt to be the real goal: enabling good programming.[98] These members signed a "minority report" (drafted by Dijkstra)[99] and formed a new Working Group, whose remit was "Programming methodology." WG 2.3 was something of an "elite member's club," with gate-kept entry and a deliberately relaxed atmosphere.[100] The initial chair, Mike Woodger, noted that "programmers needed tools other than bigger and better programming languages";[101] and rather than focusing on languages, proving program correctness was one of the group's chief concerns.

Around the same time, IFIP WG 2.2 on language description went on hiatus following a period of identity crisis.[102] One émigré was Cliff Jones, who moved to WG 2.3 in line with his own changing priorities;[103] Hoare, Strachey, and others moved at the same time. Summaries of the first decade of WG 2.3 meetings show a number of similar talking points to WG 2.2, including formal semantics.[104] These were, however, discussed as general *concepts* rather than in the context of particular programming languages. Even as the focus among formalists changed from languages to programs, the emphasis on abstraction and mathematics had not gone away.[105]

Conclusions

The history of formal semantics is a key part of the history of programming languages and program verification. Formal semantics offered a way to understand programming languages and programs that had been decoupled from machines and stood as a chief component in a theory of computation. Crucially, this is a European narrative.[106] In the United States, complexity theory dominated the academic discourse, and Mc-Carthy's work on semantics did not impress his American colleagues.[107] Formal semantics work fits instead within the ALGOL research program, a peculiarly European initiative.[108] This was noticed at the time: McIlroy recalls a stark difference between the people he met at FLDL and his American colleagues: "It was real computer science, in Europe. There were no real computer scientists in the US at the time."[109] The national computing cultures in Europe, with their focus on formal logics and mathematics, shaped a different interpretation of the science of computing, challenging the US-dominated narrative.[110]

Semantics work was also advanced as a tool with practical implications: compiler creation and language design. However, the complexity of embodying the abstractions of formal semantics in full-scale programming languages prevented immediate use. Instead, as Hoare outlined, there was a slow bridging of practice from formal semantics and verification: over-engineering, bug-checking, and the associated mindset of structured programming.[111]

The 1970s saw a change amongst semanticists, who came to realize that programming languages were "difficult things" which were "difficult to describe." Instead, a new emphasis came on applying formalism at the program level, a smaller-scale challenge: in a kind of academic sleight of hand, specifying languages was substituted for specifying programs. Although the focus changed, the researchers involved continued to claim the same end results: mathematized computing and more reliable programs.

The relationship between formal semantics and verification had in a sense switched. While the ability to write proofs was once a goal of a language description, now the role of formal semantics had become instead a part of verification: the fundamental basis for some formal proofs.[112] The intellectual (or even moral) commitment to provability as a form of knowability shows a continuity in the ontology of programming shaped initially by formal semantics work, and the fervor of narratives like "Grand Challenges in Verification" persists well into the twenty-first century.[113]

Acknowledgments

Many thanks to Janet Abbate and Stephanie Dick for helpful editorial remarks; Cliff Jones and David Dunning for feedback on drafts; and Margaret Gray for advice and copy editing. The chapter emerged from a talk at the 5th History and Philosophy of Computing Conference; thanks to the attendees who made comments. This research was supported by an EPSRC studentship and writing this chapter by a Leverhulme Trust grant RPG-2019-020.

Notes

1. Friedrich L. Bauer, "From Scientific Computation to Computer Science," in *The Skyline of Information Processing: Proceedings of the Tenth Anniversary Celebration of the IFIP*, ed. Heinz Zemanek (North-Holland, 1972).

2. Thomas Haigh and Paul E. Ceruzzi, *A New History of Modern Computing* (Cambridge, MA: MIT Press 2021), 43.

3. R. E. Utman, "Minutes of the 3rd Meeting of IFIP WG 2.1," March 1964, 19–20, Ershov Archive, http://ershov.iis.nsk.su/en/node/778087.

4. Peter Wegner, "Research Paradigms in Computer Science," in *Proceedings of the 2nd International Conference on Software Engineering* (IEEE Computer Society Press, 1976), 322–30.

5. See, for example, Donald MacKenzie, *Mechanizing Proof: Computing, Risk, and Trust* (MIT Press, 2001), and Matti Tedre, *The Science of Computing: Shaping a Discipline* (Chapman & Hall, 2014).

6. R. W. Floyd, "Assigning Meanings to Programs," in *Mathematical Aspects of Computer Science*, ed. J. T. Schwartz, vol. 19, Proc. of Symposia in Applied Mathematics (American Mathematical Society, 1967), 19–32; C. A. R. Hoare, "An Axiomatic Basis for Computer Programming," *Communications of the ACM* 12, no. 10 (1969): 576–80.

7. Mark Priestley, *A Science of Operations: Machines, Logic and the Invention of Programming* (Springer Science & Business Media, 2011), 231. "Bridging" was coined in Andrew Pickering, *The Mangle of Practice: Time, Agency, and Science* (University of Chicago Press, 1995).

8. C. A. R. Hoare, in Kurt Walk, "Minutes of the 3rd Meeting of IFIP WG 2.2 on Formal Language Description Languages," April 1969.

9. Donald MacKenzie, "A View from the Sonnenbichl: On the Historical Sociology of Software and System Dependability," in *History of Computing: Software Issues*, ed. Ulf Hashagen, Reinhard Keil-Slawik, and Arthur Norberg (Springer-Verlag, 2002), 97–122.

10. Peter Naur and Brian Randell, eds., "Software Engineering: Report on a Conference Sponsored by the NATO Science Committee, Garmisch, Germany, 7th to 11th October 1968" (1969).

11. Michael S. Mahoney, "Computers and Mathematics: The Search for a Discipline of Computer Science," in *The Space of Mathematics: Philosophical, Epistemological, and Historical Explorations*, ed. Javier Echeverria, Andoni Ibarra, and Thomas Mormann (De Gruyter, 1992), 352.

12. Zachary Loeb, "Waiting for Midnight: Risk Perception and the Millennium Bug," this volume.

13. Thomas Haigh, "Assembling a Prehistory for Formal Methods: A Personal View," *Formal Aspects of Computing* 31, no. 6 (2019): 667.

14. E. W. Dijkstra, "The Humble Programmer," *Communications of the ACM* 15, no. 10 (1972): 862.

15. María Eloína Peláez Valdez, "A Gift from Pandora's Box: The Software Crisis." (PhD thesis, University of Edinburgh, 1988), 4.

16. Troy K. Astarte and Cliff B. Jones, "Formal Semantics of ALGOL 60: Four Descriptions in Their Historical Context," in *Reflections on Programming Systems—Historical and Philosophical Aspects*, ed. Liesbeth De Mol and Giuseppe Primiero (Springer Philosophical Studies Series, 2018), 83–152.

17. Priestley, *A Science of Operations*, 229.

18. Stephanie Dick, "Of Models and Machines: Implementing Bounded Rationality," *Isis* 106, no. 3 (2015): 623–34.

19. Ekaterina Babintseva, "Engineering the Lay Mind: Lev Landa's Algo-Heuristic Theory and Artificial Intelligence," this volume.

20. Cliff B. Jones, "The Early Search for Tractable Ways of Reasoning about Programs," *IEEE Annals of the History of Computing* 25, no. 2 (2003): 26–49; Stephanie Dick, "Computer Science," in *A Companion to the History of American Science*, ed. Georgina M. Montgomery and Mark A. Largent (John Wiley & Sons, 2015), 79–93.

21. John McCarthy, "History of LISP," in *History of Programming Languages*, ed. Richard L. Wexelblat (Academic Press, 1981), 173–97.

22. John McCarthy, "Towards a Mathematical Science of Computation," in *IFIP Congress*, vol. 62, 1962, 21–28.

23. John McCarthy, "LISP—Notes on Its Past and Future—1980," in *Proceedings of the 1980 ACM Conference on LISP and Functional Programming*, 1980, v–viii.

24. Michael S. Mahoney, "The History of Computing in the History of Technology," *Annals of the History of Computing* 10, no. 2 (1988): 113–25.

25. Michael Sean Mahoney, "Software as Science—Science as Software," in *History of Computing: Software Issues*, ed. Ulf Hashagen, Reinhard Keil-Slawik, and Arthur Norberg (Springer-Verlag, 2002), 25–48.

26. Liesbeth De Mol and Maarten Bullynck, "What's in a Name? Origins, Transpositions, and Transformations of the Triptych Algorithm–Code–Program," this volume.

27. Peter J. Landin, "Reminiscences," in *Program Verification and Semantics: The Early Work*, BCS Computer Conservation Society Seminar, Science Museum, London, UK, June 5, 2001, https://vimeo.com/8955127.

28. Peter J. Landin, "My Years with Strachey," *Higher-Order and Symbolic Computation* 13, no. 1 (2000): 75–76.

29. Richard Bornat, "Peter Landin: A Computer Scientist Who Inspired a Generation, 5th June 1930–3rd June 2009," *Formal Aspects of Computing* 21, no. 5 (October 1, 2009): 393–95.

30. Peter J. Landin, "The Mechanical Evaluation of Expressions," *The Computer Journal* 6, no. 4 (1964): 308–20.

31. C. Strachey and M. V. Wilkes, "Some Proposals for Improving the Efficiency of ALGOL 60," *Communications of the ACM* 4, no. 11 (November 1961): 488–91.

32. Martin Campbell-Kelly, "Christopher Strachey, 1916–1975: A Biographical Note," *IEEE Annals of the History of Computing* 1, no. 7 (1985): 19–42.

33. Christopher Strachey, "Curriculum Vitae," December 1971, Box 248, A.3, Christopher Strachey Collection, Bodleian Library, Oxford.

34. T. B. Steel, ed., *Formal Language Description Languages for Computer Programming* (North-Holland, 1966). Other attendees included Adriaan van Wijngaarden, Edsger Dijkstra, Peter Naur, Saul Gorn, Brian Randell, and Doug McIlroy.

35. Troy K. Astarte, "Formalising Meaning: A History of Programming Language Semantics" (PhD thesis, Newcastle University, 2019), chapter 4.

36. R. E. Utman, "Minutes of the 4th Meeting of IFIP WG 2.1," September 1964, Ershov Archive, http://ershov.iis.nsk.su/en/node/778103.

37. Peter Lucas and Hans Bekič, "Compilation of ALGOL, Part I—Organization of the Object Program," Laboratory report (IBM Laboratory Vienna, 1962).

38. Peter Lucas, "Formal Semantics of Programming Languages: VDL," *IBM Journal of Research and Development* 25, no. 5 (1981): 549–61; Heinz Zemanek, interview by William Aspray, 1987.

39. Fred Brooks, in Len Shustek, "An Interview with Fred Brooks," *Communications of the ACM* 58, no. 11 (November 2015): 40.

40. R. A. Larner and J. E. Nicholls, "Plan for Development of Formal Definition of PL/I" (IBM internal memo, September 1965).

41. ULD-III, "Formal Definition of PL/I (Universal Language Document No. 3)" (IBM Laboratory Vienna, December 1966).

42. Saul Rosen, "Programming Systems and Languages 1965–1975," *Communications of the ACM* 15, no. 7 (1972): 592. To see examples of the style, see Astarte and Jones, "Formal Semantics of ALGOL 60."

43. Lucas, "Formal Semantics of Programming Languages."

44. Peter E. Lauer, "Formal Definition of ALGOL 60" (IBM Laboratory Vienna, December 1968); Peter Lucas, Peter E. Lauer, and H. Stigleitner, "Method and Notation for the Formal Definition of Programming Languages" (IBM Laboratory Vienna, June 1968).

45. Peter Lucas, "Two Constructive Realisations of the Block Concept and Their Equivalence" (IBM Laboratory Vienna, June 1968).

46. W. Henhapl and C. B. Jones, "Some Observations on the Implementation of Reference Mechanisms for Automatic Variables" (IBM Laboratory, Vienna, May 1970); W. Henhapl and C. B. Jones, "The Block Concept and Some Possible Implementations, with Proofs of Equivalence" (IBM Laboratory Vienna, April 1970).

47. Kurt Walk, "Roots of Computing in Austria: Contributions of the IBM Vienna Laboratory and Changes of Paradigms and Priorities in Information Technology," in *Human Choice and Computers* (Springer-Verlag, 2002), 82.

48. Astarte, "Formalising Meaning," chapter 5.

49. Robert Milne and Christopher Strachey, "A Theory of Programming Language Semantics" (privately circulated, 1974).

50. C. B. Jones and A. W. Roscoe, "Insight, Inspiration and Collaboration," in *Reflections on the Work of C.A.R. Hoare,* ed. Cliff B. Jones, A. W. Roscoe, and Kenneth Wood (Springer-Verlag, 2010), 1–32.

51. Hoare, interviewed in Edgar G Daylight, "From Mathematical Logic to Programming-Language Semantics: A Discussion with Tony Hoare," *Journal of Logic and Computation* 25, no. 4 (2013): 1091–1110.

52. C. A. R. Hoare, in *Formal Language Description Languages for Computer Programming* (North-Holland, 1966), 142–43.

53. Cliff Jones and Troy Astarte, "Challenges for Semantic Description: Comparing Responses from the Main Approaches," in *Proceedings of the Third School on Engineering Trustworthy Software Systems,* ed. Jonathan P. Bowen, Zili Zhang, and Zhiming Liu (Springer-Verlag, 2018).

54. For example, C. A. R. Hoare, "The Axiomatic Method: Part I" (Unpublished, December 1967).

55. Walk, "Minutes of the 3rd Meeting of IFIP WG 2.2 on Formal Language Description Languages."

56. C. Strachey, "Towards a Formal Semantics," in *Formal Language Description Language for Computer Programming,* ed. T. B. Steel (North-Holland, 1966), 198–200.

57. Quoted in Dana S. Scott, "Some Reflections on Strachey and His Work," *Higher-Order and Symbolic Computation* 13, no. 1 (2000): 110.

58. Dana S. Scott, "Logic and Programming Languages," *Communications of the ACM* 20, no. 9 (1977): 637.

59. Dana Scott and Christopher Strachey, "Toward a Mathematical Semantics for Computer Languages" (Monograph, Oxford PRG; 1971).

60. Peter David Mosses, "The Mathematical Semantics of ALGOL 60" (Monograph, Oxford PRG, January 1974).

61. Milne and Strachey, "A Theory of Programming Language Semantics."

62. Quoted in Roger Penrose, "Reminiscences of Christopher Strachey," *Higher-Order and Symbolic Computation* 13, no. 1 (2000): 83.

63. Scott also used the phrase, in D. Scott, "Outline of a Mathematical Theory of Computation" (Monograph, Oxford PRG, November 1970).

64. Hoare, "The Axiomatic Method," 1.

65. See, for example, in Thomas B. Steel, "Preface," in *Formal Language Description Languages for Computer Programming* (North-Holland, 1966).

66. Stephanie Dick, "Machines Who Write [Think Piece]," *IEEE Annals of the History of Computing* 35, no. 2 (2013): 88.

67. Peter Lucas, "On the Formalization of Programming Languages: Early History and Main Approaches," in Dines Bjørner and Cliff Jones, *The Vienna Development Method: The Meta-Language*, vol. 61, Lecture Notes in Computer Science (Springer-Verlag, 1978), 3. See also David Nofre, Mark Priestley, and Gerard Alberts, "When Technology Became Language: The Origins of the Linguistic Conception of Computer Programming, 1950–1960," *Technology and Culture* 55, no. 1 (January 2014): 40–75.

68. Michael S. Mahoney, "Computer Science: The Search for a Mathematical Theory," in *Histories of Computing*, ed. Thomas Haigh (Harvard University Press, 2011), 128–46.

69. Priestley, *A Science of Operations*, 230.

70. Babintseva, "Engineering the Lay Mind."

71. Quoted in Tedre, *The Science of Computing*, 67. Emphasis original.

72. Walk, "Minutes of the 3rd Meeting of IFIP WG 2.2 on Formal Language Description Languages."

73. Hoare, "The Axiomatic Method."

74. P. J. Landin, "The Next 700 Programming Languages," *Communications of the ACM* 9, no. 3 (1966): 157–66.

75. Quoted in C. A. R. Hoare, "A Hard Act to Follow," *Higher-Order and Symbolic Computation* 13, no. 1 (2000): 71–72.

76. Graham White, "Hardware, Software, Humans: Truth, Fiction and Abstraction," *History and Philosophy of Logic* 36, no. 3 (2015): 278–301.

77. W. M. Turski, "Minutes of the 8th Meeting of IFIP WG 2.1," May 1967, Ershov Archive, http://ershov.iis.nsk.su/en/node/778136.

78. G. W. Bonsall et al., "ULD3 and Language Development" (IBM internal memo to J. L. Cox., February 1967).

79. Astarte, "Formalising Meaning," Sections 5.7 and 7.1.

80. Milne and Strachey, "A Theory of Programming Language Semantics," 22.

81. In Cliff B. Jones et al., "Reminiscences on Programming Languages and Their Semantics" (Panel at Mathematical Foundations of Programming Language Semantics XX, May 2004).

82. Gordon D. Plotkin, interview by Troy Astarte and Cliff Jones, April 2018.

83. Peter Mosses lists advances in the area in "The Varieties of Programming Language Semantics and Their Uses," in *International Andrei Ershov Memorial Conference on Perspectives of System Informatics*, ed. A. V. Zamulin, D. Bjørner, and M. Broy, vol. 2244, Lecture Notes in Computer Science (Springer-Verlag, 2001), 165–90; and Peter O'Hearn and Robert Tennent collect more recent work in *ALGOL-Like Languages* (Springer Science & Business Media, 2013).

84. ANSI, "Programming Language PL/I" (American National Standard, 1976).

85. John McCarthy, "Letter to A. P. Ershov," December 1965, Ershov Archive, http://ershov.iis.nsk.su/en/node/777964; John McCarthy, "Letter to A. P. Ershov," March 1968, Ershov Archive, http://ershov.iis.nsk.su/en/node/779549.

86. Tedre, *The Science of Computing*, 155.

87. Quoted in Mahoney, "Software as Science—Science as Software."

88. Astarte and Jones, "Formal Semantics of ALGOL 60," 98.

89. C. B. Jones, "The Transition from VDL to VDM," *Journal of Universal Computer Science* 7, no. 8 (2001): 631–40.

90. Hans Bekič et al., "A Formal Definition of a PL/I Subset" (IBM Laboratory Vienna, December 1974).

91. Bjørner and Jones, eds., *The Vienna Development Method: The Meta-Language.*

92. Walk, "Roots of Computing in Austria."

93. Tedre, *The Science of Computing,* chapter 4.

94. Gerard Alberts and Edgar G. Daylight, "Universality Versus Locality: The Amsterdam Style of ALGOL Implementation," *IEEE Annals of the History of Computing* 36, no. 4 (2014): 52–63.

95. E. W. Dijkstra, "On the Design of Machine Independent Programming Languages," Report (Mathematisch Centrum, October 1961).

96. E. W. Dijkstra, "Guarded Commands, Nondeterminacy and Formal Derivation of Programs," *Communications of the ACM* 18 (1975): 457.

97. Amnon H. Eden, "Three Paradigms of Computer Science," *Minds and Machines* 17, no. 2 (July 1, 2007): 135–67.

98. Peláez Valdez, "A Gift from Pandora's Box."

99. Edsger Dijkstra, "Minority Report," 1969, Ershov Archive, http://ershov.iis.nsk.su /en/node/805785.

100. Haigh, "Assembling a Prehistory for Formal Methods."

101. M. Woodger, "A History of IFIP WG 2.3: Programming Methodology," in *Programming Methodology: A Collection of Articles by Members of IFIP WG 2.3,* ed. David Gries (Berlin, Heidelberg: Springer-Verlag, 1978).

102. Astarte, "Formalising Meaning," Section 7.2.

103. Cliff Jones, personal communication, 2020.

104. Woodger, "A History of IFIP WG 2.3: Programming Methodology."

105. Peláez Valdez, "A Gift from Pandora's Box," 198.

106. Astarte, "The History of Programming Language Semantics," Section 8.

107. Dick, "Computer Science"; Mahoney, "Computers and Mathematics."

108. David Nofre, "Unraveling Algol: US, Europe, and the Creation of a Programming Language," *IEEE Annals of the History of Computing* 32, no. 2 (2010): 58–68.

109. Doug McIlroy, interview by Cliff Jones, May 2018.

110. Babintseva describes an alternative Russian view of computified thought in "Engineering the Lay Mind," this volume.

111. C. A. R. Hoare, "How Did Software Get So Reliable without Proof?" in *International Symposium of Formal Methods Europe* (Springer-Verlag, 1996), 1–17. That this process took time, Hoare argued, should not have been surprising in retrospect, since deploying brand new research methods in commercial products carried significant risk.

112. Dana Scott, interview by Tony Dale, April 1994.

113. Tony Hoare and Robin Milner, "Grand Challenges for Computing Research," *The Computer Journal* 48, no. 1 (January 2005): 49–52.

CHAPTER 7

What's in a Name?

Origins, Transpositions, and Transformations
of the Triptych Algorithm–Code–Program

Liesbeth De Mol and Maarten Bullynck

Words such as "code," "program," and "algorithm," once specific to specialist technical discourse, now belong to popular discourse. They are used every day to talk about what lies "underneath" technological products like Amazon, Facebook, or Google. "Coding" and "programming" are considered important "skills" to teach our children, and "algorithms" allegedly rule our world. But these interconnected terms are not constant in meaning; they have a complex history and impact how history is or can be written.[1] Forgetting about their historical dimensions may lead to anachronisms, wrong interpretations or Whig history. As Karine Chemla has noted: "Historians often worked under the assumption that the main components of scientific text problems, algorithms and so on are essentially ahistorical objects, which can be approached as some present-day counterparts."[2] This insight also applies to everyday usage.[3] For instance, the current usage of "algorithm" in popular *and* scientific discourse often no longer refers to some mathematical object but rather to what one used to call software or programs, e.g., the so-called Facebook algorithm. The mathematical origin of the word helps to obscure human bias (so-called algorithmic bias) and commercial interests behind the software, aiding the industry in outsourcing its accountability.[4]

This chapter follows the historical origins, transformations, and interrelations of the triptych Code, Program, and Algorithm (and the related term Software) to render these words historically transparent.[5] We show how the technologies of the twentieth century reclaimed these words and reshaped their meanings. This happened in two phases. Around 1900 early automation made its mark on the words "code" and "program" first, then, after World War II, computing adopted these words.

Asking "who uses the words" and "in what context" reveals the motives at play and the shifting objectives. The words are regimented to make new distinctions, sometimes reflecting a reality of relations on the workfloor, sometimes projecting a vision onto the workfloor or onto the division of labor and its possible automation (as in the "myth" of the coder, see section 4 of this chapter: "Shifting Frontiers in the 1950s"). The words can be reminiscent of local working practices, but also vectors of managerial phantasies, they are symptoms of how the organization of labor is negotiated between technological conditions and human work and power structures.[6] Both the computing people themselves and the management put words in the field to further their agenda, e.g., "algorithm" to further computer *science,* or "software" to make a service more tangible. As all these words later became part of general everyday discourse, it is paramount to become conscious of the economic, industrial, and scientific stakes that have been invested in those words in the past.[7]

Origins and the First Technological Transfers

The words "code," "program," and "algorithm" were already part of the English vocabulary long before the first computers. They are derived from Greek or Latin roots and became proper members of the English reservoir of words in the seventeenth century during the Renaissance.

Algorithm

The word "algorithm" probably has the most intricate history. It was originally a latinized version of Al-Khwarizmi, surname of the Arab mathematician Abu Ja'far Mohammed Ben Musa (c. 780–c. 850). His treatise on the Hindu way of reckoning is now lost in its original Arabic version but has survived in a number of medieval Latin translations of the twelfth century. Typically, a medieval, hand-copied manuscript began by *dixit* (said), *inquit* (said), or *scripsit* (wrote) followed by the name of the original author. In this case "dixit algorismus," Al-Khwarizmi said. As the first (relevant) word of the codex, the name of the author became the way to refer to this manuscript, its title in a sense. When the treatise was translated into the vernacular, the word *algorismus* lost its original meaning and with time, the word came to refer to the content of Al-Khwarizmi's treatise, the Hindu system of writing and calculating numbers (viz., our decimal positional system and its rules of computation).

From this meaning, it became used to refer to other systems of calculating, either extending the decimal positional form of computation or creating new forms analogous to it, such as the algorithm of fractions or proportions, or, later, the "algorithm of infinitesimal differentials" on the Continent or "algorithm of fluxions" in Newtonian England. In the famous priority dispute between Newton and Leibniz over the invention of the calculus, the algorithm of the calculus became a point of contention in itself, the mathematician J. C. Hauff even claiming that while Newton invented the method, Leibniz invented the algorithm, thus introducing a subtle differentiation between the conceptual solution of a problem (the method) and the material or notational implementation of that solution (the algorithm).[8] The close association of notation and algorithm only loosens up during the nineteenth century, when mathematicians will use "algorithm" to refer to a schematic computational process or of a stepwise procedure. "Algorithm" thus became one of the words, alongside "method," "procedure," "rule," or "calculus," mathematicians use to denote a (semi-)formalized (numerical) solution to a (mathematical) problem.[9]

Code

The word "code" derives from the Latin word *codex*, that is, a collection of texts, in particular law texts. In its transferred meaning, it also refers to a collection of rules to be followed, such as a code of honor or a recipe for preparing a certain medicine. In this sense, the original meaning of "code" in Europe was closer to "algorithm" in its actual understanding. Only in the nineteenth century "code" started being used for a system of signs to "encode" a certain information. It had been common practice in eighteenth-century Europe to encrypt diplomatic or important commercial messages by one long word, e.g., the short word "shamefaced" stood for "sell at current market price." These words were collected in a book or in a codex. This practice became more widespread with the telegraph, especially in the United States, these code words being both cost-efficient and secure. These codes were then "translated" into a so-called telegraphic alphabet and telegraphic language, viz., combinations of long–short signals. "Code" later shifted to refer also to this telegraphic alphabet, e.g., Morse code or Baudot-Murray code.[10] When in the early twentieth century punched cards were introduced for Hollerith machines and other unit record devices, the word "coding" was introduced to denote, by analogy, the process of translating questionnaires or data sheets into the "language of punched cards," holes, and notches.[11]

Program

Finally, the word "program" was a seventeenth-century neologism derived from the concatenation of the two Greek words προ (before) and γραφειν (writing), a pre-scription,[12] a written notice of things to come. "Program" thus describes the ordering of an activity like, for instance, the program of a concert or, more generally, a general plan or scheme of something to be done, like an itinerary, a training schedule, a production plan, etc. The word became used, along with "plan" and "schedule," by management, economists, and "planning departments" as part of a vocabulary to streamline and control industrial processes in the late nineteenth century.[13]

As "code" met up with telegraphy, so did "program" meet up with another technological advance in the late nineteenth century: "program(me) clocks" developed to "furnish a convenient and practical clock, that may be set to strike according to any required programme."[14] These devices automated time schedules or production plans, ringing at preset times on a factory workfloor, at railway stations, or in a school. With time, more complex and general "program devices" or "program machines" were invented to automate the operations of machines such as a paper-cutting machine, a washing machine, etc. "Program" came to stand for the automatic carrying out of a sequence of operations or as an automated scheduler.

Around that time, the word "program" was also picked up in the context of radio engineering. At first, a program referred to the *physical* program as transmitted within a broadcasting network. With the rapidly expanding broadcasting industry and increased network complexity, scheduling "programs" in different networks became an important problem. Any program had to be connected to the right station at the right moment at so-called switching points." Originally, this was done by an operator who had to "listen for cues indicating the end of a program, and then operate the proper keys or change connections."[15] An increase in the number of switches made this impossible and the manual switching had to be (partially) automated through (relay) switching equipment. In this context, "program" steadily transposed from radio programs to be embodied by technology itself, with terms like "program circuits," "program trunks," "program switching," "program line," "program loop," etc.[16]

Though all three words ("algorithm," "code," and "program") were items present in the language's lexicon since the seventeenth century, they were recycled in new contexts that were both specialist and technical already in the late nineteenth and early twentieth century. "Algorithm" lost its footing in notation to become a more general mathematical term as "stepwise procedure," shifting from what Kenneth O. May has called "mathematical technology" to "mathematical science." "Code," then, was taken up by the professional telegraph users to denote any form of compression of messages, be it through code words or (binary) code symbols. From there on, "coding" became the verb to refer to the, often repetitious and boring, activity of translating language into code words or symbols for telegraphy, telephony, or also card punching. Finally, "program" entered the engineering discourse in the mid-nineteenth century to talk about the automation of time schedules, production plans, or, later only, the switching of radio programs in a network and the automatic sequencing of operations. Thus, even before our three words became part and parcel of computing vocabulary, they had already migrated from their general meaning to usages in specialist and technological communities. From these, they will eventually spread to computing.

The Second Transfer of Meaning under the Sign of Digital Computing

The semantics of "code," "program," and "algorithm" shifts into computing when it encounters the large electromechanical and electronic calculators of the 1940s, mostly built to help computing ballistic tables. In these machines, the traditions of business computing, scientific computing, and military command and control meet up with the reliable relay and the pioneering electronics technologies of automation.

The mathematical field now called "numerical analysis" slowly emerged after World War II; many groups and bureaus for (manual or machine-aided) computation developing their own techniques and workflows.[17] To refer to the computational procedure and its steps to be followed, the mathematicians involved sometimes used the word "algorithm," sometimes "rule," "formula," or "procedure," but the most common word was without a doubt "method."

While each computation group had its own ways of organizing computation, most of them had some kind of tiered process. First came the mathematical analysis of the problem (including finding the right "method" in the literature), then the preparation of the problem for the human computers using a "computing plan," followed by the computation itself, taking down the results using "computing sheets" and finally checking the results. When this manual (or machine-aided) process was ported to large mechanical calculators such as the ASCC/ Harvard Mark I or the Bell machines, there was continuity with these practices, but replacing the human by a machine changed some things. Usually, mathematical analysis of the problem still came first. After that, the method had to be prepared in a form that allowed an easy setup on the machine, finally the machine computation itself, followed by error analysis and checking. In order to prepare a problem for setup on the machine, one had to "code" the computation in a form that was machine-understandable, on the other hand, one had to operate the machine during execution. This whole process from mathematical analysis (planning) to setting up the problem on the machine (coding) would eventually evolve in *one* of the definitions of programming in the 1950s. The widespread use of the word "program," however, did not originate in this context.

Instead, the germ of what our modern word "program" would become lies in the application of automatic control to calculating machines. Already in the 1930s engineers had started to build "program devices" to control the sequence of operations in calculating machines. With the new complexities of time scheduling and radio broadcasting, also the automation of transfer of control or even conditional transfer of control had been achieved. As IBM engineer James W. Bryce describes in a 1937 patent: "Such programming means will enable the operator to program the sequence of transfers and to selectively route the transfers from any selected accumulator to any other selected accumulator."[18] The IBM team that would develop the ASCC/Harvard Mark I with Howard Aiken applied this technology to the machine. As a consequence, they speak in the patent of a "program tape" (instead of the later "control tape") that "schedules the operations of the machine, selecting the functions and the sequences of their performance."[19] Further mention is made of a "print program," or a "multiplying program," but in the later ASCC/Mark I manual these have disappeared and are now referred to as, e.g., "multiplying sequences."[20] Grace Hopper and other members of the ASCC/Mark I team would later speak of "coding routines" or sequences. The routines frequently needed would be stored in a "tape library" containing "control tapes [that] are of general application."[21] Thus, whereas the engineers, who were thinking in terms of the automation of control via program devices spoke of "programs," Aiken and his team, who were coming from the mathematical and human side of the computation problem, spoke of "coding routines" or, more frequently, "sequences."

From a certain viewpoint, coding the ASCC/Mark I is the automation of a machine setup. That development can be found in other contemporary machines too. Samuel Caldwell and Vannevar Bush devised a control mechanism to automate the setup of their improved differential analyzer in the 1940s.[22] Before, the machine had to be set up manually connecting all the right parts of the machine—a process that took several hours. Now, the connections, empirical data, and initial conditions were all coded on tape, which then controlled the automatic setup of the analogue analyzer. The Bell machine relay calculators model III to V all had a battery of Baudot-coded tapes with both data and instructions to automatically set up the right calculation to be executed. All these machines, however, could only follow a sequence of coded instructions. Coded iterations and full automatic conditional transfers were, at least initially, not possible.[23]

These control structures were, however, present on ENIAC once it was ready for service (1946). That machine differed fundamentally from its mechanical and analog contemporaries because it was electronic, providing a much higher speed of computation. However, in its original form, it had to be manually set up, not unlike the original differential analyzer. It is here where the most momentous transfer of meaning to "program" would happen. Mauchly's original short 1943 proposal for an "electronic computer" already referred to a "program device," which later became a "program control unit" and evolved into the ENIAC's "master programmer," which centrally controlled the local programming circuits for sequencing loops and conditionals. It is from there that the term in ENIAC developed, playing over different semantic extensions, referring both to individual (control) units (as in "program switches"); smaller pieces of an entire program (as in "program sequences"); or the complete schedule that organizes program sequences (as in a "complete program [for which] it is necessary to put [the] elements together and to assign equipment in detail").[24] "Program" in this context always refers to how automatic control, locally or globally, is organized.

The practice of putting computations on the ENIAC, however, rapidly made the term "program" drift further away from hardware to be transferred to the organization of a computation. In an appendix to a 1945 report entitled *Remarks on Programming the ENIAC*, Eckert, Mauchly, and others wrote about the entire "computational program" and how to "link the elementary programming sequences into a complex whole" using an "elaborate hierarchy of program sequences" that could be built up with the master programmer and which relied on, what they call, "sub-routines."[25] In another report written by Haskell B. Curry and Willa Wyatt planning for ENIAC to do ballistic calculations, the problem "is studied with reference to the programming on the Eniac as a problem in its own right."[26] They develop a method where each computation is "broken into pieces, called stages" which are defined as "a program sequence with an input and one or more outputs." Linking together the inputs and outputs, smaller program sequences could be combined into more complex programs and a general "schedule" of programs could be translated into wiring diagrams that indicate how ENIAC should be set up. The semantics of the "program device" discourse is still at play here, but it generalizes from the sequencing of operations to include also the scheduling of sequences of operations.

The intricacy and time-consuming process of manually setting up the ENIAC for computation, and the sheer speed of the machine, led to ENIAC being rewired, resulting in an automation of the set-up process inspired by the relay machines. In this new configuration, a sequence of coded instructions could be introduced on ENIAC, through the setting of switches or the reading of punched cards.[27] Now the logic of linking elements together in a program that automatically controlled the machine could be done through symbolic encoding. As a result, the word "program" slowly transferred even further from hardware toward what is now called software.

It is commonly known that the word "program" in its current meaning comes from ENIAC.[28] But it is the encounter of automating sequences and scheduling with calculation that put "program" there in the first place. Actual practice then made "program" shift further, from actual hardware to the configuration of that hardware. It is in that sense that Douglas Hartree defines "program" as "the process of drawing up a schedule of the sequence of individual operations required to carry out the calculation."[29] Compared to the Harvard Mark I team, Hartree is less focusing on planning and coding sequences (viz., the work of the human computer and its translation to the machine) than on how to translate the plan into a configuration of the program device that will start the required operations, either in the form of wiring diagrams or in the form of symbolic coded instructions.

Shifting Frontiers in the 1950s

The work on computing in the 1950s toward reliability, mass production, and standardization is also reflected in the attempts to define basic terms in glossaries and shape common practices. Looking at usages and definitions of "program" and "code" in different professional and local contexts brings out how fluid their meanings still are in the 1950s. Especially the development of automatic coding or programming in the mid-1950s impacted on the semantic envelope of these words.

Looking at the influential work by Goldstine and von Neumann,[30] they do not use the word "programming" but instead differentiate between planning and coding of a problem. After the mathematical preparation of a problem, a flow-diagram is introduced to plan a coded sequence which is then used to derive the coded sequence. This distinction was often picked up but with the important shift that "programming" would then be used too, referring to the whole process of planning, flowcharting and coding, whereas "coding" would now be reduced in meaning to the machine coding part only.

This is the distinction one finds in the 1954 ACM Glossary, viz.,

Code (verb): to prepare problems in computer code or in pseudocode for a specific computer.
Program (verb): to plan a computation or process from the asking of a question to the delivery of the results, including the integration of the operation into an existing system. Thus programming consists of planning and coding.[31]

Some definitions restrict "programming" further and contrast it with "coding"; e.g., the Bureau of Standards in 1948 defined "program" as a "general verbal description of the method of solving a particular problem on a computer." This is even more pronounced in the IBM glossaries of the 1950s; for the IBM 650, one author even writes: "Programming and flow charting are synonymous—the remainder is mere coding."[32] This shows that "programming" has shifted once more in the 1950s. From structuring automatic control, in some contexts it can now generalize to the whole process of planning and coding or, in some cases, even be restricted to one aspect of planning, viz., flowcharting.

While a strict differentiation was made between flowcharting and coding, usually, coding was subsumed under programming. Or, put differently, coding was just another task of the programmer besides flowcharting. This becomes explicit in a 1951 discussion:

> *L. A. Ohlinger (Northrop Aircraft Company):* I would like to ask how many programmers and coders are employed in order to keep UNIVAC busy full time?
> *J. L. McPherson:* We do not distinguish between programmers and coders. We have operators and programmers.[33]

Indeed, while both activities of flowcharting and coding can be distinguished in theory, they cannot in practice; the same person has to do both.

However, classic computing history has it that a distinction between the jobs of "programmer" and "coder" existed already in the 1950s, the first being occupied with problem analysis and flowcharting, the latter with porting the flowchart to the machine.[34] In practice, however, no such distinction appears, and even a report from the US Department of Labor describing the *Occupations in Electronic Data-Processing Systems,* identifying no fewer than 13 different occupations in electronic data processing, only uses "coder" in reference to "coder clerks."[35] These are people who "convert items of information obtained from reports and records to codes for processing by automatic machines," viz., the people who already coded information in the beginning of the twentieth century. The coding of instructions, however, is considered to be part of the job of the programmer. What did exist was a distinction between the chief programmer or systems analyst and junior programmers. Evidently, it is reasonable to assume that flowcharting was more in the hands of the former and coding more in the latter category, even though a neat separation in practice was impossible.

It thus seems a fair question to ask where the "myth" of the coder comes from? The answer lies in Grace Hopper's influential talks on automatic coding or programming.[36] Hopper remarks that "the analyst, programmer, coder, operator and maintenance man were separated," though, admittedly, the "distinction between a programmer and a coder has never been clearly made." Her distinction is, a programmer "prepares a plan for the solution of a problem" (viz., a flowchart), while a coder has "to reduce this flow chart to coding, to a list in computer code." The motivation for this, self-admitted, artificial separation becomes clear in the rest of the paper: "It is this function, that of the coder, . . . that is the first human operation to be replaced by the computer itself." Thus it is the introduction of "automatic coding" (sometimes also unfortunately called "automatic programming") that accounts for a retrospective, artificial distinction.

This observation helps to explain the different ways "coding" and "programming" are used. In UNIVAC and IBM circles usage was mostly according to the definitions quoted above, "coding" was translation into machine code, "programming" was either everything from planning to coding, or the part before coding, viz., planning and flowcharting. This is in part a reflection of the hierarchy on the workingfloor of a commercial computer installation. In other places, especially at universities, both words are used interchangeably or coding is subsumed under programming. The ambition there was often to make the computer accessible to every user, in particular scientific users not necessarily versed in engineering and machine details. This contrasts with commercial computer installations, where the user had to pass through the programmers and operators to get his work done on the machine.

A symbolic "readable" way of programming the machine directly after problem analysis was first championed by Hartree or Wilkes in the United Kingdom[37] or Zuse and Rutishauser in Germany and

Switzerland. On the EDSAC a symbolic assembler-like code was developed so "the machine may be said to understand the same language as a computor [*sic*]."[38] Or, quoting Aleck Glennie on his Autocode system for the Manchester Mark I, "We must make coding comprehensible, [t]his may be done only by improving the notation of programming."[39] Similarly, at MIT's Whirlwind it was decided early on to make the computer available to the "casual user" through "automatic standard subroutines" that "can be used almost as easily as an equivalent built-in order, with resultant saving in the programmer's time." Eventually this resulted in one of the first "automatic coding systems," "a comprehensive system of service routines . . . to simplify the process of coding."[40]

With "automatic coding systems" of the 1950s, "coding" becomes less of an issue and "programming" increasingly is the focus of human effort. This is tangible in the definitions of "program." In November 1952 a "program" on the Whirlwind is defined as "a sequence of actions by which a computer handles a problem," a definition still close to Hartree or the EDSAC team, and a "coded program" is a "set of instructions that will enable a computer to execute a program."[41] In December 1952 then, now with automatic coding, it becomes a "program is an ordered sequence of words, written with the intention of having it typed on paper tape in the (new) Flexocode and inserted in [Whirlwind I] by the intermediary of the Comprehensive Conversion Program."[42] This shift suggests the one that will happen later, at the end of the 1950s, when so-called programming languages would become used and "program" will become a "text" in those languages.

Commercial computer firms investing in automation of the programming process such as IBM or UNIVAC undergo the same evolution. The IBM FORTRAN system is an "IBM 704 program which accepts a source program in a language . . . resembling the ordinary language of mathematics, and which produces an object program in 704 machine language, ready to be run." Thus, "a FORTRAN source program consists of a sequence of source statements, of which there are 32 different types."[43] Equally, UNIVAC's FLOW-MATIC is described as shifting "the programming effort from detailed coding to problem definition and system analysis," and MATH-MATIC "describes the problem from the user's standpoint, rather than the program required by the hardware of the computer." As a consequence writing FLOW-MATIC or MATH-MATIC "programs" amounts to writing "sentences," and the "conversion of the problem, expressed in pseudo-code, into the necessary program, in machine code, is performed entirely automatically and internally."[44] This more "linguistic" or even "syntactic" definition of "program"[45] would later be confirmed by ALGOL's definition of "program": "sequences of statements and declarations, when appropriately combined, are called programs,"[46] and, one year later, "A program is a self-contained compound statement, i.e., a compound statement which is not contained within another compound statement and which makes no use of other compound statements not contained within it."[47]

Business and Science, or, Software and Algorithm?

Around 1960 there is the simultaneous surge of the computer service industry and the slow establishment of computing as an academic discipline. Like canaries in the coal mine, the usage of two small words, viz., "algorithm" and "software," tells us about both. As the statistics show, the rare word "algorithm" starts to spread from 1958 onward, the neologism "software" from 1961 onward (see fig. 7.1a).[48] Their appearance and fast dissemination is indicative of the self-consciousness of new professional groups. As the prevalence of the term "software" in the trade magazine *Datamation* shows, it is mainly used by the computer service industry (later: software industry). The term "algorithm," on the contrary, is most present in the publications of the Association for Computing Machinery (ACM), a professional society of computer scientists. Both words were "launched" into professional and public discourse purposefully.

"Algorithm" had been used by numerical analysts and computing professionals occasionally before 1958, but the choice to name the new international scientific programming language ALGOL, acronym of "ALGOrithmic Language," made the word a household term. As is well known, the origins of the ALGOL-language date back to a meeting of German and Swiss mathematicians and it was allegedly Heinz

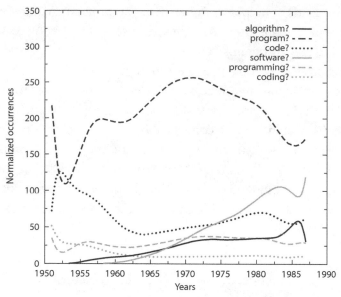

Figure 7.1a. Occurrence of various computing terms in AFIPS Proceedings, 1951–1987.

Rutishauser who repeatedly used the term "algorithmic notation" since 1955, but Herman Bottenbruch who coined the phrase "algorithmic language."[49] The German mathematicians had used "algorithmischer Programm" to distinguish between a mathematical solution to a problem and the program in machine code: "Such algorithmic notations, as we shall call them, have the appearance of classical mathematical notation but include certain dynamic elements which remind one of ordinary programming."[50]

While the first description of ALGOL in 1958 was still couched in mathematical terms and spoke of an "International Algebraic Language,"[51] the 1959 description moved to "algorithmic." With the buzz created around ALGOL in some circles, say the ACM and universities, the term "algorithm" gained currency, while words such as "method" or "rule" faded away. In particular, the specially created "Algorithm section" in the *Communications of the ACM,* first edited by J. H. Wegstein, to publish "algorithms consisting of 'procedures' and programs in the ALGOL language"[52] helped to establish it (see fig. 7.1b). It also shows that "algorithm" is situated at the boundary between "method" and

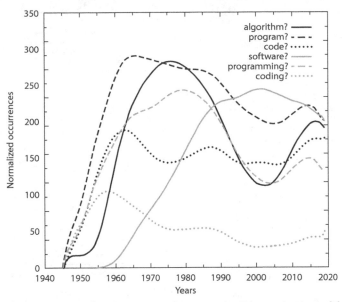

Figure 7.1b. Occurrence of various computing terms in *Communications of the ACM*, 1944–2020.

"program" and may refer to both. So ALGOL became the "internationally accepted method of describing numerical calculations in most journals devoted to computation."[53]

When Donald E. Knuth in his *Art of Computer Programming* series (1962–today) chose to talk about "analysis of algorithms" rather than "non-numerical analysis" as the topic of the series,[54] the word "algorithmic" became even more cemented as a key word, certainly for those who would call themselves computer scientists.[55] This also shows up in the statistics. While "'algorithm" surges around 1960 in all three publications (the ALGOL effect), it remains a rare word in the trade journal *Datamation*, whereas it features prominently in the *Communications of the ACM* since 1960, even if subject to some waves of fashion.

The neologism "software" seems to have started as a joke in the 1950s, the other side of the more common "hardware" of the military or the computer industry.[56] The statistics convincingly show that "software" is first used in print in 1960–1961 and then becomes a common term. It is mainly used by the computer business and service industry

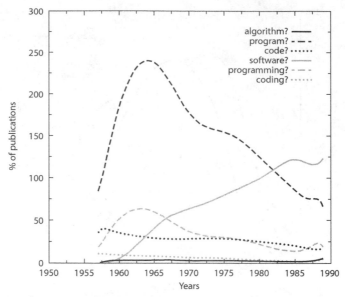

Figure 7.1c. Occurrence of various computing terms in *Datamation*, 1957–1989.

as represented by *Datamation* (see fig. 7.1c). The term also appears in publications of the ACM or the IEEE, but it never reaches the same prevalence among computer scientists and engineers.

In fact, the word "software" starts its course in advertising. The first occurrences in journals on the West Coast are all in job advertisements, at first those of Ramo-Woolridge, later of other companies too.[57] On the East Coast, the word was used first by the computer service firm C-E-I-R.[58] In both cases, the word may have been chosen for accounting reasons. Ramo-Woolridge, a primary consultant for the US space program, was under a "hardware ban" (viz., forbidden to sell hardware to the military, to avoid monopoly). Calling their services "software," even if that meant microprogramming computers, might have been a decoy tactic. In a similar vein, C-E-I-R had had problems capitalizing the costs incurred over training programmers or developing programs[59] and might thus have come up with a new term to make this investment more tangible.

The term certainly stuck quickly. It surfaces in the development of COBOL and, at first, is mainly used to refer to programs that help to program, such as compilers, assemblers, utility or monitor programs.[60] But soon "software" became the generic term used by the programming services industry for talking about programs as a commodity, as a commercial product. The prominent role of C-E-I-R's president H. W. Robinson at the Association of Data Processing Service Organizations (ADAPSO) may have played an important role in establishing the word. IBM, which had had a quarrel with C-E-I-R around a rental of its STRETCH computer, at first dismissively defines "software" as a "slang term for programming system" but will adapt the term eventually in its programming service bureaus. As the graphs of *Datamation* show, the spread of "software" correlates with a slow but steady decline of the word "program," substituting the "neutral" program for the "commercial" software, turning the "programming service industry" into "software industry."

Discussion and Outlook

The semantic "tectonics" still goes on. The further evolution of "program" and "code" may be gleaned from the statistical graphs. As "automatic coding" and "programming languages" gain currency, the frequency of "coding" goes down, while "programming" goes up. "Coding" in most cases now mainly refers to coding data or encoding or decoding practices. In the 1980s and, more recently in the 2010s, "coding" and "coder" became slightly popular again. "Coding" now has become a colloquial term for "programming," emphasizing the recreative side of programming, closer to "hacking," contrasting it with the professional business of programming applications.[61] One can also observe from the graphs that the popularity of "programming" has fallen since the economic crisis of 1973, while at the same time "software" continued its upward trend. If the ACM statistics (the only ones going beyond the 1980s) are representative, it would seem this evolution was returned around 2000, when "software" lost ground to "program(ming)" and to "code(ing)" again. "Algorithm" seems subject to some "seasonal" variation. After its ALGOL-popularity around 1960, it is barely mentioned in the trade magazine *Datamation*, whereas proceedings of the American Federation of Information Processing Societies (AFIPS) show steady use, but the ACM publications feature it most prominently. It peaks around 1975 with a downward trend afterward until 2000, when it goes up again. The heydays of structured programming might play a role here, but there is also a hint from the fact that in 1975 the Algorithms section of the *Communications of the ACM* became a journal in its own right, *Transactions on Mathematical Software,* changed the focus from algorithms to programs or even software again.[62] More research would be needed to correctly interpret these data.

It is important, finally, to note that nowadays the word "algorithm" has begun to be used in a broader sense still.[63] Examples are Google's algorithm, Facebook's algorithm, algorithmic trading, or even algorithmocracy, a state form where political decisions are influenced or even formed by algorithms. This signification of "algorithm" extends from the stepwise recipe-like numerical or non-numerical procedure to the complex or system of programs and parameters that underly parts of technological infrastructure. This recent evolution of the word would certainly merit detailed historical and political scrutiny, especially given its boost in popularity with the latest wave of artificial intelligence. A superficial browsing of our data seems to suggest that its roots are to be found in the late 1960s or early 1970s, when engineers began to speak of "scheduling algorithms" in operating systems.[64] It further gained traction in the 1970s and 1980s, when some communities started using the word in a more general way, viz., for any method that is programmed within a software project. Such usage can be found in research on predictive, genetic, backtracking, etc. algorithms in artificial intelligence (e.g., David Rummelhart or David Marr), research in query and update algorithms for database structures, or in economists' programs for automatic trading on the stock market.[65] As it becomes invested with economic or political interests, as it did for "algorithm" in the twenty-first century, it silently but not innocuously becomes part of the often subconscious set of values and distinctions that characterize everyday vocabulary. Looking beyond the screen of words is a necessity, and history helps to identify and critically engage with the forces that drive words today.

Notes

1. R. Koselleck, *Begriffsgeschichte* (Frankfurt-am-Main: Suhrkamp, 2006).
2. Karine Chemla, "On Mathematical Problems as Historically Determined Artifacts: Reflections Inspired by Sources from Ancient China," *Historia Mathematica* 36, no. 3 (2009): 213–46, here 213.
3. See Kushner's chapter on "lurking" in this volume.
4. On algorithmic bias, see Cathy O'Neil, *Weapons of Math Destruction: How Big Data Increases Inequality and Threatens Democracy* (New York: Broadway Books, 2016); Virginia Eubanks, *Automating Inequality: How High-Tech Tools Profile, Police, and Punish the Poor* (New York: Macmillan, 2017); Safiya Noble, *Algorithms of Oppression: How Search Engines Reinforce Racism* (New York: New York University Press, 2018).
5. Prior work on the origins of "program" includes David Alan Grier, "The ENIAC, the Verb 'to Program' and the Emergence of Digital Computers," *IEEE Annals of the History of Computing* 18, no. 1 (1996): 51–55; Grier, "Programming and Planning," *IEEE Annals of the History of Computing* 33, no. 1 (2011): 85–87; Thomas Haigh and Mark Priestley, "Where Code Comes From: Architectures of Automatic Control from Babbage to Algol," *Communications of the ACM* 59, no. 1 (2016): 39–44.
6. See, for instance, Jean-Louis Peaucelle, *Adam Smith et la division du travail, La naissance d'une idée fausse* (Paris: L'Harmattan, 2007); David F. Noble, *Forces of Production: A Social History of Industrial Automation* (New York: Knopf, 1984).
7. Both authors are supported by the ANR PROGRAMme project ANR-17-CE38-0003-01.
8. Lazare Carnot, *Betrachtungen über die Theorie der Infinitesimalrechnung*, translated and augmented by J. C. Hauff (Frankfurt: Jäger, 1800), 84.
9. See, e.g., Alonzo Church, "An Unsolvable Problem of Elementary Number Theory," *American Journal of mathematics* 58, no. 2 (1936): 345–63.
10. William Friedman, *The History of the Use of Codes and Code Language* (Washington, DC: US Government Printing Office, 1928).
11. Herman Hollerith, "The Electronic Tabulating Machine," *Journal of the Royal Statistical Society* 57, no. 4 (December 1894): 678–89, here 684. See also Haigh and Priestley, "Where Code Comes From."
12. Martin Carlé analyzed "pro-gram" in terms of its Greek origins in his talk *Literate Programming, Containerisation and the Future of Digital Humanities* at the Autumn 2018 meeting of the PROGRAMme project in Bertinoro.
13. See Grier, "Programming and Planning"; also Devin Kennedy, "Virtual Capital: Computers and the Making of Modern Finance, 1929–1975" (PhD diss., Harvard University, 2019), chapter 3.
14. S. F. Estell, Improvement in programme-clocks, US patent nr. 98678, patented January 11, 1870.
15. P. B. Murphey, Broadcast switching system, US patent nr. 2238070, filed May 10, 1940, granted April 15, 1941.
16. For more detail, see De Mol and Bullynck, "Roots of '*Program*' Revisited," in *Communications of the ACM* 64, no. 4 (2021): 35–37.

17. David Alan Grier, *When Computers Were Human* (Princeton, NJ: Princeton University Press, 2005).

18. J. W. Bryce, Cross-adding accounting machines and programming means therefor, US patent nr. 2,244,241, filed October 1, 1937, granted June 3, 1941.

19. Claire Lake, Zero eliminating means, US patent 2,240,563, filed August 31, 1938, granted May 6, 1941.

20. Grace Hopper et al, "A Manual of Operation for the Automatic Sequence Controlled Calculator," *The Annals of the Computation Laboratory of Harvard University*, vol. 1 (London: Harvard University Press, 1946).

21. The coders of ASCC/Mark I developed a systematic practice of storing correct pieces of code in Notebooks, which were then consulted to copy code when needed again, thus a manual way of subroutining. Hopper explained how this affected her later compiler work; cf. Computer Oral History Collection, Grace Murray Hopper (1906–1992).

22. Vannevar Bush and Samuel H. Caldwell, "A New Type of Differential Analyzer," *Journal of the Franklin Institute* 240, no. 4 (1945): 255–326.

23. The ASCC/Harvard Mark I would later have a conditional stop through a special device called the "subsidiary sequence unit" which allowed the moderator to move the tape to another position; for loops, they glued tapes together. For the Bell machines, iteration was present in Model III, conditional jumps only on Model V.

24. H. B. Curry and W. Wyatt, *A Study of Inverse Interpolation on the Eniac*, Aberdeen Proving Ground, Maryland, Report No. 615 (August 19, 1946).

25. John Presper Eckert, John W. Mauchly, Hermann Goldstine, and J. G. Brainerd, *Description of the ENIAC and Comments on Electronic Digital Computing Machines*, Contract W 670 ORD 4926 (Moore School of electrical engineering, University of Pennsylvania, November 30, 1945): 3–7.

26. Curry and Wyatt, *A Study of Inverse Interpolation on the Eniac*, 6.

27. Thomas Haigh, Mark Priestley, and Crispin Rope, *ENIAC in Action: Making and Remaking the Modern Computer* (Cambridge, MA: MIT Press, 2016).

28. Grier, "The ENIAC, the Verb 'to Program,'" 1996; Haigh, Priestley, and Rope, *ENIAC in Action*.

29. Douglas Hartree, *Calculating Instruments and Machines* (Urbana: University of Illinois Press, 1949), 111–12.

30. Herman Goldstine and John von Neumann, *Planning and Coding of Problems for an Electronic Computing Instrument*, vol. 2 of *Report on the Mathematical and Logical Aspects of an Electronic Computing Instrument* (Report prepared for US Army Ord. Dept. under Contract W-36-034-ORD-7481, 1947–48).

31. First ACM Glossary, prepared by a committee chaired by Grace Murray Hopper, 1954.

32. R. V. Andree, *Programming the IBM 650* (1958): 81.

33. J. Presper Eckert, James R. Weiner, H. Frazer Welsh, Herbert F. Mitchell, *The UNIVAC System*, AIEE–IRE '51: *Papers and Discussions* presented at the December 10–12, 1951, *Joint AIEE–IRE Computer Conference* (1951): 6–16.

34. The claim is prominent, e.g., in Nathan Ensmenger, *The Computer Boys Take Over: Computers, Programmers, and the Politics of Technical Expertise* (Cambridge, MA: MIT Press, 2010).

35. US Department of Labor, *Occupational Analysis Branch of United States Employment Service, Occupations in Electronic Data-Processing Systems* (1959).

36. Grace Hopper, *Automatic Programming—Definitions, Symposium on Automatic Programming for Digital Computers* (Office of Naval Research, Department of the Navy, Washington, DC, May 13–14, 1954): 1–5.

37. M. V. Wilkes, D. J. Wheeler, and S. Gill, *The Preparation of Programs for an Electronic Digital Computer* (Reading, MA: Addison-Wesley, 1957 [1951]).

38. M. V. Wilkes, "Programme Design for a High-Speed Automatic Calculating Machine," *Journal of Scientific Instruments* 26, no. 6 (1949): 217–20.

39. A. Glennie, "The Automatic Coding of an Electronic Computer," lecture notes (1952).

40. Project Whirlwind Summary Report no. 22 first quarter 1950 p. 24 and no. 31 third quarter (1952), 12.

41. MIT Computation Laboratory, Memorandum M-1624-1, 1.

42. MIT Computation Laboratory, Engineering Note E-516, 3.

43. J. W. Backus et al., *The FORTRAN Automatic Coding System for the IBM 704* (Poughkeepsie, NY: IBM, 1956), 7.

44. R. Ash et al., *Preliminary Manual for MATH-MATIC and ARITH-MATIC Systems for ALGEBRAIC TRANSLATION and COMPILATION for UNIVAC I and II* (Automatic Programming Development, Remington Rand UNIVAC, April 19, 1957).

45. Compare also with David Nofre, Mark Priestley, and Gerard Alberts, "When Technology Became Language: The Origins of Linguistic Conception of Computer Programming, 1950–1960," *Technology and Culture* 55, no. 1 (2014): 40–75.

46. A. J. Perlis and K. Samelson, "Preliminary Report—International Algebraic Language," *Communications of the ACM* 1, no. 12 (1958): 8–22.

47. P. Naur, ed., "Report on the Algorithmic Language ALGOL 60," *Communications of the ACM* 8 (1960): 299–314, and Numer. Math. 2 (1960), 106–36.

48. The statistics of figure 7.1 were produced with GNUplot. The data for *Datamation* and AFIPS were obtained by automatically searching with grep through the OCR-ed issues of the journals in question. The ACM data were obtained by using the search engine of the ACM Digital Library.

49. H. Durnova and G. Alberts, "Was Algol 60 the First Algorithmic Language?" *IEEE Annals of the History of Computing* 26, no. 4 (2014): 104–6.

50. H. C. Schwarz, "An Introduction to ALGOL," *Communications of the ACM* 5 (February 1962): 82–95.

51. Perlis and Samelson, "Preliminary Report—International Algebraic Language."

52. *Communications of the ACM* 3, no. 2 (February 1960).

53. R. E. Grench and H. C. Thatcher, eds., *Collected Algorithms 1960–1963 from the Communications of the Association for Computing Machinery* (Argonne National Laboratory, report ANL-7054), iii.

54. Donald Knuth, *The Art of Computer Programming: Fundamental algorithms* (Reading, MA: Addison-Wesley, 1967): vi–vii.

55. See also Astarte's chapter in this volume on how formal semantics acted as a marker of scientism.

56. Many claims to first uses exist, e.g., Paul Niquette, Grace Hopper, or the RAND Corporation, but they all have in common that it was originally used as a joking designation of things not hardware.

57. E.g., "Senior Programmers Are Urgently Needed to Help Develop a Large 'Software' Package for Commercial and Military Applications for R-W Stored Logic Computers," *Datamation* (1961): 1.

58. E.g., "large systems of programs have now reached such a degree of complexity and power as to rival the machines. This 'software,' as it is currently called . . . ," *Datamation* (1960): 9.

59. Interview with H. W. Robinson by Bruemmer, July 13, 1988, *Oral History Collection,* Charles Babbage Institute.

60. Thomas Haigh, "Software in the 1960s as Concept, Service, and Product," *IEEE Annals of the History of Computing* 24, no. 1 (2002): 5–13.

61. According to Haigh and Ceruzzi's forthcoming new edition of *A History of Modern Computing,* it might have been the inflation of job titles such as "analyst," "software engineer," "program architect," etc. that made the less pretentious "coder" seem more appealing. Comment on SIGCIS discussion list, May 5, 2020.

62. John R. Ryce, "Purpose and Scope" in *ACM Transactions on Mathematical Software* 1, no. 1 (1975): 1–3.

63. See also Babintseva's chapter in this volume, which shows how the word "algorithmic" changes in meaning in a Soviet and, later, an American financial context.

64. "Scheduling algorithm" and the like account for a large part of the increase in using the word "algorithm" in the AFIPS proceedings.

65. The economists knew the word through "linear programming," a mathematical theory developed in the late 1940s. See Kennedy, "Virtual Capital," chapter 4 for more details.

CHAPTER 8

The Lurking Problem

Scott Kushner

In the summer of 2005, on The Admin Zone, an online forum fre-
quented by people who run online forums, a short post showed how
lurking—the practices of not generating content where the generation
of content is expected—was understood to violate normative notions
of proper conduct in online social spaces. A forum operator with the
screen name DataHunter2009 sought counsel in a post titled "Huge
lurker problem!" After adjusting the forum's access settings, Data-
Hunter2009 was baffled: "Since this change, every new member has
become a lurker . . . and i mean EVERY new member. It used to be at
43 members, but in 2 months, my member count jumped to 120. But,
77 of the members looked at the messages for a few minutes after join-
ing and then left, never to be seen again."[1] Lurking frustrates expectations, small and large. DataHunter2009's personal expectations were
disappointed: for all the work invested in the site, these users offered
nothing in return. But there was also a sense that this imbalance reflected broader expectations of what a user should do.

An episode on BLADES-L, an early-2000s genealogy listserv, makes
these expectations explicit. The list manager, Peter Blades, was frustrated by a "large number of long term hangers-on" who contributed
nothing to discussions. Blades wrote that he would "unsubscrib[e] any
subscriber who has not contributed" within two weeks. Moreover,

Blades wrote that he would "give all new subscribers one month to get acquainted with the List and contribute."[2] This amounted to a probationary period for new workers to complete onboarding. The allusions to work are intentional here: lurking is inextricable from labor.

The link to labor emerges from the expectation that users generate content, that they "contribute," in Blades's terms. In today's social web, users are assigned tasks that generate data. First, they must allow the platform to record their clicks, scrollings, and lingerings, yielding to what Shoshana Zuboff calls "surveillance capitalism."[3] Under this arrangement, social platforms provide services to users at no monetary cost, settling instead for what Eric Posner and Glen Weyl call "all the upside of the data they generate."[4] That upside becomes significant by the 2010s and '20s,[5] though it was likely less tangible to hobbyists like DataHunter2009 and Blades. Second, users are summoned to post, comment, rate, like, respond, or otherwise generate content for others to read, watch, look at, or listen to. Lurkers meet the first demand but refuse the second: they don't create content, which means that they don't do their jobs.[6] By framing lurking as problematic—and pursuing solutions to the problem thus framed—DataHunter2009 and Blades both attempt to configure their users, to "[set] constraints on their likely future actions."[7] But their disappointment lingers.

At the heart of their complaints are questions of surveillance, differential visibility, and participation. DataHunter2009 and Blades can see their users' activity (or lack thereof) in server logs and membership rolls. Lurkers' offense consists in not making themselves visible to other users. Simply put, lurkers don't participate in the right ways, and lurking is therefore often understood to be, as DataHunter2009 put it, a problem.

This chapter traces the history of the problematization of lurking in networked computing. To problematize something is to subject it to "an analytic practice that takes place at the level of deep-seated presuppositions—necessary meanings antecedent to an argument— and assumptions about the world."[8] Unpacking lurking's presuppositions will reveal assumptions about how social media environments— and society—should work. When system administrators, platform owners, scholars, and users problematize lurking, they implicitly make claims about what behaviors are appropriate in particular situations: lurking "becom[es] an object of concern, an element for reflection, and a material for stylization."[9]

Those who build, maintain, operate, and frequent online social spaces have long understood lurking to be a problem in need of a solution, but the valences of that problem shifted as commercialized social platforms eclipsed amateur sites in the early 2000s. From the 1980s through the early 2000s, system administrators constructed lurking as a frustration, a drag on a community, and a deviant form of online behavior—but one with limited consequences. DataHunter2009 wrote that "the problem bugs me everytime I log in,"[10] but that appears to have been the extent of the damage. (The only substantive reply was a virtual shrug: "This is normal.")[11] Non-administrators had different relationships with lurking, often understanding it as unremarkable or a stage through which users pass, as in the case of a 1996 post in a newsgroup for recovering addicts: "I've been in recovery since Nov of 94. . . . and I've been lurking in this group most of the time."[12] With the rise of the commercialized social web in the mid-2000s, however, lurking posed an existential threat to firms, such as Facebook, that counted on users to generate content.[13]

Although archival evidence dates at least to the early 1980s, the scholarly literature on lurking emerges in the mid-1990s, gathering momentum in the early 2000s. Scholars in human–computer interaction (HCI), sociology, and media studies offered various definitions for lurking.[14] Most center either on metrics that lurkers fail to meet (some number of contributions over a given time period)[15] or concepts such as "free riding"[16] or "listening."[17] Much of the scholarship aims to rescue lurking, casting it as somehow productive[18] (or at least harmless)[19] and devising schemes to reduce its prevalence.[20] Such rescue efforts reckon with lurking's most salient fact: it's what most people are doing most of the time. The proportion of lurkers in online social spaces has been pegged as high as 90%, with only 10% of users producing any content, and a mere 1% generating most of it.[21] Some scholars have portrayed lurking as normal,[22] nondeviant behavior: "The default user," wrote Geert Lovink in the aptly titled *Zero Comments*, "is the lurker."[23] Or, as Barry Wellman put it: "Lurking is normal, it is the people who post who are abnormal."[24]

In what follows, I will elaborate an understanding of lurking as the practices of not generating content where the generation of content is expected. This definition echoes earlier attempts to delimit lurking but with one substantial adjustment: it accounts for context, because lurking makes sense only where users are both able and expected to contribute content. Not all networked computing contexts admit of user-generated content, and not all such contexts demand it. The specifics of the contexts establish the norms that construct lurking.

To construct lurking in normative terms is to imagine it as a historical construct, wrapped in the technological, economic, and social fabric of networked computing. Even though sociology, HCI, and media studies have attended to lurking, no history of the practice has yet been written. This is not entirely surprising, as there are major methodological challenges to overcome. Lurking, a practice of *not* doing something, leaves few traces and is maddeningly difficult to pin down. All reception research poses challenges, leading fields like the history of computing to focus mostly on extant records held by corporations, research institutes, and academic libraries. Historians of lurking seek evidence that something is missing, such as the user contributions that failed to materialize for DataHunter2009. In what follows, primary source evidence takes the form of complaints by sysadmins, forum operators exchanging advice, and confessions of having lurked. Accumulating such evidence entails lurking through archived message boards and usenet groups, untangling musty threads of online sociality in search of interactions that point to something that was not said or a person who said nothing.

Most conversations where evidence of lurking surfaces are unremarkable: an admission that one has spoken up for the first time, counsel offered to those wishing to draw other users into conversation, a recommendation to newcomers. Lurking does not manifest as a blockbuster event that leaves extensive documentation of motivations, causes, and consequences. There are no great lurking events, nor have any people become famous for lurking and left records behind. Lurking's lived history is instead impressionistic, comprising small moments, most of them similar to DataHunter2009's complaint or Blades's decree.

The history I'll recover in this chapter is largely compressed in the last 40 years, but I'll start much deeper in the past by opening up the word *lurking* itself. Other terms could be used to describe the practices of not generating content,[25] but the word *lurk* is itself instructive, because it frames lurking as socially deviant. Indeed, an archaic meaning, lurking in the word's history, suggests not doing what one is supposed to do. With this etymology unpacked, I'll explore the ways that the term was used in the 1980s, 1990s, and early 2000s by system administrators and users. Then, I'll elaborate a definition of *lurking* that accounts for historical and contextual difference. I'll conclude by pushing from computing to the street, exploring how lurking interacts with broader notions of participatory politics.

Lurking Language

People have been lurking—and talking about lurking—for quite some time. Tracing the word's history illuminates the consequences of using it to describe online behavior. *To lurk* is among English's oldest words, dating to the early fourteenth century.[26] It derives from an older verb, *to lour*, meaning "to frown" or "scowl" but also "to threaten." A frequentative suffix, *-k*, was added, leaving *lurk* to indicate habitual, sustained frowning or threatening, as the common verb *to talk* suggests the habitual, sustained telling of tales. Related languages had similar words: Low German's *lurken* (to shuffle along) and Norwegian's *lurka* (to sneak away). *Lurka* also appears in Swedish, where it suggests slowness in work, a sense that later resurfaces in English.

All of *lurk*'s historical connotations are negative, and many concern seeing without being seen. In addition to current meanings, centered on "remain[ing] furtively or unobserved about one spot," there is the rarer meaning of "mov[ing] about in a secret and furtive manner." In early usage, the person who lurks is often evading authority (as did a king's foes in one Middle English romance)[27] or engaging in unsanctioned movements (as in a Middle English poem where a man lies in wait for a maiden he covets).[28] Lurking is thus either stationary or ambulatory but always marked by invisibility, concealment, and the hint of a threat. In one obsolete meaning, *to lurk* is "to peer furtively or slyly": the subterfuge remains, joined now by an explicit invocation of looking. In all of these cases, the habitual nature of lurking is an integral component of the threat it poses: lurking also means "to lie in ambush." The danger of lurking, etymologically speaking, is not simply associated with an unseen seer but also with impending attack.

Most interesting, however, is an obsolete meaning, "to shirk work; to idle," which echoes the Swedish *lurka*, "to be slow in one's work." Instances of this disused meaning include a 1551 critique of fathers who raise lazy children[29] and, more to the point, a 1570 warning against employing rascals and thieves: "slouthfull to wurke," they "loue for to lurke."[30] The imputation here is not lying in wait but simply lying. Lurking in this sense is both unauthorized inactivity and the concealment of that inactivity from others. Similar to the word's other meanings, this lurking always threatens to explode into view when the boss checks the laborer's progress.

Lurking has thus always been bound up in social relations. Hiding, watching, and ambushing are all actions taken in relation to other people: one hides from others, watching them, ready to pounce. If one shirks labor, one frustrates others' expectations. Lurking is a matter of hiding one's movements, stillnesses, and machinations from others.

The notion that lurking has to do with labor, obsolete in modern English, is essential to making sense of lurking in networked computing. The word was used in relation to email lists, discussion boards, and social platforms to invoke the tangled threats of lying in wait and failing to complete a task. When *lurking* entered the computing lexicon,[31] it carried all of these connotations.

". . . So you know they're there"

Fittingly enough, when *lurking* enters the realm of networked computing, it arrives quietly. The earliest instance I've found of the term in relation to computing comes in a 1983 *New York Times* article, surely capturing use that predated publication. The concept of lurking is deployed to add color near the end. Lurking is framed, oddly enough, as a polite gesture, which shrouds more sinister connotations.

The article explains that, on electronic bulletin boards, "strangers debate and harangue; shy people lose their shyness; and many people invent fantasy lives about themselves." However, the article notes, not everyone is so talkative: "Some computer owners . . . observe others' conversations, eavesdropping but not participating." Indeed, the *Times* quotes a man from Phoenix, Arizona, named Kael Smith, "who uses his computer nightly," and who reports: "Sometimes they even type 'lurking' so you know they're there."[32]

What Smith described to the *Times* shows one way that BBS (bulletin board system) users navigated the asymmetrical visibility that characterizes lurking. Aware that they can see without being seen,[33] these lurkers make themselves visible, even if only to announce their invisibility. (A paradox: are lurkers who announce they're lurking still lurkers?)[34] More importantly, the anecdote shows lurking's cultural valence. To announce oneself offers a courtesy to those who speak: it makes them more fully aware of their audience and the extent to which their words circulate. Smith's remark suggests an expectation of polite visibility on BBSs, where the number of users was often countable in a manner unimaginable in the context of twenty-first-century commercial social media.

As polite as these lurkers may have been, the *Times* frames lurking in problematic terms. The paragraph about lurking—two sentences near the end of a four-column story with a photograph—sits among a string of examples of deviant behavior. The short mention of lurking follows Smith's account of the "weirdos" and "goofy people" one could meet on a BBS, some of whom "are flat-out nuts"; it precedes discussions of aliases (framed as textual devices that conceal identity), "abusive" behavior on the boards, sexually explicit content (which "stirred concern" of state intervention), and users admonished for "filling the exchanges with obscenities."[35] Even as lurking shows deference to more talkative users, the *Times* places it alongside questionable company.

Elsewhere, on the message boards and forums of the noncommercial web, lurking was seen as a natural part of the landscape (lurking is "allowed" and "desirable") that served a purpose ("They are a quiet but uncritical audience") and was often a waystation preceding regular participation.[36] Emerging from lurkerdom often assumed a ceremonial sheen, as if the delurker performed a transition from apprentice to practitioner. In 2007, Mel, a user on a beekeeping usenet group, debuted by writing: "normally I make a point to lurk in a newsgroup for a few weeks before I post, but I need an answer fairly quickly. So I've read back through the archives for a few weeks instead. You seem like a helpful bunch."[37] Lurking is presented here as proper etiquette—speaking privileges must be earned—and, in this first contribution, the user seeks to compensate for not having observed silently for an adequate time by performatively reviewing the group's history. Indeed, lurking is sometimes even recommended, as when one user on an automobile enthusiasts' board suggested that another user "might want to lurk over at the pelican parts forum for a while and chime in if you don't see an answer to your question."[38]

Frequent users often applauded delurking. After a first post, it was not uncommon for regular users to fête the newcomer, sometimes winkingly, as on a motorcycle board in 2011 ("'Lurker' status is now revoked!!!"),[39] and sometimes mock-flirtatiously, as when a user with the screenname "The Cutest Atheist" delurked in 2003 ("Hello Cutie. Welcome to the party").[40] On a board dedicated to R.E.M., the American popular music group, lurking was invoked when a user tried to summon discussion: "So where is everybody? Time to come out of lurker mode."[41] In these contexts, lurking is tolerated, but the unspoken assumption is that fully formed users contribute. Among users, the notion that delurking is a desirable sign of growth—and, by extension, that lurking is a behavior that users should move beyond—is received wisdom.

The question for users, then, is how to graduate from lurking to visibility, from silence to participation. At times, this is a matter of users reflecting on their own habits, as in the 2011 case of a blogger admitting that he lurked on others' blogs—and promising to stop: "Knowing there's a problem is half the battle, right?"[42] This blogger's confession is also a search for causes ("I get nervous," "technical difficulties"), and he emplots his transition away from lurking as a form of self-improvement.

From other perspectives, the question is how to motivate others to delurk. As the social spaces of the internet became workplaces, the problem narrative became louder. Such efforts are framed not as self-improvement but rather in terms of improving an interaction, a community, or a profit margin. Heidi Medina, a user engagement expert, offered advice in 2018 to social media managers about "enticing [lurkers] into action." The goal is to "turn that lurker into a buyer." In the meantime, "lurkers aren't really a problem," because they "are still learning about the brand."[43] Virtual sex workers see lurking in similar terms. In 2019, Taja Ethereal, a webcam model, dished advice from "the Camgirl Book Of Hustles" to colleagues confronting "the abject fuckery of a chatroom lurker." Ethereal implies that, in a space designed for looking, visitors should speak up. Lurkers could be converted by "creat[ing] or fak[ing] chatroom engagement" that would elicit participation: "If talking dirty isn't your shtick, then go off into a long story about the best concert you've ever been to. Eventually someone will speak up." The appearance of engagement begets engagement, which can jumpstart a quiet crowd.[44]

Medina and Ethereal operate in unambiguously commercial spaces that are seemingly far removed from the chatrooms, BBSs, and forums that were more prominent in the 1980s, '90s, and early 2000s. Yet, their concerns echo those of not-for-profit operators like DataHunter2009 and Blades: the questions of visibility and participation persist. For both the social engagement manager and the webcam model, lurkers are visible as they log in. Their efforts are organized around compelling lurkers to make themselves visible to other users, to engage with content that might "create leads" or "make money" for a brand[45] or "convert lurkers into spenders."[46] Like so many other strands of computing history, the visible importance of money grew over time.

Today's massive social media corporations recognize that they are "highly dependent on [their] users and the engagements they make."[47] Lurking becomes a constant threat to the bottom line: a thriving social scene is a prerequisite to a healthy balance sheet. This isn't new. A former AOL forum moderator reports that he "would frequently reward first-time participants with free time (at that time worth $5 an hour) if they asked good questions or offered valuable input." He also notes that CompuServe "[told] users how many responses they'd received to a new comment when they logged back in," in order to encourage ongoing participation,[48] foreshadowing visible Like, comment, and retweet counts on platforms such as Facebook or Twitter. Such public displays of participation not only create a "Like economy,"[49] casting deep, sociality-drenched links across the entire web. They also create a participation economy, where users learn to savor the responses their content provokes and to produce content that will push other users to engage.[50]

As BBSs, mailing lists, and forums met the glare of twenty-first-century corporate social media, platform managers wanted lurkers to do more than type "lurking." While noncommercial lurking in the 1980s and '90s could be understood as a matter of good manners, the financial pressures of advertising-driven social media demand that more content be generated more often.[51] As the context changed, the expectations grew.

Not Generating Content Where the Generation of Content Is Expected

Lurking is a combination of practice and context. Different contexts demand different sorts of participation, and it is only in specific contexts that specific behaviors can be understood as a problem and integrated in a larger social field. *Lurking refers to the practices of not generating content where the generation of content is expected.* This definition comprises three elements.

First, lurking refers to *practices.* Lurking is not a non-event but a specific kind of activity: a thing that someone does. As demonstrated by the care with which platforms like Facebook monitor user activity,[52] lurkers' movements are visible to those with access to the backstage portions of social spaces. Some contemporary platforms even make public the presence of those who read or watch, as in the view counts on YouTube videos.[53] Lurking is not a complete absence of action.

It does, however, denote the absence of a specific action: lurking is the practice of *not generating content.* This element descends directly from existing scholarship, which coalesces around the notion that lurking consists in not contributing to discussion; it also carries the accumulated etymological weight of the verb *to lurk.* It's by not generating content—by not making one's presence felt to others—that one lurks. This is where the two meanings of the verb *to lurk* meet: lurkers are not only hidden, but they explicitly do "not fulfill their roles." Social media users are workers whose wages are access to human sociality. Their role is precisely to create content: to contribute to bottomless, algorithmically curated piles of user content or to respond to the stimuli of Likes, retweets, and comment counts. As they watch furtively, lurkers generate content that is visible to platforms but which platforms cannot profitably make visible to other users. The problem is lurkers' invisibility to other front-end users.

Third, lurking occurs only in contexts *where the generation of content is expected.* Different environments have different participant structures, "the types of roles that are available in a situation."[54] Some sites' participant structures allow for lurking, but others' do not. For example, in June 2020, the three most visited sites on the World Wide Web were google.com, youtube.com, and tmall.com.[55] Although each leverages user-generated content, a brief review shows that none of them constitutes a context where lurking makes sense.

While it is owned by a large company that offers a wide range of services, the google.com search utility does not solicit content directly from its users. Google generates search results using an algorithm that feeds on user activity across the web: Google ranks sites as a function of user behavior, an example of usability guru Jakob Nielsen's suggestion to "let users participate with zero effort."[56] But lurking is nonsensical in the context of google.com, because users are never made visible to one another.

YouTube, a Google subsidiary, is more ambiguous. Much of the content on YouTube is created by users, and comments threads and chat boxes dangle below videos and livestreams, respectively. But YouTube presents visually as a place to watch videos; features inviting active participation cling to the margins. Comments are literally placed off-screen (one must scroll down even to see them, never mind to reply), and the upload link is relegated to a corner. At YouTube, the generation of content is possible—encouraged, perhaps—but not expected.

TMall is a Chinese e-commerce site. As on amazon.com (the thirteenth-most-visited site in June 2020, likely more familiar to Anglophone readers), TMall's product pages include purchaser reviews. These reviews are valuable because, as Joseph Reagle notes, they allow e-commerce sites to deputize customers to evaluate products' quality, which drives sales.[57] Again, on TMall and Amazon, participation is encouraged but not demanded, a distinction made clear by user-generated content's placement, as in Reagle's title, at "the bottom of the web."

It is the fourth website on the list, facebook.com, where lurking begins to compute. On Facebook, the generation of content is expected. This expectation is made explicit by a textbox at the top of Facebook's screen asking, "What's on your mind?" (Or, in the mid-2010s, offering clear instruction: "Write something.")[58] On Google's search pages, a box often appears, near the top, with the title "People also ask," under which sit a handful of related questions. These boxes "us[e] machine learning [to] identify what questions . . . users have next before they even ask,"[59] and they suggest which concepts belong together in the collective consciousness of Google users. When one enters the terms "lurking facebook," Google provides such a box: the top item is "What is lurking on Facebook?" No box appears when querying "lurking youtube" or "lurking google." This absence implies that Google finds little semantic connection between the terms. Lurking is possible in Facebook's participant structure (or, again according to the "People also ask" box, in Instagram's or Snapchat's), but not in YouTube's or Google's. It is on sites like Facebook that lurkers fail to meet the participatory expectations that formats use.

Generating content in the social web is fundamentally a question of format, of doing something in a proscribed way. Christopher Kelty documents the century-long transformation of participation from a process that can "reveal ethical intuitions, make sense of different collective forms of life, and produce an experience beyond that of individual opinion, interest, or responsibility" to one that is "a formatted procedure" whereby "autonomous individuals . . . experience an attenuated, temporary feeling of personal contribution."[60] For the twenty-first-century platforms that facilitate participation, format matters. "To participate is to live stories,"[61] but to participate on social platforms is to live stories in a formatted way—via the textboxes, webforms, and social buttons (such as Facebook's ubiquitous Like)[62] that abstract the embodied experience of everyday life into data that can be processed and reallocated for other users' consumption.

The problem with lurking is that it leaves the wrong kinds of traces in the wrong kinds of places. Platforms push users to generate content, in order both to attract other users' attention and to mine it for actionable data.[63] Lurkers don't make themselves visible in a way that platforms find profitable, a refusal—intentional or otherwise—to be properly visible. If participation is structured according to a historically specific grammar, then lurking is the misconjugated verb of contemporary social media. It is the participation that does not parse.

Post-Script: "No elegant form of solidarity"

Structuring participation and lurking around platforms' balance sheets has broader consequences for the practices of political action. I write this essay as protests around the world object to policing practices that injure and kill Black people disproportionately. Calls for antiracist reforms fill streets and social platforms, two essential media of political activism in the early twenty-first century.

But these two media—the street and the screen—have different participation structures. As the protests marched on, Jamila Thomas and Brianna Agyemang, two Black women working in the music industry, led an online protest among record labels and recording artists "to show solidarity for the Black Lives Matter movement."[64] Participants would place black squares on their social profiles and refrain from posting on Tuesday, June 2, 2020. In effect, Thomas and Agyemang asked people to commandeer the logics of participation in order to lurk visibly, to be seen not generating content where the generation of content is expected. The idea gained momentum and an endorsement from Patrisse Cullors, one of the co-founders of the Black Lives Matter movement,[65] and it spread beyond the music industry.

But Blackout Tuesday had mixed results: some saw empty posturing. Michael Harriot downplayed the notion of white celebrities stepping away from their social feeds for a day as another way to have a "conversation about race" that offers absolution in exchange for an unburdensome demonstration.[66] Another writer declared: "just posting a black square and then logging off gives both brands and nonblack people a way of signaling support on social media without providing any real help."[67] Indeed, others suggested it was counterproductive: "my instagram feed this morning is just a wall of white people posting black screens. like . . . that isn't muting yourself, babe, that's actually kind of the opposite! it's taking up an absolutely WILD amount of space and does nothing!"[68] Moreover, many participants labeled their blacked-out status updates with the #BlackLivesMatter hashtags (despite organizers' instructions to the contrary[69] and a plea from Alicia Garza, another co-founder of Black Lives Matter, to use a different hashtag),[70] and the blackout effectively overwhelmed attempts to organize and support protests in the streets on June 2—the very protests the blackout was meant to encourage.[71] The performance of not generating content (possible on platforms like Twitter and Instagram only by generating content)[72] is an ineffective form of political action. Its risks are too low, its gestures too disembodied, and its focus too neatly placed on the individual who posts the square on the screen. What possibilities exist for politically engaged silence in a media context that registers only active participation? What might it mean to reconfigure use to serve interests of parties beyond platform owners?

A reporter for *The Atlantic* summarized the trouble with Blackout Tuesday: "The insta/twitter blackout controversy captures so much of what makes the platforms awful for politics: There is no elegant form of solidarity #onhere; of skeptical, sympathetic, or silent active listening; and every conversational contribution looks like an assertion of identity."[73] The format of online participation transforms collective efforts into individual statements. Gestures of solidarity blur into virtue signaling.

At protests and marches, political action takes the embodied form of physical co-presence. A protest's power lies in a mass of people showing up and clogging streets, a corporeal denial of service attack that snarls traffic and makes a demand on public space. A street protest embodies "politics understood as the *disruption* of power," forceful changes to the flows of bodies, objects, ideas, and decision-making, rather than "the *distribution* of power" through preexisting channels.[74] Marchers risk their physical safety by massing their bodies in public spaces, by reconfiguring the street, whose logics mandate that "only certain forms of access/use are encouraged"[75]—and this potentially in opposition to state and police power, a fact underscored by injuries sustained by some protesters during the spring 2020 Black Lives Matters actions.[76] As protesters' individual identities melt into the crowd, the potential of the street as a medium of political action outshines that of the information superhighway.

Are lurking and political action compatible? Because "contemporary participation is resolutely focused on the individual participant,"[77] the refusal to generate content online—especially where nonparticipation is invisible to others—has little political salience.[78] At a street protest, one might maintain distance, observe while remaining unseen. But it is also possible to join the crowd, to march alongside, to blend one's voice into the chorus: to contribute to collective action without foregrounding one's own contributions. This was the Blackout Tuesday organizers' intent, frustrated though it was by participants' execution and the affordances of social platforms. Indeed, even as participants performed their temporary media abstention, they fed content to platforms, neither disrupting nor redistributing their power.

A mass movement makes participants visible as a collective, not as individuals. This is why protests can have political impact, if the circumstances are right, contributing to the reshaping of policy, law, society, and culture. Online lurking disrupts social platforms by refusing the demand of atomized, individual visibility. But as it withholds content, lurking fails to communicate much of anything; it is ill suited to focusing political energy, and its meanings are difficult to parse for fellow travelers and social platforms alike.[79] It is a form of resistance that can testify as easily to deeply held convictions as it can to distractedness or laziness. A problem, indeed.

Notes

1. DataHunter2009, "Huge lurker problem," The Admin Zone, July 23, 2005, https://www.theadminzone.com/threads/huge-lurker-problem.12300/#post-87924.
2. Peter Blades, "Announcement," BLADES-L, February 14, 2000, https://lists.rootsweb.com/hyperkitty/list/blades.rootsweb.com/thread/31876044/.
3. Shoshana Zuboff, *Surveillance Capitalism* (New York: Public Affairs, 2019).
4. Eric A. Posner and E. Glen Weyl, *Radical Markets: Uprooting Capitalism and Democracy for a Just Society* (Princeton, NJ: Princeton University Press, 2018), 231.
5. Dan Bouk, "The History and Political Economy of Personal Data over the Last Two Centuries in Three Acts," *Osiris* 32 (2017): 103, https://doi.org/10.1086/693400.
6. Scott Kushner, "The Instrumentalised User: Human, Computer System," *Internet Histories* 5, no. 2 (2021), https://doi.org/10.1080/24701475.2020.1810395.
7. Steve Woolgar, "Configuring the User: The Case of Usability Trials," in *A Sociology of Monsters: Essays on Power, Technology and Domination,* ed. John Law (London: Routledge, 1991), 57–99.
8. Carol Bacchi, "The Turn to Problematization: Political Implications of Contrasting Interpretive and Poststructural Adaptations," *Open Journal of Political Science* 5, no. 1 (January 2015): 2, http://doi.org/10.4236/ojps.2015.51001.
9. Michel Foucault, *The Use of Pleasure,* vol. 2: *The History of Sexuality,* trans. Robert Hurley (New York: Vintage, 1990), 23–24.
10. DataHunter2009, "Huge lurker problem."
11. TheViper, reply to "Huge lurker problem," The Admin Zone, July 24, 2005, https://www.theadminzone.com/threads/huge-lurker-problem.12300/#post-87941.
12. tunaman, "A lurker comes to the surface," alt.recovery.na, September 15, 1996, https://groups.google.com/g/alt.recovery.na/c/oxym2fNFafc/m/onw-NBekT-MJ.
13. For examples of other computing terms whose meanings have shifted over time, see Liesbeth De Mol and Maarten Bullynck, "What's in a Name? Origins, Transpositions, and Transformations of the Triptych Algorithm–Code–Program," this volume.
14. Christian Stegbauer and Alexander Rausch, "Lurkers in Mailing Lists," in *Online Social Sciences,* ed. Bernad Batinic, Ulf-Dietrich Reips, and Michael Bosnjak (Seattle: Hogrefe & Huber, 2002), 264.
15. Blair Nonnecke and Jenny Preece, "Lurker Demographics: Counting the Silent," *CHI '00: Proceedings of the SIGCHI Conference on Human Factors in Computing Systems* (2000): 76, http://doi.org/10.1145/332040.332409; Jenny Preece, Blair

Nonnecke, and Dorine Andrews, "The Top Five Reasons for Lurking: Improving Community Experiences for Everyone," *Computers in Human Behavior* 20, no. 2 (March 2004): 202, https://doi.org/10.1016/j.chb.2003.10.015.

16. Peter Kollock and Marc Smith, "Managing the Virtual Commons: Cooperation and Conflict in Computer Communities," in *Computer-Mediated Communication: Linguistic, Social and Cross-Cultural Perspectives*, ed. Susan C. Herring (Amsterdam: John Benjamins, 1996), 109–28.

17. Kate Crawford, "Listening, Not Lurking: The Neglected Form of Participation," in *Cultures of Participation*, ed. Hajo Grief, Larissa Hjorth, and Amparo Lasén (Berlin: Peter Lang, 2011), 63–77.

18. Blair Nonnecke and Jenny Preece, "Silent Participants: Getting to Know Lurkers Better," in *From Usenet to CoWebs: Interacting with Social Information Spaces*, ed. Christopher Lueg and Danyel Fisher (Berlin: Springer, 2003), 126–27; Judd Antin and Coye Cheshire, "Readers Are Not Free-Riders: Reading as a Form of Participation in Wikipedia," *CSCW '10: Proceedings of the 2010 ACM Conference on Computer Supported Cooperative Work* (2010): 128, https://doi.org/10.1145/1718918.1718942; Masamichi Takahashi, Masakazu Fujimoto, and Nobuhiro Yamasaki, "The Active Lurker: Influence of an In-house Online Community on Its Outside Environment," *GROUP '03: Proceedings of the 2003 International ACM SIGGROUP Conference on Supporting Group Work* (2003): 1, https://doi.org/10.1145/958160.958162.

19. Barry Wellman and Milena Gulia, "Virtual Communities as Communities: Net Surfers Don't Ride Alone," in *Communities in Cyberspace*, ed. Marc A. Smith and Peter Kollock (London: Routledge, 1999), 180.

20. Rosta Farzan, Joan M. DiMicco, and Beth Brownholtz, "Mobilizing Lurkers with a Targeted Task," *Proceedings of the Fourth International AAAI Conference on Weblogs and Social Media* (2010): 235, https://www.aaai.org/ocs/index.php/ICWSM/ICWSM10/paper/view/1483.

21. Nielsen, "The 90-9-1 Rule for Participation Inequality in Social Media and Online Communities," Nielsen Norman Group, October 8, 2006, https://www.nngroup.com/articles/participation-inequality/. On the significance of these figures and platforms' responses, see Scott Kushner, "Read Only: The Persistence of Lurking in Web 2.0," *First Monday* 21, no. 6 (June 2016), https://doi.org/10.5210/fm.v21i6.6789.

22. Noella Edelmann, "Reviewing the Definitions of 'Lurkers' and Some Implications for Online Research," *Cyberpsychology, Behavior, and Social Networking* 16, no. 9 (September 2013): 647, http://doi.org/10.1089/cyber.2012.0362.

23. Geert Lovink, *Zero Comments: Blogging and Critical Internet Culture* (New York: Routledge, 2008), 241. Speaking of a "default user" may obscure the ways that markers of social difference intersect with lurking. On gender and lurking, see Stacy Horn, *Cyberville: Clicks, Culture, and the Creation of an Online Town* (New York: Warner Books, 1998), 100–101. On race and lurking, see Lorna Roth, "Reflections on the Colour of the Internet," in *Human Rights and the Internet*, ed. Steven Hick, Edward F. Halpin, and Eric Hoskins (Hampshire, UK: MacMillan, 2000), 176–77. On sexuality and lurking, see Michael W. Ross, "Typing, Doing, and Being: Sexuality and the Internet," *Journal of Sex Research* 42, no. 4 (November 2005): 348.

24. Quoted in Nonnecke and Preece, "Silent Participants," 125. Fittingly, Wellman didn't publish this claim; it arrived to Nonnecke and Preece as a "personal communication," a back-channel message in the lurking literature. Wellman was lurking the scholarly conversation about lurking.

25. Meredith McGill once remarked to me that lurking was what we used to call reading.

26. Definitions and etymological data in this section are drawn from the Oxford English Dictionary.

27. *The Lay of Havelok the Dane,* 1.68.

28. John Gower, *Confessio Amantis,* 1.6746.

29. Robert Crowley, *The Select Works of Robert Crowley, Printer* (London: Early English Text Society, 1872), 117.

30. Thomas Tusser, *Five Hundred Pointes of Good Husbandrie,* chap. 10, quatrain 54.

31. A 2001 draft addition to the OED definition of *lurk* concerning the word's use as "computing slang" has not yet been fully revised and incorporated into the dictionary.

32. Robert Lindsey, "Computer as Letterbox, Singles Bar and Seminar," *New York Times,* December 2, 1983, A18.

33. Though note that, in some environments, users can enter a command that reveals which users are logged in. See Horn, *Cyberville,* 22–23.

34. Cf. fictualities, "The sound of one hand clapping: a lurker's manifesto," LiveJournal, January 16, 2007, https://fictualities.livejournal.com/51776.html.

35. Lindsey, "Computer as Letterbox," A18.

36. Horn, *Cyberville,* 30.

37. Mel Rimmer, "Moving a WBC Hive," sci.agriculture.beekeeping, May 30, 2007, https://www.usenetarchives.com/view.php?id=sci.agriculture.beekeeping&mid=VoNIRFRJYnpuUjQ.

38. Nate Nagel, reply to "Pontiac Dead," alt.autos.gm, February 23, 2009, https://usenetarchives.com/view.php?id=alt.autos.gm&g=74958&p=0.

39. Kenny Lukenbill, reply to "Lurker," Bluebird510 (Google Groups), December 14, 2011, https://groups.google.com/g/bluebird510/c/5FiqsDRBJNI/m/bvo9_s3KH2MJ.

40. Woden, reply to "Why, Hello There. I'm An AA Newbie," alt.atheism, October 11, 2003, https://usenetarchives.com/view.php?id=alt.atheism&g=72306&p=0.

41. Andrew Palka, "Lurkers Unite!" rec.music.rem, March 16, 2008, https://usenetarchives.com/view.php?id=rec.music.rem&g=10127&p=0.

42. paul, "Comments? A Lurker Confesses," *The Ossington Kitchen* (blog), December 14, 2011, http://theossingtonkitchen.blogspot.com/2011/12/comments-lurker-confesses.html.

43. Heidi Medina, "Are Social Media Lurkers Really a Problem?" LinkedIn, October 22, 2018, https://www.linkedin.com/pulse/social-media-lurkers-really-problem-heidi-medina/.

44. Taja Ethereal, "How to Deal with Chatroom Lurkers," *BoleynModels* (blog), June 27, 2019, https://boleynmodels.com/blog/dealing-with-lurkers-and-a-quiet-camroom/.

45. Medina, "Are Social Media Lurkers."

46. Ethereal, "How to Deal."

47. Tero Karppi, *Disconnect: Facebook's Affective Bonds* (Minneapolis: University of Minnesota Press, 2018), 7.

48. Christopher Allen, "Community by the Numbers, Part III: Power Laws," Life with Alacrity, March 19, 2009, http://www.lifewithalacrity.com/2009/03/power-laws.html.

49. Carolin Gerlitz and Anne Helmond, "The Like Economy: Social Buttons and the Data-Intensive Web," *New Media & Society* 15, no. 8 (December 2013): 1348–65, http://doi.org/10.1177/1461444812472322.

50. Kushner, "Read Only."

51. Mark Andrejevic, "Surveillance and Alienation in the Online Economy," *Surveillance & Society* 8, no. 3 (January 2011): 280, https://doi.org/10.24908/ss.v8i3.4164.

52. Karppi, *Disconnect*, 7, 11.

53. Karin van Es, "YouTube's Operational Logic: 'The View' as Pervasive Category," *Television & New Media* 31, no. 3 (March 2020): 223–39, https://doi.org/10.1177/1527476418818986.

54. Ilana Gershon, *Down and Out in the New Economy: How People Find (or Don't Find) Work Today* (Chicago: University of Chicago Press, 2017), 47.

55. "The Top 500 Sites on the Web," Alexa, updated June 24, 2020, https://www.alexa.com/topsites.

56. Nielsen, "The 90-9-1 Rule."

57. Joseph M. Reagle Jr., *Reading the Comments: Likers, Haters, and Manipulators at the Bottom of the Web* (Cambridge, MA: MIT Press, 2015), 44–45.

58. Aimée Morrison, "Facebook and Coaxed Affordances," in *Identity Technologies: Constructing the Self Online*, ed. Anna Poletti and Julie Rak (Madison: University of Wisconsin Press, 2014), 122.

59. Courtney Cox Wakefield, "Achieving Position 0: Optimising Your Content to Rank in Google's Answer Box," *Journal of Brand Strategy* 7, no. 4 (Spring 2019): 331, https://www.ingentaconnect.com/content/hsp/jbs/2019/00000007/00000004/art00005.

60. Christopher M. Kelty, *The Participant: A Century of Participation in Four Stories* (Chicago: University of Chicago Press, 2019), 1.

61. Kelty, *The Participant*, 3.

62. Gerlitz and Helmond, "The Like Economy."

63. Kushner, "Read Only."

64. J'Na Jefferson, "What Is Blackout Tuesday? Industries, Brands and More Go Black in Solidarity of Black Lives Matter," *The Root*, June 2, 2020, https://www.theroot.com/what-is-blackout-tuesday-industries-brands-and-more-g-1843852383.

65. Jem Aswad, "Black Lives Matter Cofounder Patrisse Cullors on Blackout Tuesday and How the Music Community Can Help," *Variety*, June 2, 2020, https://variety.com/2020/music/news/black-lives-matter-patrisse-cullors-blackout-tuesday-1234622767/.

66. Michael Harriot, "Top 5 Ways to Erase Racism (According to White People)," *The Root*, June 16, 2020, https://www.theroot.com/top-5-ways-to-erase-racism-according-to-white-people-1844054873.

67. Rebecca Heilweil, "Why People Are Posting Black Squares on Instagram," *Vox,* June 2, 2020, https://www.vox.com/recode/2020/6/2/21278051/instagram-blackout -tuesday-black-lives-matter.

68. Jeanna Kaldec (@jeannakadlec), Twitter, June 2, 2020, https://twitter.com/jeanna kadlec/status/1267799619699957760.

69. Brianna Agyemang (bri_anna), "We are tired and can't change things alone. This is a call to action for those of us who work in music/entertainment/show business to pause on Tuesday, June . . . ," Instagram, May 29, 2020, https://www.instagram .com/p/CAyqPSxAHOt/.

70. Alicia Garza (@aliciagarza), "Please remove the #BlackLivesMatter hashtag from the #blackoutday action. Today we need to be MORE connected than ever," Twitter, June 2, 2020, https://twitter.com/aliciagarza/status/1267820821201825804.

71. Zoe Haylock, "How Did #BlackOutTuesday Go So Wrong So Fast?" *Vulture,* June 2, 2020, https://www.vulture.com/2020/06/blackout-tuesday-guide.html. For further discussion of the complications of horizontal organizing, see Zeynep Tuekci, *Twitter and Tear Gas: The Power and Fragility of Networked Protest* (New Haven, CT: Yale University Press, 2017), ch. 3, esp. p. 53.

72. Cf. Laura Portwood-Stacer, "Media Refusal and Conspicuous Non-Consumption: The Performative and Political Dimensions of Facebook Abstention," *New Media and Society* 15, no. 7 (November 2013): 1046, https://doi.org/10.1177/1461444812465139.

73. Robinson Meyer (@yayitsrob), Twitter, June 2, 2020, https://twitter.com/yayitsrob /status/1267892672884285440.

74. Darin Barney, "'We Shall Not Be Moved': On the Politics of Immobility," in *Theories of the Mobile Internet: Materialities and Imaginaries,* ed. Andrew Herman, Jan Hadlaw, and Thom Swiss (London: Routledge, 2015), 15, italics in the original.

75. Woolgar, "Configuring the User," 89.

76. Knvul Sheikh and David Montgomery, "Rubber Bullets and Beanbag Rounds Can Cause Devastating Injuries," *New York Times,* June 12, 2020, https://www.nytimes .com/2020/06/12/health/protests-rubber-bullets-beanbag.html.

77. Christopher M. Kelty, "Too Much Democracy in All the Wrong Places: Toward a Grammar of Participation," *Current Anthropology* 58, suppl. 15 (February 2017): S88.

78. On race and the political salience of computing resources (including decisions about which software to use), see André Brock, "Beyond the Pale: The Blackbird Web Browser's Critical Reception," this volume.

79. On the internet's inherent political frustrations, owing to its decentralized nature, see Marc Aidinoff, "Centrists against the Center: The Jeffersonian Politics of a Decentralized Internet," this volume.

CHAPTER 9

The Help Desk

Changing Images of Product Support
in Personal Computing, 1975–1990

Michael J. Halvorson

This chapter explores some of the ways that PC users received customer service and technical support from software companies during the early years of personal computing, c. 1975–1990. My main historiographic concern is to understand how technical competence was constructed and developed in user communities influenced by personal computing and to explore how new forms of labor were valued, gendered, and made visible or invisible by emerging software companies. A preliminary review of the evidence indicates that by the mid-1980s, many PC firms were significantly overrun by the needs of their consumers. Although it was possible to support early microcomputers through person-to-person interactions and creative documentation, the surging demand for PC products meant that new infrastructures were needed to support users and the message that computers were easy to use for a wide range of consumers. The new mechanisms included novel business structures, enhanced communication systems, and new types of workers. Taking center stage during this period of transition were customer service and technical support employees, who joined the ranks of expanding PC companies in significant numbers. These "help desk" workers were the front-line laborers who answered support calls and responded to inquiries, often working long hours for modest financial rewards. Several trends soon developed in the PC industry. Among help desk workers, most of the front-line consumer response and customer service representatives were women, and they were usually paid hourly wages and denied the benefits of salaried employees. The technical support specialists were usually men, and they often had more direct career paths into IT jobs and software development roles if they were successful at their work. These gendered expectations fit the employment patterns noted in several recent studies about labor in computer-related organizations.[1] The consequence of this gendering influenced the career path and benefits of many support workers and preserved the status of certain kinds of technical knowledge in companies while demoting others.

My examination of product support practices also opens a new window on the historiographic question of "how users are made" and what types of technical knowledge are valued and impressed upon emerging user communities.[2] I explore some of the diverse experiences that lie beneath common terms such as "technical knowledge," "help desk labor," and "customer service." My suggestion is that PC users and technical support personnel exercised a mutual influence on each other through their interactions. In the 1980s, technical support workers used a range of tools adapted from diverse settings, including touch-tone phones, microfiche, prewritten scripts, paper engineering diagrams, time-sharing terminals, and more. Yet their customers were often calling about the newest versions of PC software and sharing feedback and concerns that were important for the company to hear. As a result, help desk workers did not just help to make users/consumers, but they transmitted important knowledge that would shape corporate policies and products.

To survey the rise of help desk services, I analyze the offerings of several early PC firms, including MITS, Apple Computer, VisiCorp, and Microsoft Corporation. As part of the presentation, I will use interviews and a selection of customer service documents from Microsoft, where I worked as a technical editor, writer, and localization project manager from 1985 to 1993. Using company records, I chart the growth of product support services to understand how user communities were built around key products, what the typical questions of customers were, and the organizational challenges facing companies during a period of rapid growth and restructuring. The "hidden workers" examined at Microsoft include consumer response agents, order entry personnel, customer service representatives, corporate accounts representatives, and product support technicians.

Why was there such uncertainty about how to support customers and products in the early PC industry? The context of this question takes us back to the business assumptions of early microcomputer manufacturers in the 1970s. Coming from mostly engineering backgrounds, the entrepreneurs who created the first microcomputer kits assumed that their buyers could figure out how to use the circuit boards and other materials sold to customers. These products included rudimentary hobbyist systems like the MITS Altair 8800 and Apple I microcomputers, but also early versions of shareware or commercial software, including Tiny BASIC and Tiny C. In each of these computing scenarios, it was assumed that if you had some experience with similar products, you could probably learn to use the new systems on your own, even if you didn't know precisely what computers could do.[3] Between 1975 and 1990, however, expectations about the "tacit knowledge" of hobbyists shifted dramatically, from the interpersonal modes of learning that were typical of user group meetings and small trade shows, to the sprawling technical communities shaped by nationwide registration agendas, formal support mechanisms, and the rising tide of computer books and magazines.

Early Customer Support for Microcomputers

The earliest microcomputer systems were component-based products that were assembled by small firms with little in the way of dedicated support systems. When the MITS Altair 8800 microcomputer debuted in early 1975, it was sold primarily via mail order transactions. The hobbyists and tinkerers who purchased the device were assumed to be familiar with radio electronics, integrated circuits, and the construction of kit-based products. Thomas Haigh and Paul Ceruzzi have described how the first users assembled these components and developed problem-solving techniques by working together and reading magazines such as *Radio-Electronics* and *Popular Electronics*.[4] Although a step-by-step instruction manual was included with the Altair 8800 kit, the consumers who purchased the device were encouraged to call the company's main office if they encountered problems. Soon after the product started shipping, callers began asking for help from owner Ed Roberts and his small staff in Albuquerque. But reports indicate that the phone lines were soon jammed with callers, and customers were forced to use more traditional methods. These included consulting technical newsletters, computer books, and user groups, the latter of which allowed for rich, person-to-person contact and the transfer of information through social mechanisms. User groups had a long history in computing communities. For example, IBM mainframe customers had routinely sought help from the SHARE user group, and DEC customers were able to consult the DECUS user society for new information about hardware and software products. Attempting to construct a user community in this way, MITS advertised that each customer who purchased an Altair 8800 would be supported by The Altair User's Group, a loosely organized social organization.

The story of interpersonal interaction and learning was similar for owners of the original Apple I microcomputer, released in July 1976. Hobbyists were able to purchase the bare-bones motherboard at The Byte Shop in San Rafael, California. They then needed to add a power supply, keyboard, and display device to the unit and make the appropriate connections. If they encountered problems, they were encouraged to contact Apple directly, or they could seek help from local user communities, such as The Homebrew Computer Club. During the limited release of the product, it was not uncommon for customers to call Apple and speak personally with Steve Wozniak or Steve Jobs.

Computer book author Mitchell Waite recalls how he had a problem with his Apple I motherboard and asked Wozniak for help in this way. Waite brought the motherboard to Apple and Wozniak carefully soldered a new socket into the breadboard area, then connected it to the device's peripheral interface adapter.[5] Jobs also visited Waite on his sailboat and learned how Waite had customized his Apple I to track local weather data. Interpersonal episodes like this indicate that during the start-up phase of the PC industry, it was not uncommon for users to seek tacit knowledge about computing systems directly from the personnel who represented the products. Microcomputer companies also learned what users were *doing* with their hardware and software, shaping expectations about what systems might accomplish in the future. In many respects, the microcomputer community was *defined* by its ability to learn in face-to-face ways.

VisiCalc and the West Coast Computer Faire

A fascinating nexus where sales, support, and customers met face to face were early computing trade shows in the United States, which gained momentum in the late 1970s. An interesting example of this is the West Coast Computer Faire, which held its first convention in San Francisco in April 1977. Over 12,000 people attended the three-day event and learned about microcomputer hardware, software, and accessories. Attendees had the opportunity to buy components, take classes, and visit the booths of over 180 computer exhibitors.[6] Unlike later commercial computing shows, which grew into massive events dominated by industry luminaries, the first microcomputer shows allowed consumers to meet the designers of their systems and interact with them. Buyers could ask questions, register complaints, and advocate for improvements—all in simple pipe-and-drape booths adorned with six-foot tables and humble signage.

A firm that announced its products in this context was Personal Software Inc. (later VisiCorp), the distributors of the spreadsheet VisiCalc. Software engineer Dan Bricklin introduced VisiCalc at the Computer Faire in May 1979, where he demonstrated the program to users of the Apple II platform. VisiCalc sales took off in 1980, achieving a run rate of 12,000 copies per month by the end of the year. New versions of the innovative product soon arrived for the Tandy TRS-80 and the IBM Personal Computer. By 1983, the spreadsheet had sold some 700,000 copies, showing how large the market for PC software could be.

How did Personal Software shape its user community and provide support for its customers? The organization first changed its name to VisiCorp, aligning its branding with the name of the popular spreadsheet. The company then expanded the line by publishing ten complementary products, including VisiWord and VisiFile. This was a corporate consolidation strategy designed to increase brand loyalty and shape a user community around a unified platform. Recognizing the need to support their diverse collection of applications, VisiCorp organized a new computer book division named VisiPress, formed to publish comprehensive "how to" guides for the company's products. VisiCalc also created a support division named VisiCare, designed to assist customers with bug fixes, replacement disks, and other types of technical support. Finally, VisiCorp created a seminar series for business executives called VisiTraining, which the firm believed would inspire decision makers to further adopt their products.

Despite VisiCorp's comprehensive growth strategy, most of its initiatives would fail within the first 12 to 18 months.[7] Rather than focusing on social, technical, or business competencies, the company focused on platform dominance. Not long after, VisiCorp's revenues were eclipsed by Lotus Development Corp., a new competitor that released a rival spreadsheet, Lotus 1-2-3. Part of the problem for VisiCorp was strategic—it shifted from one system to the next in an awkward way and confused users with a raft of new tools. However, it also failed in creating a functioning user community that could share technical knowledge and welcome new demographics and skill levels. Its main emphasis was on selling add-on products and seminars to *customers* rather than shaping a robust VisiCalc community.

The book publishing division, VisiPress, closed its doors in September 1984, due to a lackluster demand for VisiCorp's new products. The VisiTraining seminars were also a flop. The seminars were priced at between $650 and $750 per person, and they were designed primarily for executives, decision makers, and Wall Street analysts—the community imagined as key users of the product.[8] But the groundswell of support for VisiCalc came from small business owners, accountants, secretaries, and home users who hoped to manage more typical financial transactions. The training program fell flat when Lotus 1-2-3 cut into the market share of VisiCalc and reduced interest in VisiCorp's line of compatible products.

Were there ways that VisiCorp successfully created new users for its innovative spreadsheet? The organization's most successful strategy may have been its effort to support users through the relatively "hidden" influence of software documentation. Recognizing that user manuals were widely disparaged in the technical products industry, VisiCorp did an admirable job of focusing on the needs of early adopters who were learning about spreadsheets on their own and had limited access to face-to-face instruction. In 1979, Dan Bricklin wrote the first VisiCalc manual himself, a guidebook that included an innovative reference card to teach the program's major commands and functions. Each new software release included an improved user manual, and the materials were often prepared by writers with corporate experience. When VisiCorp released the IBM PC version of VisiCalc in 1981, the user guide was written by Van Wolverton, a technical editor with years of experience leading documentation teams at IBM and Intel. *PC Magazine* wrote a glowing review of Wolverton's manual in a feature article that also evaluated the components of the first IBM PC systems.[9] Wolverton went on to write the *Running MS-DOS* book series that further shaped the community that operated IBM PCs and compatibles, teaching them what they should learn about operating systems and hardware.

Microsoft's Customer Service and Product Support

An analysis of Microsoft's support infrastructure offers additional insight into how growing PC companies worked to create product-based communities and address customer concerns when things went wrong. The mid-1980s is a useful vantage point for Microsoft, because at this time the company expanded its offerings into new categories and dramatically expanded its customer base. When IBM released the IBM PC in 1981, Microsoft supplied an operating system for the device, and the company gradually developed a suite of programming tools, operating systems, applications, and hardware accessories. In 1987, Microsoft was among the top revenue producers in the burgeoning PC software industry, at a time when four out of the top five PC companies had grossed in excess of $100 million in a year.[10] This expansion meant that the market leaders needed to adopt more systematic measures to track and support their customers. These mechanisms included prioritizing high-level accounts, organizing call centers, enhancing communication systems, and training support personnel in new ways.

Although the earlier face-to-face approach to support continued in smaller companies, the larger firms pursued the integrated support practices they had learned about from mainframe and minicomputer companies, as well as retail sales giants like Sears or JC Penney. When I took an editorial position at Microsoft in 1985, I observed some of these strategies firsthand in the sales, marketing, and product development groups. Like most of the company's employees, I was constantly working to learn new computing systems, and this process continually shaped my understanding of what hardware and software could do for customers. It was part of my job at Microsoft Press to communicate with users about how they could extend our products or build new systems of their own. Since software revision cycles were short and the competition was fierce, we continually cultivated feedback from registered users to improve our offerings. Companies that designed applications for Microsoft's operating systems regularly sent us feedback through beta testing programs and forums like Microsoft Developer Network (MSDN). This criticism shaped company policies and influenced new software releases. The company was particularly sensitive to feedback from large accounts, such as Fortune 500 companies with profitable site licenses. But individual "users" could also exercise some agency by choosing which Microsoft products to buy (or not), and they regularly wrote detailed "letters to the editor" in computer magazines that the company took seriously. Along with these mechanisms, customers attempted to shape corporate attitudes by interacting with support services.

In the mid-1980s, Microsoft's support services were organized into several functional groups within the US Sales & Marketing Division. Although US-based calls to the company were "free," most customers still had to pay long-distance charges. The "toll free" number was a closely guarded secret, only shared with software dealers, value-added resellers, and major accounts. Microsoft's phone lines were usually open for calls 12 hours a day, from 9:00 a.m. EST to 6:00 p.m. PST. In its 1986 annual report, Microsoft noted that it had ramped up its product support capacity to receive up to 35,000 calls per month.[11] Two years later, the company reported that the demand for support had risen to a monthly average of 62,500 calls, or approximately 750,000 calls a year.[12]

One strategy for managing call volumes was flowcharting, a process adapted from engineering that presented common user questions and solutions on paper diagrams. Microsoft's teams utilized these call routing charts to quickly discern what customers were asking about, so that they could direct them to the proper resources. (Essentially, it was an application of the same principles that were foundational to early programming, applying the algorithmic language of software to the people who supported it.) There were several entry points into the system. When users dialed Microsoft's retail sales number, they were greeted by a team known as Consumer Response. According to company records, there were fifteen employees working in Consumer Response in late 1986, including twelve women and three men. These workers were paid an hourly wage for their labor, and they were not eligible for the benefits offered to salaried employees. It was considered an entry-level position, and many of the workers did not have two- or four-year college degrees. However, the agents were trained to manage a wide variety of customer inquiries, and it was their responsibility to transfer customers to the appropriate area of Microsoft as soon as it could be determined what they needed.

The Consumer Response team worked largely from printed sources, and it recorded call statistics by hand, using computers later in the day for email and basic record keeping. (The company had a Xenix-based email system connected to a DEC PDP-11, but external customers were unable to access these resources.) The team had several technical tools to help it route calls, including prewritten scripts which it could read from to assist customers. Many agents also used large, foldable diagrams that indicated visually where the incoming calls should be routed. Customer inquiries were passed from one person to the next via standard touch-tone phones with speed dial functions. The agents worked in cubicles and wore headsets to improve audibility. It was necessary for them to be on the phones promptly when their shifts began, and their call times and logs were reviewed regularly for communication statistics. Although Microsoft tended to describe this work as "non-technical," it is clear that significant technical skills *were* required to do this work, a theme that I will return to several times in this chapter. For organizational reasons, however, a subtle line was being drawn in relation to what was considered "technical knowledge" in the company and how this classification could be related to status, compensation, and gender. In an effort to build a cost-effective system to manage its support calls, the company raised the status of certain kinds of knowledge (programming and an understanding of software internals) while demoting other types of knowledge (communication systems, how to use flowcharts, etc.).

Although some support queries could be dealt with immediately by experienced agents, other calls needed to be routed to specialized groups within US Sales & Marketing. These support teams included Telemarketing (for people who wanted to buy Microsoft products), End User Customer Service (for manufacturing problems with existing products), Customer Service Major Accounts (for corporate, government, and education partnerships), and Product Support (for hardware or software functionality issues, also called Technical Support). Supply and stocking questions about goods in the warehouse could be directed to employees at Kamber/Fulfillment in Bellevue, Washington.

In September 1986, Microsoft created and distributed an internal manual for End User Customer Service representatives, which I reviewed for a more detailed look at the company's systems when the organization employed about 1,400 people. The unpublished binder includes scripts for customer service representatives; a list of supported products; data entry procedures; order forms; contact information for corporate, government, and education partners; and basic organizational structures within the company. When the manual was distributed, the

End User Customer Service team was staffed by thirteen women and one man. The group was led by one manager and one supervisor (both female). Although there was no stated preference for female customer service representatives in the materials, the gender imbalance indicates that certain types of technical labor were conceived of as ideal roles for women at Microsoft, especially candidates who had worked in retail customer service environments. According to representatives, job candidates who had worked at Nordstrom were particularly sought after in the 1980s.[13] Female employees were also believed to be more comfortable with interpersonal communication tasks, especially phone work.

Microsoft's Customer Service manual begins with a reminder that staff members should be speedy and polite as they interact with customers. They are also encouraged to be gracious if "rude" product owners called, a relatively common occurrence. Kim Ullom, a customer service representative in 1986–1987, remembers that people were often angry because they had waited for some time in the phone queue before they could get through. "Customer wait times were 30 minutes or more as our customer base expanded. Often callers were anxious because they were unsure that we could solve their problems."[14] Although interpersonal skills were valued in all parts of the company, customer service put a premium on these abilities.

An experienced Microsoft End User Customer Service representative could handle 30 to 40 calls per day. In addition to time on the phones, the representatives researched new products, wrote up orders, prepared mailings, and conferred with colleagues about policy issues. Each of these tasks required technical skill, although it wasn't described as such in company materials. The following is an excerpt from Microsoft's training manual about how to handle difficult callers. A number of technical abilities are implied in the description. Notice also that, in several places, the caller is assumed to be male.

Call Handling
Always stick completely with the phone script trees provided. If you find resistance from a customer at your direction, use the appropriate response scripts you have been given.

> Try to transfer all calls as quickly as possible. Don't let the customer wander while he's talking to you. Use the phone screening tree you were given to direct the caller. If necessary, to get him back on track, gently break in by saying "Excuse me, but do I understand you to need . . . ?" But at the same time remember that you are the first MS voice that the customer hears, and the image you convey is important. Always make some transition statement such as "one moment please, and I will transfer you to" when forwarding a call . . .
>
> Always give the customer the benefit of doubt. If you've had three rude callers, don't expect the fourth one to be the same—it will reflect in the manner you speak to him.[15]

Like many advertisements and user scenarios in the early PC industry, this calling script uses a subtle form of gendering to construct who a typical "user" of the software might be and what technical "paths" they should be led on. Quite typically, the caller was assumed to be male and the representative female.

But why did Microsoft's customers call customer service? They typically made contact because they were current product owners, but the software they had purchased was defective, did not meet expectations, or needed updating.[16] At the time, all of Microsoft's packaged products were purchased by consumers in retail stores or bought via mail order. When the customer opened the software box, there might be a physical problem with the software (the wrong program or manual was included), or the computer disks might be defective or contain a version of the software that was different than advertised. Sometimes, customers damaged their software disks inadvertently after purchase. For example, they stepped on diskettes or ran them over with a desk chair. These programs needed to be replaced.

Before software downloads were widely available, all PC software needed to be shipped to customers via physical media. At Microsoft, this might include diskettes that included bug fixes, software updates, or new versions of the product. Before a representative could send this media to customers, they needed to verify that the customer was a registered owner. Although customers could accomplish this by filling out the registration cards that came in their software, many did not send their cards in, or they were still being processed when the support call was made. In 1986, the registration information was consolidated and distributed internally on microfiche forms, which representatives could read using older fiche readers at their desks. This rather "low-tech" solution to managing customer records was an embarrassment to managers (because the information was not available in an electronic database), but it is an indication of the hybrid information systems that were typical in the 1980s. If a caller claimed to be a Microsoft customer but had not submitted a registration card, the staff was trained to ask them to read a part number from their software box or send in something *physical* related to their purchase—either the software disks, a manual, or part of the physical box. "One customer was so angry that they included a cigarette butt and ash in their mailing to me along with the diskettes," Ullom recalled. "Customers didn't understand why we had to verify their purchases, and sometimes they would lash out."[17] In addition to providing a glimpse at consumer frustration, the registration process offers another view of the process of "user formation," in which customers were registered in corporate systems and linked together via US Mail and other processes. Customers often preferred anonymity and resisted becoming "Microsoft users" but gradually relented when it was the only way to get product support or a low-cost upgrade. More subtly, the formation process was a two-way street, in which support workers were also "constructed" by the expectations of callers and the technical tools and procedures used to support them.

Microsoft was not a direct retail sales business. It was not possible for customers to visit the Redmond campus in person and interact with customer service personnel. All product-related issues had to be managed by phone transactions and the era's physical delivery systems, including US Mail, UPS, and the recent option of FedEx. If there was a charge associated with an upgrade or replacement, representatives wrote down the credit card numbers by hand and delivered them to the Order Entry staff in nearby cubicles. In 1986, Order Entry was a group of five non-exempt female employees who processed paperwork related to customer transactions. Their jobs straddled the boundary between information technology work and traditional business functions, and they were related in some respects to earlier positions in secretarial pools.

The Microsoft Product Support (or Technical Support) group responded to hardware- or software-related questions about Microsoft's products. This support group went through a dramatic period of expansion in the late 1980s, as Microsoft's product lineup expanded into a range of operating systems, applications, development tools, and hardware accessories. The platforms supported included CP/M, Apple DOS, MS-DOS, Xenix, Apple Macintosh, Windows, and OS/2. The company supported operating systems through Original Equipment Manufacturers (OEMs) when possible, a strategy designed to redirect support calls and utilize industry partners. Internal documents suggest that the primary reasons for contacting Product Support included learning more about the features of a product, getting help with installation, learning how to make back-up disks, interpreting error messages, reporting bugs, and resolving compatibility issues. In the early years, Product Support also helped to manage the beta testing for programs like QuickBASIC and Flight Simulator.

When Ken Boyer was hired as a member of the Product Support team in 1984, he was employee #495 at the company. "There were 14 people on the team worldwide when I was hired," Boyer recalls.[18] A big part of the team's identity, according to Boyer, was to manage a large volume of calls within a relatively small group. There were usually just two to three people per product category to handle all the questions about professional languages, business languages, MS-DOS applications, Macintosh applications, and hardware. In fact, by late 1986, there were still just 20–25 product support employees working on the entire Redmond campus.[19] Although the group hoped that registered customers would call only if they had encountered a troublesome glitch in the software or hardware, some users called with relatively simple questions that seemed more appropriate for a classroom setting or perhaps a computer book or magazine tutorial. However, in the early days of personal computing it remained an open question what users should figure out on their own and what kinds of things users were expected to ask a manufacturer about.

Microsoft occasionally posted new product support jobs in *MicroNews,* the company's weekly newsletter. These notices sought "product support technicians," Microsoft's preferred term for its help desk workers. Employee lists in the customer service manual indicate that the technicians were predominantly male, but there were also experienced female workers in the group, including Delores Bergstrom, a well-known lead who was quoted in industry publications.[20] But more typically, technical support employees were male because that was the typical gender profile of the company's engineers, as well as the users of many of Microsoft's systems and language products. By gendering technical support positions as male, even in subtle ways, it opened up successful employees to professional advancement in the groups that were higher status and higher paying, including the teams involved with product development, testing, documentation, and management.

Product support technicians were a core part of the US Sales & Marketing Division in the early to mid-1980s. It was not until 1987 that the company decided to consolidate its various support functions and separate them from the retail sales group. This organizational revision took shape in March 1987, when the company combined retail and OEM support services into a single entity called Product Support Services. Soon after, a dramatic physical expansion began. In December 1988, the company leased office space in the Lincoln Plaza complex in Bellevue, opening a facility for 250 support technicians and staff. The expanded footprint allowed the group to receive 110,000 support calls per month, and its communication systems were enhanced with advanced database tools and a powerful DEC VAXcluster.[21] In September 1990, the company opened a second support center in Charlotte, North Carolina. This location was initially staffed by 85 workers. The group continued to grow in 1991 by opening a third call center in Dallas, Texas. *Computerworld* magazine subtly critiqued the expansion by noting that the organization was "swamped with customers phoning in with DOS 5.0 and Windows 3.0 questions."[22] In retrospect, however, it seems to have been a deliberate effort at constructing a virtual user community through technical means. How else could valuable information be for-malized and transferred to millions of customers?

What happened to promote the sudden surge in product support calls? From a consumer products standpoint, 1990–1991 was the period in which the PC industry finally distributed systems to a significant de-mographic of the American public. Microsoft released Windows 3.0 in 1990, followed by DOS 5.0 in 1991. DOS 5.0 achieved an installed base of some 50 million users. A midrange IBM PC or compatible running these programs likely featured an Intel 486 microprocessor, 4MB of RAM, a 200MB hard drive, one 3.5" diskette drive, and an optional CD-ROM drive—a configuration that vastly surpassed the power and ca-pacity of earlier PCs. A substantial user community took shape around this platform, and it looked to Microsoft to provide software and sup-port via an amended corporate infrastructure.

When asked about the rapid growth of product support, Microsoft general manager Patty Stonesifer described the meteoric rise of a team that had expanded from a few dozen employees in 1986 to over 1,400 in 1991.[23] The organization now processed approximately 3 million calls a year, and the inquiries ran the gamut from Flight Simulator to Win-dows to SQL Server. An average call took 10 minutes, but some com-plex requests took days to resolve and involved people and resources from across the company. Stonesifer summarized why product support had become so important: "Sometimes I'm asked why we need so much product support. Can't we just make the products easier to use? But if you listen to the customers on the phone, you'll discover that people are doing more with our products than ever before, mostly because the graphical user interface makes those features so much more accessi-ble. In fact, Bill Gates came over and listened in on the Windows lines a couple of months ago. He said that even he was amazed by the in-credible ways people were pushing the software."[24]

Stonesifer was quick to defend the quality of Microsoft's products, which she linked to the enticing user interface of Windows. It is con-structive to see this as a way of shaping an emerging user community around a specific set of attributes, such as "usability," "accessibility," "productivity," etc. But Stonesifer also argued for a new way of envisioning the company's phone-support technicians. These workers were not just responding to bug reports and helping users troubleshoot their systems. Instead, Stonesifer described her team as "engineers," and she recognized their work as an extension of the company's software development community. This valuing seemed strategic because Microsoft very much needed the group's assistance with its customers. The hidden work of technicians and other support employees at Microsoft had contributed to making the company's software a viable consumer product for a wide range of users. Employing sophisticated database tools, Microsoft's support teams carefully coded each call, consolidated the information received, and reported it to management and the development teams. In many cases, this feedback helped shape future products, including training materials that could be sent to customers.

In retrospect, Microsoft had gone through a substantial period of growth very quickly. Near the end of the transition, Microsoft director Jan Claesson announced optimistically, "Every sixth Microsoft employee is somehow involved with product support worldwide."[25] Claesson's announcement was meant to reassure customers that their demands for improved services were being taken seriously. But within the company, a subtle dividing line between "nontechnical" and "technical" employees had developed, often with gender-related consequences. Although it was common for high-achieving programmers to be lionized in the popular press, there were few comparable stories about successful consumer response or customer service workers. This sense of invisibility increased when the employees of Product Support Services were moved "off campus" to secluded buildings in Bellevue or other locations.

The isolation of support workers has continued as a general trend in the US software industry. In a 1999 report on Information Technology (IT) workers, the authors concluded that support and customer service personnel seemed remarkably invisible in the companies that they worked in.[26] For example, only one contemporary definition of IT worker in the United States includes the job descriptor "supporter" or "call consultants." In comparable studies, "help desk" workers are typically lumped in with other categories and are rarely studied as unique groups with distinctive jobs, skills, and functions. Greg Downey supported this assessment by observing that most studies of high-technology work spaces fail to include a range of workers in the mix, focusing instead on just a few millionaires.[27]

Despite their hidden status, the constellation of workers referred to as "help desk" employees contributed greatly to the success of many companies in the early PC industry. They helped to construct user communities and became an important conduit for customers to influence support teams and the products they ultimately represented. It is important for historians to recognize the wide range of technical skills and knowledge that product support and administrative employees possessed and used in their daily work. A broader definition of what constitutes technical knowledge is important as the PC industry works to hire, promote, and compensate employees in an equitable manner.

Notes

1. See Laine Nooney, "The Uncredited: Work, Women, and the Making of the US Computer Game Industry," *Feminist Media Histories* 6, no. 1 (2020): 119–46; Mar Hicks, *Programmed Inequality: How Britain Discarded Women Technologists and Lost Its Edge in Computing* (Cambridge, MA: MIT Press, 2017); Lisa Nakamura, "Indigenous Circuits: Navajo Women and the Racialization of Early Electronic Manufacture" (2014, repr. this volume); Janet Abbate, *Recoding Gender: Women's Changing Participation in Computing* (Cambridge, MA: MIT Press, 2012); Nathan Ensmenger, *The Computer Boys Take Over: Computers, Programmers, and the Politics of Technical Expertise* (Cambridge, MA: MIT Press, 2010); and Jennifer S. Light, "When Computers Were Women," *Technology and Culture* 40, no. 3 (1999): 455–83.
2. The question of "how users are made" is also creatively explored in this volume by Scott Kushner ("The Lurking Problem") and Laine Nooney ("'Have Any Remedies for Tired Eyes?': Computer Pain as Computer History").
3. For a glimpse at how PCs were first marketed to American consumers, see Otto Friedrich, "The Computer Moves In," *Time*, January 3, 1983. On the ontological uncertainty of what microcomputers were for, see Bryan Pfaffenberger, "The Social Meaning of the Personal Computer: Or, Why the Personal Computer Revolution Was No Revolution," *Anthropological Quarterly* 61, no. 1 (1988): 39–47.
4. Thomas Haigh and Paul E. Ceruzzi, *A New History of Modern Computing* (Cambridge, MA: MIT Press, 2021), 173–175.
5. A longer version of Waite's story is recounted in Michael J. Halvorson, *Code Nation: Personal Computing and the Learn to Program Movement in America* (New York: ACM Books, 2020), 192–200.

6. For a firsthand account of the show, see David H. Ahl, "The First West Coast Computer Faire," *Creative Computing,* July–August, 1977, 98–102.

7. Burton Grad, "The Creation and the Demise of VisiCalc," *IEEE Annals of the History of Computing* 29, no. 3 (2007): 20–31.

8. For VisiCalc's reputation on Wall Street, see William Deringer, "Michael Milken's Spreadsheets: Computation and Charisma in Finance in the Go-Go '80s," *IEEE Annals of the History of Computing* 42, no. 3 (July–September 2020): 53–69. For the influence of managerial labor on computing communities, see Thomas Haigh, "Technology, Information and Power: Managerial Technicians in Corporate America, 1917–2000" (PhD diss., University of Pennsylvania, 2003).

9. Jeremy Joan Hewes, "A Glimpse at Two PC Manuals," *PC Magazine,* February–March 1982, 116–18.

10. Martin Campbell-Kelly, *From Airline Reservations to Sonic the Hedgehog: A History of the Software Industry* (Cambridge, MA: MIT Press, 2003), 235.

11. Microsoft Corporation, 1986 Annual Report (Redmond, WA), 13.

12. Microsoft Corporation, 1988 Annual Report (Redmond, WA), 9.

13. Kim Ullom, email correspondence with the author, April 20, 2020.

14. Ullom, email correspondence.

15. Unpublished Customer Service Manual, Microsoft Corporation (1986), 10.

16. Unpublished Customer Service Manual, 12.

17. Ullom, email correspondence.

18. Ken Boyer, email correspondence with the author, May 24, 2020.

19. Clay Jackson, phone interview with the author, May 25, 2020.

20. Elizabeth Bibb, "Thirteen Going Professional," *PC Magazine,* January 24, 1984, 49–52.

21. Microsoft Corporation, 1989 Annual Report (Redmond, WA), 4.

22. "Microsoft Plans Dallas Center," *Computerworld,* July 22, 1991, 100.

23. Microsoft, 1991 Annual Report, 11.

24. Patty Stonesifer, quoted in Microsoft, 1991 Annual Report, 11.

25. Stuart J. Johnston, "Microsoft Expands Its Support Services," *InfoWorld,* December 10, 1990, 45.

26. Peter Freeman and William Aspray, *The Supply of Information Technology Workers in the United States* (Washington, DC: Computing Research Association, 1999), 32.

27. Greg Downey, "Virtual Webs, Physical Technologies, and Hidden Workers: The Spaces of Labor in Information Internetworks," *Technology and Culture* 42 (2001): 209–35.

CHAPTER 10

Power to the Clones

Hardware and Software Bricolage on the Periphery

Jaroslav Švelch

In recent decades, historians of both computing and video games have criticized the unhealthy obsession with origin stories and genius inventors. Joy Lisi Rankin has written of a "Silicon Valley mythology" that celebrates American "hackers, geniuses, and geeks" but obscures the role of government funding, educational programs, and the whole non-Western world.[1] Erkki Huhtamo has likewise lambasted video game histories "built around the same landmarks, breakthroughs, and founding fathers (not a word about mothers!)."[2] The contributions of "mothers," as Laine Nooney has pointed out, often go uncredited.[3] According to video game historian Jaakko Suominen, the enthusiast style of writing histories tends to be guilty of "fetishizing video games," putting certain developers on a pedestal, creating canons, and supporting the master narrative of innovation and progress.[4] Collectors are willing to pay exorbitant prices for fetishes such as the Apple I home computer or the so-called Nintendo PlayStation prototype. The market value of these objects, as well as the attractiveness of the stories told about them, derives from their "firstness" and "originality." But histories of Apple or Nintendo tell us little about the practices of computing outside the Western world and Japan. Any truly international history must dispel the myths, overthrow the fetishes, and open its doors to the myriads of clones and other nonoriginal artifacts that have had an enormous impact on the practices and cultures of computing.[5]

Originally from the domain of genetics, the word "clone" is often used in its figurative meaning, defined by the Merriam-Webster dictionary as "one that seems to be a copy of the original form." Interestingly, the dictionary lists "a clone of a personal computer" as a prototypical use, hinting at the proliferation of such clones.[6] But despite their abundance, clones had long been relegated to the footnotes of computing history. This oversight can be attributed to the idea of the "primacy of the origin," explored by Ivan da Costa Marques. In his view, this idea "invokes the precedence, priority, predominance, preference, prerogative, privilege, right-of-way, seniority, supremacy of the original over the copy, of the model over the imitated."[7] The status of the original is, however, not inherent but constructed through legal litigation, advertising discourses, and origin stories. According to da Costa Marques, "laboratories and courtrooms of the West emerge as places that concentrate resources to carry out the work of division, classification, and purification in the modern world."[8] Embodied in the institutions of patenting and intellectual property rights, primacy of the origin thus perpetuates the dominance of the richer and more developed countries in historical narratives.

Histories of computing in non-Western contexts have been previously explored by authors such as Eden Medina (writing on Chile), Benjamin Peters (on the Soviet Union), or Maria B. Garda (on the People's Republic of Poland); Chan and Stevens write of Singapore later in this volume.[9] This chapter demonstrates how a focus on clones in particular can enrich and diversify historiography and give voice to the periphery. It builds on the material I gathered for my recent monograph, *Gaming the Iron Curtain*, and focuses on the examples of mainframe, microcomputer, and computer game cloning in Communist-era Czechoslovakia.[10] To counter the power of origin stories, it forefronts the process of bricolage—the creative recombination of existing materials and ideas—and shows the need for computing histories that are transnational, material, and nonhierarchical.

Researching Clones

To survey the state of the art of clone histories, I complemented a traditional open-ended literature review with a small mixed method probe of texts published in the *Annals of the History of Computing,* the longest-serving computing history journal. Using full-text search, I identified 39 articles that contain the word "clone" or "clones."[11] Within this corpus, the term is never explicitly defined but refers to both hardware and software, as well as computer networks such as ARPANET.[12] The original whose clones are most often mentioned is the IBM PC, occurring in sixteen articles, followed by VisiCalc in four articles, and the Apple II, IBM System/360, and WordStar in three articles each. Interestingly, the clones of the IBM PC and Apple II machines are, with one exception, never named within the corpus and are instead referred to generically as "IBM PC *and clones,*" while VisiCalc and WordStar clones (such as the Lotus 1-2-3) are mentioned by name. In part, this may be due to differences in the perception of clones between hardware and software production, as cloning is portrayed as more common and permissible in the latter. Perhaps more likely, the difference reflects the fact that *Annals* is a US-based journal and most of the articles in the corpus (22) focus on US developments. While the VisiCalc and WordStar clones were produced in the United States, many of the IBM PC and Apple II clones were manufactured in other countries, especially in Asia. The few articles that engage with clones on a deeper level focus on peripheral countries associated with widespread cloning, such as Taiwan or the (former) Soviet Union.[13]

Histories written from the perspective of the original producer tend to view clones as an uncontrollable and unpredictable threat. IBM's decision to base its personal computer on widely available components and to allow Microsoft to resell the machine's operating system has resulted in ceding control of the market, and eventually losing against clone manufacturers, usually smaller companies with faster management processes and lower administrative and research costs.[14] Therefore, IBM PC clones could "flood the market" as a massive, sudden wave.[15] Apple defended itself against cloning through litigation and political pressure in various regions, as meticulously explored by da Costa Marques on the case of Unitron, a Brazilian clone of Apple Macintosh.[16] Other manufacturers, like Atari or Commodore, curtailed cloning by utilizing proprietary custom chips.

Things look very different from the users' perspective. To explore this perspective, I step out of the *Annals* corpus and encompass literature on regional computing and game histories, which tends to pay more attention to local consumption. In Taiwan, there were reportedly up to 100 different companies manufacturing Apple II clones in 1982, providing affordable machines for the domestic market.[17] In Brazil, importing Western machines was prohibited, and people instead programmed and played on the TK90X, a local clone of the British ZX Spectrum, which was also exported to Uruguay.[18] The South Korean Zemmix console was a clone of the MSX standard, as was the Kuwaiti Sakhr computer, popular among middle-class families in the Middle East.[19] In the Soviet Union and its successor countries, a confounding multitude of Sinclair ZX Spectrum clones were produced in the late 1980s and early 1990s, some of them branded after the cities and regions of manufacture, such as "Spectrum St. Petersburg."[20] Each was housed in a different, often crude, chassis, and might have featured different circuitry, but all of them could run Spectrum software. In many peripheral contexts, clones thus provided access to technology that was otherwise unavailable or unaffordable.

Very often, cloning was a result of *bricolage*. In Claude Lévi-Strauss's original conceptualization of bricolage, the bricoleur is introduced alongside the engineer as two divergent types of creative practice.[21] While the engineer starts from a conceptual blueprint to procure required materials, the bricoleur "addresses himself to a collection of oddments left over from human endeavors" and has to "make do with 'whatever is at hand.'"[22] The two types of practice often overlap. Bricolage has played an important role in computing, a field that has time and again repurposed devices and components from other domains (for example, the teletype from telegraphy). Bricolage is, however, especially typical of the periphery, where various constraints—such as the lack of formalized know-how and components—create opportunities for creative bricoleurs who excel at workarounds and clever hacks. Clones produced on the geographical peripheries drew from existing foreign designs but also incorporated serendipitous combinations of locally available resources.

The study of clones requires specific methodological considerations, which include *transnationality*, *materiality*, and *variantology*. First, the study of clones needs to be *transnational*. From the dictionary definition, the central feature of a clone would seem to be its similitude to the original. However, both lay and academic usage of the term also imply separation and distance. If the same company produces a new, similar machine, it is likely to be called a new model, new version, or an upgrade. IBM's PCjr model from 1984, for example, would hardly be called a clone of its IBM PC. A clone tends to be produced by a different entity, often in a different country. The geographical distance often enables clones to be produced, as the legal and discursive power of the original owner may not reach the site of cloning.

Research into hardware clones, specifically, should also take into account the *materiality* of computing, and specifically its manufacturing. When studying computing platforms, the focus is often on the platform as an abstract system and design template. At the inception of the platform studies book series, Nick Montfort and Ian Bogost proposed that "a platform in its purest form is an abstraction, a particular standard or specification before any particular implementation of it."[23] Such focus supports contemporary tech industry's efforts to disconnect its products from both manufacturing (which often entails labor exploitation) and disposal (which is an enormous burden on the environment). In her work on laptop manufacturing in China, Ling-Fei Lin points out that the resulting statements such as "Designed by Apple in California, Assembled in China" suggest that the relation between design and production is hierarchical and disconnected.[24] To quote Nathan Ensmenger, "The dirty and dangerous work of industrial manufacturing . . . happens in other parts of the world, and by (and to) other kinds of people."[25] Given that clones reimplement existing specifications, their historiography must, instead, closely engage with the material aspects of their production, be it individual bricolage or larger-scale manufacturing.

Third, the study of clones should eschew the hierarchies imposed by historical canons. Studying Italian computer game clones, Riccardo Fassone adapts Siegfried Zielinski's concept of *variantology*[26] into a historiographical method that explores variants of software in a nonhierarchical fashion, focusing on "lower intensity processes and practices and [on] non-exceptional, often banal, cases such as adapted clones."[27] It is only when we stop considering clones shadows of the original that we start to understand the bricoleur practices of marginal and marginalized historical actors (also documented by André Brock's chapter reprinted in this volume). Unlike the acclaimed authors of "originals," these actors have often not had their voice heard, and it is important to let them—and their work—speak. The following sections, in turn, explore the transnational, material, and variantological aspects of hardware and software clones.

The Soviet Way of Cloning

During the Cold War era, an Iron Curtain of trade embargoes and travel restrictions separated the Western and Soviet bloc computer industries. The Committee for the Multilateral Export Controls (CoCom), whose members included most Western countries and Japan, restricted the export of technology with potential military use into the Soviet bloc.[28] Export was not banned altogether, but it was tightly controlled. Ivan Malec, a Czechoslovak computer expert from the Federal Statistical Office, remembers that the Control Data mainframe he purchased to process the 1970 census data came along with two US agents who oversaw the handover of the machine. The end user agreement reportedly contained a clause that the computer "would not be misused against U.S. interests." On top of that, he had to ask the Czechoslovak secret police to secure nine permanent visas for individuals who would observe whether the obligations were followed.[29]

These limitations, compounded by the lack of convertible Western currency, contributed to the demand for domestic production of computers, which were sorely needed for critical tasks in the industry, administration, and research. In the 1950s and 1960s, Prague's Research

Institute for Mathematical Machines launched two systems of its own design—SAPO (launched in 1956) and EPOS (1963).[30] Foreshadowing the bricoleur practices of the next generations of microcomputer hobbyists, the Institute's designers came up with innovative fault-tolerance features to compensate for error-prone relays produced by the domestic electronics industry.[31] However, by the mid-1960s, it was becoming clearer that autonomous breakthroughs in individual Soviet bloc countries were unlikely.[32]

In the late 1960s and 1970s, the bloc's computer industries started to rely on industrial espionage, as discussed on the Polish case by Mirosław Sikora. Following the Soviet example, the Polish intelligence service was tasked with smuggling blueprints and products from across the Iron Curtain. These operations were elaborate and costly, requiring hundreds of agents and support staff as well as ample funds to bribe Western officials and set up fake companies. The Polish leadership, however, considered these expenses justified, as they were likely to be lower than the cost of original research. The resulting intelligence was then "legalized" by filing it as local patents and shared with other Soviet bloc countries, who partook in parallel operations.[33] As a result, members of the COMECON (the Soviet bloc's economic treaty) could produce Western components and hardware, albeit with a significant lag.

The two Soviet bloc computer manufacturing initiatives of the 1970s and 1980s accordingly focused on cloning major Western platforms. The Unified System of Electronic Computers (also known as RYAD) comprised a series of models based on the IBM System/360.[34] It launched in the Soviet Union in 1967 but was subsequently joined by other countries that contributed components, machines, and peripherals, namely, East Germany, Czechoslovakia, Poland, Hungary, and Bulgaria. Choosing to adopt the System/360 architecture might seem an admission of the Soviet failure to develop a competing system, but it was also—as James Cortada puts it—a "convenient shortcut" at a time when the IBM architecture was a de facto standard in most of the world and plenty of software was available. The Unified System soon became a huge multinational cloning operation, employing close to

300,000 workers, and over 46,000 engineers and scientists.[35] Local experts, however, showed some resistance to implementing the system. The Czechoslovaks at the Institute for Mathematical Machines already had another major project in the works, and they only reluctantly agreed to scrap it and work on a midrange model in the 360 clone series, which came to be known as the EC 1021. Still, they managed to sneak parts of their scrapped project into the EC 1021, making the machine more powerful but partially incompatible with the rest of the series.[36] The second COMECON initiative, entitled System of Small Electronic Computers, started in the mid-1970s and featured various clones of the DEC PDP-11 minis, including the Soviet Elektronika 60, famously used by Alexei Pajitnov to write *Tetris*.

These clones have a peculiar position in computing historiography. Local historians approach them with less enthusiasm than they do "national" models like SAPO, which are unique and more idiosyncratic.[37] We should not, however, imagine the clones as simply identical to the Western machines. Besides differences in exterior and build quality, they used different operating systems and peripherals, and Western software required substantial adaptation and creative programming to work on the Soviet bloc systems.[38] Compared to national models, the clones had an outsized impact on computing practices in the Soviet bloc and alleviated the precarious dependence on imports. In Czechoslovakia, the share of Western mainframes and minis dropped from 35% in 1974 to 12% in 1989; and out of the 2,226 machines captured by government statistics in 1989, 871 and 779 belonged to the Unified and Small systems, respectively.[39] Such an extensive degree of top-down cloning was in part enabled by the bipolar division of the world, demonstrating that clones imply not only similitude but also disconnection and rupture. The West-imposed embargos made legal sales to the Soviet bloc extremely difficult, but illegal cloning could not be prevented, as IBM's and DEC's patents and copyrights had no protection on the Eastern side of the Iron Curtain.[40] Eventually, the two initiatives normalized cloning as a technological policy and reverse-engineering as a design practice, laying down the path for future bricoleurs.

How to Manufacture a Micro-Clone

Microcomputers presented a new challenge to the Soviet bloc computer industries. They appeared less critical for the industry and administration, and their production within COMECON was never coordinated to the same extent as in the case of mainframes and minis. In the early 1980s, however, the demand from consumers and educational institutions was rising. Instead of a centralized initiative, dozens of factories and institutions in different countries of the Soviet bloc set out to build their own machines based on COMECON-produced clones of standard Western CPUs, such as the Intel 8080, Zilog Z80, or MOS 6502. These were often bottom-up initiatives that lacked solid support in the higher levels of central planning, leaving the hardware designers to scramble for resources and employ bricoleur tactics.

Czechoslovakia's best-known microcomputer designer was Eduard Smutný, whose articulate and outspoken support for the popular availability of computers made him something of a media personality. Smutný worked at a Prague-based research and development branch of TESLA, the state-owned concern that produced almost all the country's electronics. Here, he created several computers and was active in the local computer hobby scene, encouraging hands-on experimentation with computing among kids and young people. In various interviews over time, Smutný maintained that designing and building a prototype of a working computer was not an issue for local engineers. Before 1987, micros could not be bought in Czechoslovak retail, and the local enthusiasts relied on individually imported or smuggled computers from the West, with the British Sinclair ZX Spectrum becoming the most popular. Faced with a lack of available hardware, Czechoslovak hobbyists had become proficient bricoleurs, able to repair and build their own computers or assemble a joystick from mechanical switches and a bicycle handlebar. Instead, the biggest challenge of the local computer industry was to produce hardware in sufficient quantities for the consumer market. Besides the persistent shortages of electronic components, Smutný bemoaned the lack of quality laminate for circuit boards and even of the polymers required for green solder mask, which would improve both reliability and production efficiency.[41] In a 1986 interview, he joked: "When the authorities said they would greenlight electronics, I said that the green solder mask would be enough to make me happy!"[42] He also complained about chronic supply chain problems that resulted in late deliveries of components. Additionally, there was a strong pressure to manufacture exclusively from COMECON-produced components, as importing from elsewhere would deplete the precious hard currency reserves.

Several local hardware designers took up the challenge of making a consumer-oriented computer. While nominally engineers, they operated in the bricoleur mode, building computers from scraps that fell under the table of the centrally planned economy. In Poland, for example, the Elwro 800 Junior computer—a clone of the ZX Spectrum—was housed in a chassis originally meant for a toy piano and included a folding music stand. Smutný's own project, the Ondra, was prompted by the discovery that another branch of TESLA was manufacturing a membrane keyboard for a foreign customer. It had only 37 keys and no dedicated number row, and therefore required a complex system of shift keys, but the fact that it was cheap and already approved by planners trumped all its shortcomings. In the end, however, not more than 2,000 machines were manufactured due to the indifference of TESLA's management.[43] In another division of TESLA, a team led by Roman Kišš came up with the PMD 85 microcomputer, notably designed to use faulty RAMs that would have otherwise been discarded.[44] In this case, the keyboard was assembled from keyphone buttons, which were extremely rigid and hard to type on. More successful than the Ondra, the PMD 85 and its successor models accounted for over 14,000 units but did not make it into retail and were sold directly to institutions, mostly schools. Despite using cloned chips, neither of the two computers was compatible with a Western standard.

This changed with 1987's Didaktik Gama, produced by the Didaktik Manufacturing Cooperative, a factory based in the backwater town of Skalica that specialized in producing school equipment. As opposed to the state-owned TESLA, the cooperative enjoyed a degree of autonomy from the government and managed to bypass the restriction on the use of Western components. This allowed it to acquire the Ferranti ULA chips, which handled input/output and graphics on the ZX Spectrum, enabling the Gama to run Spectrum software. But, according to the chief engineer Ľudovít Barát, the real secret of the machine's success was Didaktik's manufacturing capability. The cooperative designed and printed its own circuit boards, soldered them using the wave soldering process, molded plastic parts, and even manufactured retail boxes. In Barát's words, they had "everything under one roof," avoiding many of the supply chain issues that plagued centrally planned economies. While more comprehensive and "engineered" than the previous projects, the Didaktik operation can also be seen as taking bricolage to the next level by devising in-house manufacturing processes to produce a local stockpile of material. Unlike in the cases of the Ondra and the PMD 85, Gama's keyboard was manufactured in-house, and much more ergonomic (fig. 10.1).[45] Didaktik went on to produce over 100,000 units of ZX Spectrum clones, including the Gama and its successor models.[46] Despite the enormous impact of Didaktik machines on Czechoslovak hobby computing and gaming cultures, the achievements of Smutný and Kišš are often celebrated among retro computing enthusiasts, while Barát's seems to fly under the radar. This echoes Lin's concern that de-

Figure 10.1. The 1987 model of Didaktik Gama, a Czechoslovak clone of the Sinclair ZX Spectrum.
Photo by Vítězslav Turčín.

signing an efficient manufacturing process is valued less than designing an "original" hardware artifact, although both are equally critical
to its proliferation.

The Gama's status in the local computing discourse shifted over the
years, reminding us of the constructed nature of the link between a
clone and an original. Some of the early coverage focused on the differences and portrayed the Czechoslovak machine as something of a
pretender. Its logo and the horizontal grooves along the chassis mim-
icked the sleek design of the Sinclair's ZX Spectrum+ model, but the
Gama machine looked cruder and more angular. Its circuitry closely followed the Spectrum's cheap and simple design, and the Gama boasted
"software compatibility" on the packaging, promising to run the vast
library of Spectrum software that was already circulating in the country. Local enthusiasts, however, soon reported that a good portion of
Spectrum programs crashed on the Gama because of differences in its
ROM.[47] While Didaktik addressed these issues in later revisions, some
work was also performed by Czechoslovak hobbyists, who diligently
documented the ROM, allowing users to fix the crashing software.[48]
Thanks to joint efforts of the users and the manufacturer, these quirks
were eventually fixed or forgotten, and the local hobbyist press started
to consider Didaktik a part of the ZX Spectrum family. A 1994 survey
by a local ZX Spectrum magazine found that about two-thirds of its
readers used a Didaktik machine.[49] Nevertheless, the local community
kept identifying as Spectrum users or "Spectrists," in part out of tradition, in part because of the coolness associated with Western—as opposed to domestic—products. This case illustrates that a piece of hardware does not automatically become a clone at the point of production,
but rather *becomes* one as it is used and discursively framed as one.

A Variantology of *Boulder Dash*

In the early home microcomputer era of the late 1970s and early 1980s,
software cloning tended to complement hardware cloning. Many different, often mutually incompatible, computer models existed alongside
each other, and their users entertained themselves with homegrown

variations on mainframe and mini games (such as *Lunar Lander*) or contemporary arcade or console titles (*PacMan* or *Space Invaders*).[50] Amateur clones were especially appealing to the users of less popular computer platforms, which were underserved by the software industry. In Czechoslovakia, the owners of ZX Spectrum and Didaktik machines could run illegitimate copies of abundant British software, but the users of some other platforms—including the domestic PMD 85 or the imported Sharp MZ-800—could not. Much of the Czechoslovak production for the latter platforms, including both games and other software, therefore consisted of unlicensed clones, made by amateurs who convened in state-supported computer clubs. In the gaming discourse, the term "clone" indicates "an undesirable extent of imitation"[51] or a copy of "earlier hit games or recognized innovations."[52] Reflecting the idea of the primacy of the original, it is often used in opposition to the term "classic." This relationship between a classic and a clone is, however, far from straightforward, as I will show in a brief variantology of the Czechoslovak versions of the US hit game *Boulder Dash*.

Written by the Canadian programmers Peter Liepa and Chris Gray and published by First Star Software, *Boulder Dash* featured gameplay that consisted of exploring caverns and collecting diamonds while avoiding monsters and falling rocks.[53] Its main appeal stemmed from the interlocking mechanics and simulated physics of the game world, which gave players freedom to improvise their own strategies. Although it was originally released in North America in 1984 for the Atari 8-bit computers, many Czechoslovak players experienced it through a licensed British conversion for the ZX Spectrum, released later that year. Far from the reach of the copyright holders, local bricoleurs used various elements of the game—code, graphics, or mechanics—as building blocks for their own projects.

In 1987, the Czechoslovak hobbyist Antonín Spurný ported it to the Sharp MZ-800, a Japanese machine with next to no commercial software that was imported into Czechoslovakia to sate the demand for home micros. While essentially incompatible with the Spectrum, the Sharp had the same CPU, allowing Spurný to reuse most of the Spectrum machine code for the Sharp version.[54] He disassembled the code, identified all input and output calls, and rewrote them to work with the Sharp hardware. His version of *Boulder Dash* was an improvement on the Spectrum version, as it took advantage of the Sharp's higher display resolution to show more of the game world (fig. 10.2). While not a game design achievement, it was certainly a coding feat, and it enjoyed warm reception by the MZ-800 community.[55] Many more ports followed suit, and by the early 1990s, around 200 unofficial ports of Spectrum games were made for the MZ-800 in Czechoslovakia, making it into a formidable gaming platform.[56]

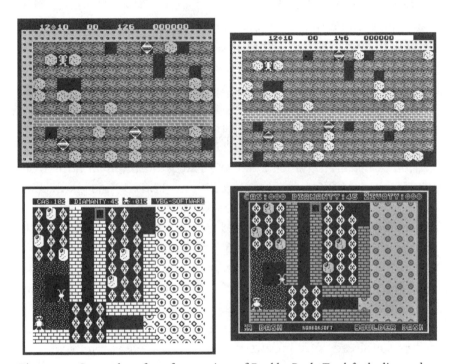

Figure 10.2. Screenshots from four versions of *Boulder Dash*. *Top left,* the licensed version for the Sinclair ZX Spectrum (1984); *top right,* the same level in the Sharp MZ-800 conversion by Antonín Spurný (1987); *bottom left,* the PMD 85 variant by Ladislav Gavar and VBG Software (1987); *bottom right,* the same level in Roman Bórik's conversion of the PMD 85 game for the Sinclair ZX Spectrum (1993).

The same year, another game called *Boulder Dash* was released by the VBG Software collective for the PMD 85 machine, popular at schools and computer clubs. The Ostrava-based trio specialized in remaking games from other machines, some of which were faithful and others quite loose. The latter tended to be written by Ladislav Gavar, who—according to his collaborator Vlastimil Veselý—enjoyed "combining elements of other games with his own ideas" and "gradually moved away from copying foreign templates."[57] *Boulder Dash* was no exception. First, all the levels were completely different. While the original contained 16 sprawling levels, the 68 levels of the PMD 85 version was composed of just 14 × 10 tiles to fit onto the screen at once, likely because PMD 85 was too slow to scroll the playfield. As a result, Gavar's levels are each built around one puzzle or challenge rather than relying on the more open-ended gameplay of the original. On the other hand, individual levels were accessed through a stand-alone labyrinth, introducing the possibility to progress through them in a nonlinear fashion.

For many Czechoslovak schoolkids, this was the truly canonical version of *Boulder Dash*. One of these young fans, Roman Bórik from the Slovak city of Košice, first played it on a PMD 85 in a youth computer club and found it immediately "spellbinding." Not even his later encounters with properly licensed versions of *Boulder Dash* could match it.[58] In 1993—as at the tail end of the 8-bit gaming era in Czechoslovakia—he converted the VBG version back to the ZX Spectrum (and Didaktik), keeping all of Gavar's levels and mechanics intact while improving the graphics.[59] Confusingly enough, even this piece of software bore the title *Boulder Dash*. Although a software industry was already forming after the switch to a free market economy, the program was released as freeware similarly to the previous two.

The case of *Boulder Dash* shows how unauthorized clones have provided access to digital entertainment in peripheral settings and on marginal platforms.[60] The concept of bricolage allows for a nuanced and context-sensitive look at what could otherwise be dismissed as simple piracy of code and assets. Each of the three bricoleurs made substantial contributions to the respective versions of the game and helped it proliferate within the local user base. If we consider copyright and piracy modern inventions, then the hobbyists—keeping with Lévi-Strauss's conceptualization—operated in a premodern manner.[61] In the absence of a commercial software market, they disregarded intellectual property and encouraged hacking and adapting Western software as a way of learning programming and participating in the community. The variantological approach reveals a history messier and more complex than a history of "classics," as the line between an original and a clone was often fuzzy and arbitrary, and one's clone could be another one's classic.[62] Embracing these messy histories of cloning and game modifications can help us appreciate local computing practices and creative work.

Conclusions

Researching clones poses several methodological challenges. Clones were often produced by semilegal or ephemeral entities that may have not kept records; and the primary sources about them may be dispersed and difficult to access. Additional challenges, especially for preservation and exhibition, stem from the clones' ontological impurity and questionable legal status. Faced with four different software artifacts named *Boulder Dash*, for example, a curator may be tempted to fall back on the canon. Even in peripheral contexts, Western or Japanese hardware may be displayed more prominently than local clones despite the latter's larger impact in the region.

As the Czechoslovak examples have shown, clones have had significant impact on computing practices, providing access to hardware as well as outlets for local tinkering. Despite the frequent belief that cloning is just a "stage" in the development of hardware and software industries, clones should be studied on their own terms, independently of the teleology of industrial growth.[63] Acknowledging them will not only paint a fuller picture of the history of computing but can also contribute to historiography in at least three interrelated ways. First, and most obviously, focusing on clones will make histories more geographically inclusive. Whole regions used to run on cloned hardware and/or software, and dismissing their stories would be a great disservice to historiography. Along with the transnational connections established through cloning, we should also keep in mind the ruptures and barriers that helped cloning flourish. Second, clones help us appreciate the material aspects of hardware, including its manufacturing. Rather than mere implementations of existing standards, we can see them as products of bricoleur tactics and idiosyncratic manufacturing processes of local factories. If we think of platforms only as abstractions, we might forget how it felt to type on their keyboards. Third, clones invite a historiography that is nonhierarchical and speaks of variants rather than originals and copies. To paraphrase Walter Benjamin, clones are unburdened by the "aura" of the original and help us step beyond origin stories.[64] They make technology profane and return it to the pedestrian world of practical politics, manufacturing plants, and decentralized tinkering.

Notes

1. Joy Lisi Rankin, *A People's History of Computing in the United States* (Cambridge, MA: Harvard University Press, 2018), 2.
2. Erkki Huhtamo, "Slots of Fun, Slots of Trouble: An Archaeology of Arcade Gaming," in *Handbook of Computer Game Studies,* ed. Joost Raessens and Jeffrey H. Goldstein (Cambridge, MA: MIT Press, 2005), 4.
3. Laine Nooney, "The Uncredited: Work, Women, and the Making of the U.S. Computer Game Industry," *Feminist Media Histories* 6, no. 1 (January 1, 2020): 119–46, doi:10.1525/fmh.2020.6.1.119.
4. Jaakko Suominen, "How to Present the History of Digital Games: Enthusiast, Emancipatory, Genealogical, and Pathological Approaches," *Games and Culture* 12, no. 6 (June 20, 2016): 551, doi:10.1177/1555412016653341.
5. Research for this chapter was supported by Charles University project PRIMUS/21/HUM/005—*Developing Theories and Methods for Game Industry Research, Applied to the Czech Case.*
6. Merriam-Webster, "Clone," *Merriam-Webster.com dictionary,* March 29, 2020, https://www.merriam-webster.com/dictionary/clone.
7. Ivan da Costa Marques, "Cloning Computers: From Rights of Possession to Rights of Creation," *Science as Culture* 14, no. 2 (June 2005): 139, doi:10.1080/09505430500110887.
8. Marques, "Cloning Computers," 148.
9. Eden Medina, *Cybernetic Revolutionaries: Technology and Politics in Allende's Chile* (Cambridge, MA: MIT Press, 2011); Benjamin Peters, *How Not to Network a Nation: The Uneasy History of the Soviet Internet* (Cambridge, MA: MIT Press, 2016); Maria B.

Garda, "Microcomputing Revolution in the Polish People's Republic in the 1980s," in *New Media Behind the Iron Curtain: Cultural History of Video, Microcomputers and Satellite Television in Communist Poland,* ed. Piotr Sitarski, Maria B. Garda, and Krzysztof Jajko (Łódź: Łódź University Press; Kraków: Jagiellonian University Press, 2020), 111–70.

10. Jaroslav Švelch, *Gaming the Iron Curtain: How Teenagers and Amateurs in Communist Czechoslovakia Claimed the Medium of Computer Games,* Game Histories (Cambridge, MA: MIT Press, 2018).

11. The search was conducted on March 4, 2020, on the *Annals* website, and initially yielded 55 results. I removed wrong hits and editorial material such as issue introductions. Interviews and practitioner recollections are included in the corpus.

12. In some cases, it is difficult to distinguish between the cloning of hardware and software. "Cloning Macintosh," for example, can mean both cloning the hardware and cloning the operating system and the system ROM.

13. Ling-Fei Lin, "Design Engineering or Factory Capability? Building Laptop Contract Manufacturing in Taiwan," *IEEE Annals of the History of Computing* 38, no. 2 (April 2016): 22–39, doi:10.1109/MAHC.2015.73; Honghong Tinn, "From DIY Computers to Illegal Copies: The Controversy over Tinkering with Microcomputers in Taiwan, 1980–1984," *IEEE Annals of the History of Computing* 33, no. 2 (February 2011): 75–88, doi:10.1109/MAHC.2011.38; Zbigniew Stachniak, "Red Clones: The Soviet Computer Hobby Movements of the 1980s," *IEEE Annals of the History of Computing* 37, no. 1 (January 2015): 12–23, doi:10.1109/MAHC.2015.11.

14. James W. Cortada, *IBM: The Rise and Fall and Reinvention of a Global Icon,* History of Computing (Cambridge, MA: MIT Press, 2019).

15. Petri Saarikoski and Jaakko Suominen, "Computer Hobbyists and the Gaming Industry in Finland," *IEEE Annals of the History of Computing* 31, no. 3 (2009): 20–33, doi:10.1109/MAHC.2009.39.

16. Marques, "Cloning Computers."

17. Tinn, "From DIY Computers to Illegal Copies."

18. Gonzalo Frasca, "Uruguay," in *Video Games Around the World,* ed. Mark J. P. Wolf (Cambridge, MA: MIT Press, 2015), 609–12.

19. Radwan Kasmiya, "Arab World," in *Video Games Around the World,* ed. Mark J. P. Wolf (Cambridge, MA: MIT Press, 2015), 29–34; Benjamin Nicoll, *Minor Platforms in Videogame History* (Amsterdam University Press, 2019).

20. Stachniak, "Red Clones."

21. Jaroslav Švelch, "Promises of the Periphery: Producing Games in the Communist- and Transformation-Era Czechoslovakia," in Olli Sotamaa and Jan Švelch, eds., *Game Production Studies* (Amsterdam University Press, 2021).

22. Claude Lévi-Strauss, *The Savage Mind* (London: Weidenfeld and Nicolson, 1966), 19, 17.

23. Nick Montfort and Ian Bogost, *Racing the Beam: The Atari Video Computer System* (Cambridge, MA: MIT Press, 2009), 2.

24. Lin, "Design Engineering or Factory Capability?" 23.

25. Nathan Ensmenger, "The Environmental History of Computing," *Technology and Culture* 59, no. 4S (2018): S7–33, doi:10.1353/tech.2018.0148.

26. Siegfried Zielinski, Silvia Wagnermaier, and Gloria Custance, eds., *Variantology 1: On Deep Time Relations of Arts, Sciences, and Technologies*, Kunstwissenschaftliche Bibliothek (Cologne: Walther König, 2005).

27. Riccardo Fassone, "Cammelli and Attack of the Mutant Camels: A Variantology of Italian Video Games of the 1980s," *Well Played Journal* 6, no. 2 (2017): 68.

28. Michael Mastanduno, *Economic Containment: CoCom and the Politics of East–West Trade*, Cornell Studies in Political Economy (Ithaca, NY: Cornell University Press, 1992).

29. Ivan Malec, interview by Jaroslav Švelch, July 8, 2015.

30. Helena Durnová, "Sovietization of Czechoslovakian Computing: The Rise and Fall of the SAPO Project," *IEEE Annals of the History of Computing* 32, no. 2 (April 2010): 21–31, doi:10.1109/MAHC.2010.7.

31. Helena Durnová, "JSEP—Jednotný systém elektronických počítačů," in *Věda a technika v Československu od normalizace k transformaci,* ed. Ivana Lorencová (Praha: Národní technické muzeum, 2012).

32. Mirosław Sikora, "Cooperating with Moscow, Stealing in California: Poland's Legal and Illicit Acquisition of Microelectronics Knowhow from 1960 to 1990," in *Histories of Computing in Eastern Europe*, ed. Christopher Leslie and Martin Schmitt, vol. 549 (Cham: Springer International Publishing, 2019), 165–95, doi:10.1007/978-3-030-29160-0_10.

33. Sikora, "Cooperating with Moscow, Stealing in California."

34. Petri Paju and Helena Durnová, "Computing Close to the Iron Curtain: Inter/national Computing Practices in Czechoslovakia and Finland, 1945–1970," *Comparative Technology Transfer and Society* 7, no. 3 (2009): 303–22, doi:10.1353/ctt.0.0039.

35. Cortada, *IBM*.

36. Durnová, "JSEP—Jednotný systém elektronických počítačů."

37. Durnová, "Sovietization of Czechoslovakian Computing"; Sikora, "Cooperating with Moscow, Stealing in California."

38. Durnová, "JSEP—Jednotný systém elektronických počítačů."

39. Petr Kovář, "Historie výpočetní techniky v Československu" (Master's thesis, Charles University in Prague, 2005), http://www.historiepocitacu.cz/o-projektu-historie-pocitacu.html; Český statistický úřad, *Stav a využití výpočetní techniky v roce 1989 v ČSR* (Prague: Český statistický úřad, 1990).

40. According to Cortada, "[IBM] complained to U.S. and Western European authorities, but they could do little to block these practices, let alone the illegal shipment of products east." Cortada, *IBM*, 368.

41. Petr Trojan, "Jsme schopni vyrábět mikropočítače?" *Informace pro uživatele mikropočítačů* 1, no. 1 (Počítač přítel člověka) (1989): 1–4.

42. Přemysl Engel, "Náš interview s ing. Eduardem Smutným," *Amatérské radio, řada A* 35, no. 3 (1986): 82.

43. Antonín Vomáčka, "Televizní klub mladých 1/1987," *Televizní klub mladých* (Praha: Československá televize, January 25, 1987).

44. *30 let osobních počítačů v Československu (1/4)—Roman Kišš: PMD-85: Jak to začalo, proběhlo a nakonec skončilo,* 2013, http://www.youtube.com/watch?v=LitYDyvJwjM.

45. Ľudovít Barát, interview by Jaroslav Švelch, October 25, 2016.

46. A figure provided in Didaktik's advertising, which seems realistic when corroborated with partial information in other sources. ZX Magazín, "Kompakt," *ZX Magazín* 6, no. 1 (1993): A2; Roman Kerekeš, "8-bitová story z moravsko-slovenského pomedzia," *Bajt* 4, no. 10 (1993).

47. Michal Bechyně, "Ještě jednou Didaktik Gama," *Mikrobáze* 5, no. 2 (1989): 25–27.

48. Jaromír Olšovský, "Didaktik Gama: Popis monitoru," *Elektronika* 2, no. 8 (1988): 32–33.

49. George K., "Anketa ZX Magazínu," *ZX Magazín* 7, no. 3–4 (1994): 48–49.

50. Petri Saarikoski, Jaakko Suominen, and Markku Reunanen, "Pac-Man for the VIC-20: Game Clones and Program Listings in the Emerging Finnish Home Computer Market," *Well Played* 6, no. 2 (June 26, 2017): 7–31; Alison Gazzard, "The Intertextual Arcade: Tracing Histories of Arcade Clones in 1980s Britain," *Reconstruction* 14, no. 1 (2014), https://web.archive.org/web/20160319130044/http://reconstruction.eserver.org/Issues/141/Gazzard.shtml.

51. Lies van Roessel and Christian Katzenbach, "Navigating the Grey Area: Game Production between Inspiration and Imitation," *Convergence: The International Journal of Research into New Media Technologies* 26, no. 2 (April 2020): 402–20, doi:10.1177/1354856518786593.

52. Saarikoski and Suominen, "Computer Hobbyists and the Gaming Industry in Finland," 31.

53. Peter Liepa and Chris Gray, *Boulder Dash*, Atari (First Star Software, 1984).

54. By "porting," I mean the creation of a version of an existing program for another platform using substantial parts of the original code. For a more detailed discussion of the ports and conversions, see Švelch, *Gaming the Iron Curtain*.

55. Antonín Spurný, interview by Jaroslav Švelch, November 18, 2019.

56. Jaroslav Švelch, "Adopting an Orphaned Platform: The Second Life of the Sharp MZ-800 in Czechoslovakia," in V. Navarro Remesal and O. Perez Latorre, eds., *Perspectives on the European Videogame* (Amsterdam University Press, 2021), https://www.aup.nl/en/book/9789463726221/perspectives-on-the-european-videogame#toc.

57. Vlastimil Veselý, personal communication, April 15, 2020. Gavar himself has not responded to interview requests.

58. Roman Bórik, personal communication, May 3, 2020.

59. Roman Bórik, *Boulder Dash*, ZX Spectrum (Romborsoft, 1993).

60. Švelch, *Gaming the Iron Curtain*; Saarikoski, Suominen, and Reunanen, "Pac-Man for the VIC-20"; Fassone, "Cammelli and Attack of the Mutant Camels: A Variantology of Italian Video Games of the 1980s."

61. Adrian Johns, *Piracy: The Intellectual Property Wars from Gutenberg to Gates* (Chicago: University of Chicago Press, 2009); Lévi-Strauss, *The Savage Mind*.

62. Nicoll, *Minor Platforms in Videogame History*; Gazzard, "The Intertextual Arcade."

63. See Tinn, "From DIY Computers to Illegal Copies."

64. Walter Benjamin, "The Work of Art in the Age of Mechanical Reproduction," in *Illuminations: Essays and Reflections,* ed. Hannah Arendt, trans. Harry Zohn (New York: Schocken Books, 1986), 217–52.

PART II

EMBODIMENTS

CHAPTER 11

Indigenous Circuits

Navajo Women and the Racialization of Early Electronic Manufacture

Lisa Nakamura

My aim in writing this article (which appeared in *American Quarterly* in 2014) about indigenous women's labor building the space program, satellite industries, and home electronics industry was to create a fuller account of where our digital devices come from and whose labor created them. I started this article in 2011 after having written two books on race, gender, and digital representation, knowing that this area needed a material turn toward the workers whose labor created computing culture, a turn that took race and gender into account as one of its preconditions. People of color's stories had been neglected, buried, or otherwise left out of existing accounts of Silicon Valley and the rise of the Internet, which tended to focus on either white male engineers and programmers or labs, like the MIT Media Lab, Bell Labs, or Xerox PARC. Cheap transistors made innovation possible in all of these spaces, for all of these people. What made cheapness possible?

My path toward answering this question took me both to the paper archive and to personal experiences from a childhood in Silicon Valley. I was raised with stories about William Shockley, my father's eccentric and famous boss, who had won the Nobel Prize for inventing the transistor, a true genius who appeared in the news and on talk shows. Shockley told his Asian employees that their children would be smarter than Black or even white children due to our genetic backgrounds. This is what he told my father.

Abridged from *American Quarterly* 66, no. 4 (2014): 919–41.

Shockley was a deeply influential and controversial figure. He was a public eugenicist who set the pattern for Valley entrepreneurs to come; he was an abusive and unreasonable technologist whose racism and egotism were tolerated and later even came to be expected as part of the price of genius. Ideas that sound completely bizarre today, like the sperm bank he donated to and promoted that only contained samples from Nobel Prize winners, all went toward legitimizing our current moment: Silicon Valley's white male entrepreneurs continue to create products that manipulate how we perceive race, state violence, and gender.

Shockley's views on the inferiority of African American intelligence were well documented and public, but I searched through Stanford University's Special Collections' Fairchild and Shockley papers looking for something else: evidence that his racist beliefs influenced his hiring decisions, and by extension, those of other CEOs in Silicon Valley's formative years. Even those who hated him admired his skill in *hiring:* almost every scientist and engineer he recruited went on to found the signature tech companies such as Intel and AMD that drove the Valley's growth. Could it be possible that the Asian "model minority" stereotype, and the scarcity of African American and Latinos in the computer industry might have something to do with hiring practices based on opinions like Shockley's? Could he have had a hand in stacking the deck for and against racial groups that he thought were inferior or superior? Might Shockley's loudly espoused racism have deepened the "digital divide" and produced one of the most monocultural white and male industries in the world?

These questions fell by the wayside when I came across a box of Fairchild corporate newsletters and annual reports for shareholders that depicted Navajo women bent over microscopes soldering together integrated circuits. Once I had seen one, I couldn't stop seeing them: they were everywhere, in multiple publications over a period of ten years. Their identities as members of an indigenous, "primitive" race who had nonetheless learned how to excel as workers within one of the most advanced assembly plants in the world were often celebrated as a triumph of modernity. At other times they were simply depicted as workers like everyone else, winning prizes at bowling, holding company rodeos and picnics, and celebrating anniversaries with the company. It became clear that between 1965 and 1975, before the plant was closed by American Indian Movement protesters and Fairchild decided to shut the plant down, Navajo women and men were some of the industry's first high-tech workers.

This isn't a story about finding something new, rare, or unviewed by others. The Navajo have been "the most studied people in the world," for many years, the American Indian Movement protests are extremely well documented, and the Fairchild protest shows up in many of these histories. The images in my essay were available as pdf's in public digital collections, free to anyone to see. But because the Fairchild "ladies," as I heard them call themselves and each other during a visit to the Shiprock chapter in 2016, did not fit into triumphal narratives of individual white and male genius, their stories were overlooked. Most of the people whose labor made digital technology cheap and ubiquitous are people and women of color and their stories are still waiting to be told.

Happily, indigenous scholarship and research on digital technology has blossomed in recent years. Marisa Duarte's brilliant *Network Sovereignty: Building the Internet across Indian Country* (Seattle: University of Washington Press, 2017) and Jodi Byrd's analyses of video games such as *Assassin's Creed III: Liberation,* which explore the complex dynamics of indigenous dispossession, empire, and settler colonialism, have established a new field of study and a new way to center digital indigenous formations. I'm grateful for their work and that of other indigenous scholars whose research tells a different story about where technology comes from and who it is for.

Donna Haraway's foundational cyberfeminist essay "A Cyborg Manifesto" is followed by an evocative subtitle: "An Ironic Dream of a Common Language for Women in the Integrated Circuit." She writes, "The nimble fingers of 'Oriental' women, the old fascination of little Anglo-Saxon Victorian girls with doll's houses, women's enforced attention to the small take on quite new dimensions in this world. There might be a cyborg Alice taking account of these new dimensions. Ironically, it might be the unnatural cyborg women making chips in Asia and spiral dancing in Santa Rita jail whose constructed unities will guide effective oppositional strategies."[1] In this passage Haraway draws our attention to the irony that some must labor invisibly for others of us to feel, if not actually *be,* free and empowered through technology use:

technoscience is, indeed, an integrated circuit, one that both separates and connects laborers and users, and while both genders benefit from cheap computers, it is the flexible labor of women of color, either outsourced or insourced, that made and continue to make this possible.[2]

. . .

Haraway's Marxian insistence on materiality rather than just virtuality in the "Cyborg Manifesto"—on the gendering and racializing of bodies as well as on computer hardware itself—anticipated many of the concerns at the center of media archaeology and platform studies in the twenty-first century. Tiziana Terranova, whose focus on the Internet as a site of digital labor brings us back to the material realm of bodies and exploitation, extends this interrogation into the way that labor is commodified and extracted, often without compensation for the laborer, within digital culture.[3] For Haraway, the women of color workers who create the material circuits and other digital components that allow content to be created are all integrated within the "circuit" of technoculture. Their bodies become part of digital platforms by providing the human labor needed to make them. Really looking at digital media, not only seeing its images but seeing *into* it, into the histories and platforms, both machinic and human, is absolutely necessary for us to understand how digital labor is configured today.

How can we take up Haraway's injunction to be guided by women of color's labor in the digital industries to form "effective oppositional strategies"? Women of color feminism's theoretical framework has much to offer digital media studies, particularly in light of the emphasis on the physical and material aspects of computing that media archaeology has brought to the field.[4] . . . When we look at the history of digital devices, it is quite clear that the burden of digital media's device production is borne disproportionately by the women of color who make them. . . .

References to "nimble fingers" as a digital resource appear in many accounts of how women of color were understood by and actively recruited to work in the electronics industry in this period. As Jefferson Cowie writes, the "nimble fingers" phrase was applied to Latino women working in maquiladoras for RCA and other electronics firms, including Fairchild. According to Karen Hossfeld, by the eighties in Silicon Valley, electronic assembly had become not just women's work but women of color's work.[5]

This essay focuses on a group of women of color who are almost never associated with electronic manufacture or the digital revolution—Navajo women. The archive of visual materials that document the history and industrial strategy of Fairchild Semiconductor, the most influential and pioneering electronics company in Silicon Valley's formative years, documents their participation through visual and discursive means, albeit never in their own voices.[6] Fairchild's internal documents, such as company newsletters, and its public ones, such as brochures, along with Bureau of Indian Affairs press releases and journalistic coverage by magazines such as *Businessweek*, paint a picture of Navajo women workers as uniquely suited by temperament, culture, and gender as ideal predigital digital workers.

My reading of these materials reveals how Fairchild produced a racial and cultural argument for recruiting young female workers in the electronics, and later digital device production industries, from among the Navajo population. As Cowie writes of young Mexican women working for RCA: "Management's standard explanation for its preference for young female workers typically rested on the idea that women's mental and physical characteristics made them peculiarly suited to the intricacies of electrical assembly work."[7] Similarly, the hundreds of Navajo women who worked at the Fairchild semiconductor plant in Shiprock, New Mexico, on Navajo land were understood through the lens of specific "mental and physical characteristics" such as docility, manual dexterity, and affective investment in native material craft. The visual rhetoric that described their unique aptitude for the work drew heavily on existing ideas of Indians as creative cultural handworkers.

A close examination of how Navajo women's labor was exploited as a visual and symbolic resource as well as a material good shows us how indigenous women's labor producing circuits in a state-of-the-art factory on an Indian reservation came to be understood as affective labor, or a "labor of love." In her work on women's affective labor in digital media usage, Kylie Jarrett uses the term *women's work* "to designate the social, reproductive work typically differentiated from productive economics of the industrial workplace."[8] A 1969 Fairchild brochure celebrates Indian women circuit makers as culture workers who produced circuits as part of the "reproductive" labor of expressing Navajo culture, rather than merely for wages.

The Anomalous Narrative of Indigenous Workers at Fairchild Semiconductor

The story of Fairchild's plant on Navajo land is not part of a narrative of development that fits comfortably into the history of the digital industries. Though documentary histories of Fairchild abound, and no history of Silicon Valley fails to mention the company, the Shiprock plant is rarely discussed in these accounts, or at best appears as a footnote or a brief mention or digression from the story of outsourcing production to Southeast Asia.[9] The company was regarded as a pioneer because of its willingness to take risks, to invent new manufacturing processes, and to venture onto foreign shores in search of cheap labor, an act that "helped to launch the PC revolution, which begot the commercial Internet, which begot everything else."[10] Fairchild's trajectory of sourcing labor domestically from female workers of color in the sixties, to outsourcing in the seventies, and eventually to offshoring in Asia was a path followed by many other electronics companies.

Since Fairchild was one of the first chip manufacturers to outsource production to Asia, this is recognized as an epochal event in the history of computing, an innovation that permitted the remarkable growth of the electronics and eventually the computing and personal digital device industry. However, the history of offshore outsourcing to Asia runs parallel with chip fabrication projects *within* and *across* US borders, specifically on Navajo land and in Mexico, respectively. In 1964 the Bracero Program officially ended, and in 1965 the Border Industrialization Program (BIP) began on the US-Mexico border. By 1973 Fairchild and other semiconductor manufacturers were operating plants in Mexico under this program, in addition to plants in Singapore, Hong Kong, and Seoul.

In 1962 Charlie Sporck, a top executive at Fairchild Semiconductor and, later, president and CEO of National Semiconductor, two of the largest and most important manufacturers of integrated circuits, knew that the industry was "running into limitations as to where we could sell the product."[11] The majority of the "product" was being sold to the military, and Sporck realized that Fairchild needed to reduce labor costs in order to break into the "vast consumer market out there" for electronic devices such as calculators, games, and eventually personal computers. In an interview recorded as part of the "Silicon Genesis: An Oral History of Semiconductor Technology" project, Sporck recalls how the quest for cheaper labor and lower overhead drove Fairchild to open a plant in Hong Kong, a move that pioneered electronics manufacture outsourcing to this and other locales in Southeast Asia, Mexico, and Southern California.

However, the interview takes an odd turn. As Sporck warms to his work of explaining how Fairchild started the "mad rush into Southeast Asia by all companies" in the sixties, the interviewer interrupts, asking, "Well, did you also go to Shiprock, New Mexico to the Indian reservation?" Sporck replies, "Yeah, that's not one of the . . ." The interviewer continues, "I noticed you didn't bring that up." Sporck replies, "No, we did, that was at the, just about the time we went to Portland, Maine. We looked elsewhere in Shiprock, looked like a possibility and we did locate down there. It never worked out, though. We were really screwing up the whole societal structure at the Indian tribe. You know, the women were making money and the guys were drinking it up and it was a failure."

Though Sporck depicts the plant as a "failure," it was depicted as a tremendous success during its years of operation. In fact, the archive of materials about the plant depicts it as doing well because it was *in line* with the "societal structure of the tribe," rather than in conflict with it.

Insourcing on the Reservation: Fairchild's Move to Indigenous Territory

Fairchild opened its state-of-the-art semiconductor assembly plant on the twenty-five-thousand-square-mile Navajo reservation in Shiprock,

New Mexico, in 1965. The plant grew from a pilot project employing fifty-five people to a thirty-three-thousand-square-foot integrated-circuit manufacturing facility where hundreds of Navajo women and some men worked on circuit assembly between 1965 and 1975; while accounts as to the exact number of Navajo employed vary, in 1966 Fairchild was the "largest of several electronics plants now located in Indian areas,"[12] and "at its height, the plant provided work for more than 1,000 Navajos. . . . Fairchild became the largest industrial employer in New Mexico and the largest employer of Indians in the country."[13] The plant, which operated twenty-four hours a day, was owned by the Navajo Tribal Council and leased by Fairchild for $6,000 a month.[14] It boasted a very low failure rate—5 percent, in contrast to rates in the twentieth percentile at other plants—and received several awards for its innovative practices.

As the historian Colleen O'Neill writes, "In 1974, prior to its closing, Fairchild employed 922 Navajos, most of whom were women. Fairchild was one of the largest employers of Navajo labor on the reservation, second only to public sector employees, including the Bureau of Indian Affairs and the Navajo nation."[15] In most histories of Silicon Valley, domestic manufacture is assumed to have given way to foreign manufacture starting in the sixties, when the first large plants in Asia and Mexico opened. Widening the perspective on outsourcing to include insourcing practices like the production of semiconductors on Navajo land provides a valuable perspective from which to view the material culture of computing.

Reservations provided spaces of exception to US laws on minimum wage; in this way they were like foreign countries, but in other ways American mythologies around Indianness gave these workers a desirable identity as culturally foreign yet familiar. Likewise, American Indian history tends to include the Fairchild plant as an example of failed economic development or as part of the history of the American Indian Movement's protests, but does not connect it to digital culture or history.

Fairchild's Shiprock plant was far more than just an outlier. Instead, the company represented it as a new and innovative model for cheap domestic electronics manufacture: insourcing rather than outsourcing. In Fairchild's promotional materials and in journalistic accounts Navajo workers were always represented as different from white workers, as possessing innate racial and cultural traits that could be enhanced or rehabilitated to produce chips accurately, quickly, and painlessly. . . . Analysis of documents from the period that describe the plant's remarkable early success and its eventual closure in 1975 reveal potent and durable claims and beliefs about gender, race, and particular labor *styles* that would quickly be appropriated to describe the Asian women workers who eventually replaced them.

How and why did the most advanced semiconductor manufacturer in the world build a state-of-the-art electronics assembly plant on a Navajo reservation in 1965? A 1969 Fairchild news release explains that the plant was "the culmination of joint efforts of the Navajo People, the U.S. Bureau of Indian Affairs (B.I.A.), and Fairchild." Though cheap, plentiful workers and tax benefits helped lure electronics companies to the reservation, Navajo leadership helped push the project forward; Raymond Nakai, chairman of the Navajo Nation from 1963 to 1971, and the self-styled first "modern" Navajo leader, was instrumental in bringing Fairchild to Shiprock. He spoke fervently about the necessity of transforming the Navajo as a "modern" Indian tribe, and what better way to do so than to put its members to work making chips, potent signs of futurity that were no bigger than a person's fingernail? The incongruity of this form of labor—the creation of the most advanced devices the world had yet known, tiny bits of matter that could tell a satellite where to point, by women who were conceived of as irredeemably primitive—was not lost on the tribes themselves.

In his address dedicating the newly built Shiprock plant, Nakai said, "It is a brilliant chapter that we write here in the dedication of this magnificent plant. It signals the real and early industrialization of the Navajo reservation. It marks the advancement of the Navajo nation from an Agrarian Nation to an Industrial Nation."[16] This attempt to rebrand the Navajo as modern through their labor within electronics manufacture seems designed to counter the notion of Indians as "suffering from a racial inability to advance," as Philip Deloria puts it.[17] This new notion of the Navajo as "Industrial" produced a complicated identity whose formation relied on the idea that the tribe could be modern, even hypermodern, precisely as a result of being distinctively Indian. Indian-identified traits and practices such as painstaking attention to craft and an affinity for metalwork and textiles were deployed to position the Navajo on the cutting edge of a technological moment precisely because of their possession of a racialized set of creative cultural skills in traditional, premodern artisanal handwork.

The building of the Shiprock plant was very much in line with the 1961 Task Force on Indian Affairs recommendations, which urged that reservations attract light industry as part of the "key to the economic and social competency program," which would "increase Indian economic self-sufficiency, and eventually terminate all services from the federal government to Native Americans."[18] As Peter Iverson writes, "The Navajo sought to lure other large-scale industry with cheap land leases, favorable construction arrangements, and a trainable work force. Two major firms accepted the Navajo's invitation: Fairchild Semiconductor and the General Dynamics corporation."[19] In turn, Fairchild benefited from a $700,000 loan from the Navajo to finance plant build-out, free equipment from the BIA supplied from "federal excess property sources," a very low hourly wage, freedom from real estate taxes, and funding for training programs supported by Department of Labor.[20] These factors all mattered, but in the end, product quality was what kept the plant in business and allowed it to expand.

Race and Gender as Digital Resource: Navajo Women as Early Creative Class Workers

Semiconductor manufacture was performed using a microscope and required painstaking attention to detail, excellent eyesight, high standards of quality, and intense focus. Not all who started to work there continued—as Jim Tutt, a Navajo process engineer who worked at Fairchild until 1974, put it, "It was tedious work under a microscope. They couldn't handle it, some of them, [because they had to spend] so many hours a day looking at it."[21] Despite these daunting conditions, the hundreds of Navajo women who stayed on excelled at this work, and the industrial discourse produced by and about the plant attributed its success to the female gender of its workers as well as Indian racial traits. At Fairchild, the preference for women assembly workers was so strong that men were effectively shut out of the vast majority of jobs at the Fairchild plant, and Nakai had to work hard to pressure the company into hiring more men at the plant.[22]

A Fairchild company newsletter published a story titled "Fairchild Shiprock: A Success Story," citing the "tremendous job" that the Navajo "ladies," pictured hovering over microscopes, were doing assembling integrated circuits. To explain the plant's success, the article equates creative cultural skills such as weaving and silversmithing with circuit building. Both Fairchild's corporate newsletter and *Businessweek* credited plant manager Paul Driscoll with discovering and exploiting the "untapped wealth of natural characteristics of the Navajo . . . the *inherent flexibility* and dexterity of the Indians": "For example, after years of rug weaving, Indians were able to visualize complicated patterns and could, therefore, memorize complex integrated circuit designs and make subjective decisions in sorting and quality control."[23]

In the days before either outsourcing or insourcing, when integrated circuits were manufactured in the same complexes or even buildings that housed the men who envisioned and designed them, immigrant women of color were hailed as the ideal workforce because they were mobile, cheap, and above all, *flexible*; they could be laid off at any time and could not move to look for alternative forms of work, while their employers could close plants and reopen them in locales with the most favorable conditions. The notion that Indians were "inherently flexible" both racializes and precedes the idea of flexible labor that informs much of the research on globalization in the information age.

As Guy Senese writes, "employee availability" was highly desired by industry, which influenced its choice to open plants on Indian reservations. The almost complete lack of other wage-based employment options in Indian country and an extremely high unemployment rate almost guaranteed a favorable environment for employers. He situates the plant as part of an ongoing project of "Indian labor exploitation," writing that both "quality and low cost of Indian labor was, along with liberal government loan and tax relief, a major attraction for industry."[24] "Quality," defined as a low failure rate, was a major issue in the industry; many parts of the chip production process required artisanal handwork. Partly because of this, failure was quite common and could have serious consequences, particularly for Fairchild's military and space program contracts, which were still a major part of its business.[25] Thus, in Fairchild's outward-facing publications, such as brochures and press releases, as well as in journalistic accounts of the Shiprock project, quality is discussed rather than cost. And it was a specific kind of quality—Indian craftsmanship.

The argument that circuit quality was a natural outcome of Indian raciocultural traits is made quite overtly in Fairchild's 1969 brochure celebrating the new Shiprock plant and its workers. The first page features a large photograph of a rectangular brown, black, and white rug, woven in a geometric pattern composed of connecting and intersecting right angles (fig. 11.1).[26] . . . The following pages depict a woman weaving the same type of rug.

The accompanying text reminds the reader that "weaving, like all Navajo arts, is done with unique imagination and craftsmanship, and it has been done that way for centuries." Just as this idyllic tribute to Navajo craft is getting started, the brochure transitions to a photograph of a Navajo woman standing over a microscope, gazing at the viewer, as a white male face gazes over her shoulder supervising and admiring her work (fig. 11.2). The text negotiates the transition from traditional artisanal cultural work to industrial wage labor by asserting that "building electronic devices, transistors and integrated circuits, also requires this same personal commitment to perfection. And so, it was very natural that when Fairchild Semiconductor needed to expand its operations, its managers looked at an area of highly skilled people living in and around Shiprock, New Mexico."

This appeal to "nature" as justification for converting "highly skilled" female cultural labor such as weaving rugs into high-tech factory labor is signaled by the following image, which depicts a Fairchild 9040 integrated circuit, "used in communications satellites like COMSAT," en-

Figure 11.1. Navajo rug.

Shiprock Dedication Commemorative Brochure, September 6, [1969], lot X5184.2009, folder 102725169, Computer History Museum, Mountain View, CA.

larged so that its geometry fills the whole page (fig. 11.3). The resemblance between the pattern of the rug depicted on the first page and the circuit is striking and uncanny. It makes the visual argument that Indian rugs are merely a different material iteration of the same pattern or aesthetic tradition found within the integrated circuit. The opposing page states, "The blending of innate Navajo skill and Semiconductor's precision assembly techniques has made the Shiprock plant one of Fairchild's best facilities—not just in terms of production but in quality as well."

Figure 11.2. Navajo semiconductor plant worker.
Shiprock Dedication Commemorative Brochure, September 6, [1969], lot X5184.2009, folder 102725169, Computer History Museum, Mountain View, CA.

Again, the notion of an "inherently flexible" laborer, a worker whose nature it is to be both adaptable and culturally suited, or hardwired, to craft circuit designs onto either yarn or metal appeals to a romantic notion of what Indians are and the role that they play in US histories of technology. The nostalgic appeal to Indian identity as a unique and valuable commodity in the world of high-tech manufacture, as both a vanishing resource and an example of a participant in the nation's unstoppable drive toward modernity, is completed on the brochure's last page. The brochure's last image is a photograph of the sun setting behind the majestic Shiprock Mountain, the namesake for the Navajo reservation, superimposed by a poem, "Song of the Earth Spirit, Origin Legend." . . . It is safe to say that poetry is not a standard convention for industrial brochures. Including it solidified claims that circuit manufacture was naturally indigenous people's work.

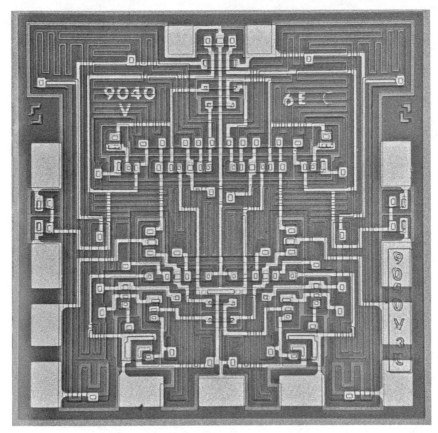

Figure 11.3. Fairchild 9040 integrated circuit.

Shiprock Dedication Commemorative Brochure, September 6, [1969], lot X5184.2009, folder 102725169, Computer History Museum, Mountain View, CA.

. . .

Depicting electronics manufacture as a high-tech version of blanket weaving performed by willing and skillful indigenous women served two goals: it permitted the incursion of factories into Indian reservations to be seen as a continuation of rather than break from "traditional" Indian activities, and it pioneered the blurring of the line between wage labor and creative-cultural labor; one seamlessly became the other. Indeed, one may have *replaced* the other: the new eight-hour workday altered many aspects of family life for the Navajo people who worked at Fairchild. However, the 1969 Fairchild brochure and other materials describing the plant assert that replacing rugs with circuits is, rather than a cultural loss or, worse yet, a form of cultural imperialism, instead an extension of an existing, indigenous cultural practice; it is culture work for the nascent information age. It posits that indigenous design informed electronic circuit design—a kind of colonialism in reverse—despite the lack of involvement of indigenous people in the company's research and development arm.

The argument that Navajo women were good at their assembly jobs because they were good blanket weavers and jewelry makers appears throughout contemporary accounts of the plant. Journalistic accounts, BIA press releases, and Fairchild internal documents alike depicted Indians as the first informationalized "creative class" workers, to use Richard Florida's influential formulation, doing what they loved well because they loved doing it.[27] Florida argues that jobs in software design, engineering, and even haircutting appeal far more to early twenty-first-century workers than jobs in the traditional industries such as manufacturing, not only because of the intrinsic pleasure involved with the act of making but also because of the personal freedom acceded to the worker. . . . The seed of this argument can be found in the Shiprock brochure that depicts naturally happy workers, expressing their creativity by creating electronic artifacts that resemble indigenous artifacts. The brochure's photographs of satisfied Navajo women busy at their looms and microscopes was especially appealing given the intense competition between states and non-US countries to attract industry by offering freedom from taxes and, most importantly, freedom from labor unions.

It was also a fortuitous moment for Fairchild to assert the connections between nature and technology, specifically electronics manufacturing: chip manufacture is a notoriously dirty business, and workers at Fairchild and other semiconductor manufacturers were falling victim to pollution-related disease and starting to blame the company. "By the mid-1970s, reports of chemical exposures among production workers had begun to surface" in San Jose, California.[28] Given the already high rates of pollution on the reservation from the extraction of resources such as uranium, gas, coal, and oil, semiconductor manufacture continued the ongoing practice of environmental degradation in a spot renowned for its natural beauty.[29] Ultimately, the Navajo nation failed to benefit economically as much as it had expected from the plant and was left to deal with the detritus and its long-term consequences.

Navajo identity had a heavy burden to bear. In the face of concerns about high-tech pollution, increasingly empowered labor organizations, and a newly politicized and visible American Indian civil rights movement, indigenous electronic workers at Shiprock were pressed into service as examples of the peaceful coexistence and integration of the past and the future, the primitive and the modern, creativity and capitalism. They were cited as evidence that digital work—the work of the hand and its digits—could be painlessly transferred from the indigenous cultural context into the world of technological commercial innovation, benefiting both in the process.

Navajos were described by their managers as having "patience, respect for private property (hence a low theft rate), lack of militancy, and pride in their work."[30] They were the ideal workforce, because in contrast to striking workers in other parts of the country, they could not relocate; Fairchild's 1969 brochure claims that "the real value of this progress lies in the creation of meaningful jobs for those who have not had jobs, jobs which will keep them in the land they love and among the people they know."

The immobility and vulnerability of the Navajo worker was rhetorically respun into an act of purposeful and care-driven cultural preservation on the part of the corporation. The original rationale for bringing industry to the reservation, which was to gradually eliminate support from the federal government to Native Americans, was represented as part of a plan to help them stay on their reservations and retain their ancestral homelands.

The benefits of a trained and seasoned indigenous labor force that was new to industrial forms of labor were not lost on managers at other factories in Shiprock. As C. J. Jameson, manager of General Dynamics's Shiprock plant, said, "they don't have the bad habits people have in more industrial areas."[31] This is an eloquent illustration of how racialization works; prior beliefs about Indians as unreliable workers unsuited for modern forms of labor are transformed into assertions of the positive *value* of "primitive" habits. This shift demonstrates the fluidity and mutability of gender and race stereotyping; Indians were described as careful, docile, and hardworking when it helped their managers to understand and explain productivity through an ethnic lens. This strategy was one of the first iterations of an exceptionally effective argument to justify digital labor exploitation by depicting it as an outlet for the expression of cultural and racial identity. Attention to detail and pattern, careful handcraft, stoicism, and flexibility are made, not born—as Cowie writes, they are invoked in response to the needs of global capital to travel, to justify manufacturing a product in the cheapest place possible.

Race and gender are themselves forms of flexible capital. When it helps create a compelling narrative that justifies, even celebrates, the yoking of corporate interests to indigenous governance, Navajo women are understood and perceived as docile, flexible, and natural electronics workers, and indigenous identities change as a result. And when it does not, they are changed back accordingly. Latinas and Asian, African American, and, later, Indian women were all viewed as having "nimble fingers and passive personalities."[32] American Indian women, as well as Mexican women working in maquiladoras, were described in much the same way as "Orientals": as ideal workers in the digital industries, because of their experience with fine crafting of jewelry and textiles. In our present day and for the past few decades, Asian fingers have been "nimble," but in the sixties and seventies, Navajo women's fingers were envisioned this way. In this case, it can be seen how racialization—the understanding of a specific population as possessing traits and behaviors that belong to a race, not an individual—is a process, not a product.

Rug weaving is the linchpin of the Shiprock brochure's visual argument that Navajos were natural circuit assembly workers. It is mentioned in every publication that attempts to explain the plant's success. Unlike silversmithing, jewelry making, and other indigenous Navajo practices that were cited as an argument for why and how Indians were so good at their work, rug weaving was a specifically female activity.

As Benny Klain discovered during his interviews with indigenous rug and blanket weavers in his documentary *Weaving Worlds* (2008), weaving was a reliable source of personal income for women during hard times as well as an important creative outlet and spiritual practice, and as one weaver explained, it "kept us fed." Yet at the same time, the low prices offered by Indian trading post owners and traveling rug buyers guaranteed that Indian women weavers' labor was not compensated fairly; it is still a potent emblem of the exploitation of indigenous women's knowledge and labor.

. . . The affinity and historical links among weaving, digital computing, and women figures centrally throughout cyberfeminist theory, most famously "A Cyborg Manifesto." Silicon Valley business discourse created an archive of materials that represented Navajo women as "natural" cyborgs, indeed, as embodying nature itself using silicon as their medium. The cyberfeminist theorist Sadie Plant completes the circuit between weaving as indigenous practice and software production: "Textiles themselves are very literally the software linings of all technology. . . . it is their microprocesses which underlie it all: the spindle and the wheel used in spinning yarn are the basis of all later axles, wheels, and rotations; the interlaced threads of the loom compose the most abstract processes of fabrication."[33]

The discourse about Fairchild's Shiprock operation described Navajo women's affinity for electronics manufacture as both reflecting and satisfying an intrinsic gendered and racialized drive toward intricacy, detail, and quality, and the women who performed this labor did so for the same reason that women have performed factory labor for centuries—to survive. The liberal discourse of the seventies assuaged its conscience in consigning vulnerable populations like Native Americans to this type of labor by *suturing the work itself to an emergent discourse of multiculturalism.* How could this type of labor be exploitative when it was already so much like the "native" cultural production that Indians had done for centuries without pay, the original "free labor," such as weaving blankets?

Thus it was semiconductor and electronics manufacture, among the most tedious of jobs in the long supply chain that produces our digital media devices and the vast array of technologies we use today, that was redefined and envisioned as *creative* labor, labor that women do to express themselves. In a BIA news release titled "Industries Turn to Indians for Precision Workers," the writer claims, "The Indian, with a natural affinity for precision work, is equally at home as a high-climbing steel structural worker and as a weaver of intricate designs. Somewhere between the two extremes lies electronic factory work, which calls for skill that is rooted in pride of workmanship."[34] Semiconductor manufacture was made to seem like an act of Navajo cultural preservation as well as a bid for economic survival.

Navajo women did not make circuits because their brains naturally "thought" in patterns of right-angle colors and shapes. They did not make them well because they had inherent Indian virtues such as stoicism, pride in craftswomanship, or an inherent and inborn manual dexterity. And Fairchild did not employ Navajo women *because* of these traits. These traits were identified after the company learned about the tax incentives available to subsidize the project, the lack of unions and other employment options in the area, and the generous donation of heavy equipment given by the US government gratis as part of an incentive to develop "light industry" as an "occupational education" for Indians.

Though in 1969 Fairchild's president and CEO Dr. C. Lester Hogan stated, "In the next several years we expect to see expansion of this nearly all Navajo operated plant, concurrent with future development of the Shiprock community and increased opportunities for all Navajos," this was not to be. . . . After 1975, when the plant was taken over by American Indian Movement members led by Russell Means and Fairchild closed it, the Navajo were no longer the digital model minority. Fairchild cited the unstable labor environment as the reason, but many suspected that this had to do with a desire to move all operations offshore, where wages were even lower than they were in Navajo country, and workers less inclined to protest conditions. In the wake of the Alcatraz Occupation (1969–71) and the Wounded Knee incident (1973), the American Indian Movement (AIM) was perceived as a militant group and certainly not one that industries wanted to tangle with directly.

The reasons stated for the occupation cited worker layoffs, but others speculated that AIM's desire to unionize a famously never-unionized industry contributed to the closure as well.[35] Two conflicting views of indigenous women—as inherently digital workers who could "see complex patterns" and effortlessly, perfectly, and "naturally" re-create them on miniature circuits, and as militant aiders and abettors of militant men or, worse yet, as themselves militant—collided in this moment. While some Navajo mourn the closure to this day, imagining a Navajo Silicon Valley, others are relieved that the reservation was saved from this fate.[36]

Race and Digital Platforms

In *Indians in Unexpected Places,* Deloria writes that the American custom of imagining Indians in terms of "primitivism, technological incompetence, physical distance, and cultural difference" has remained "familiar currency in contemporary dealings with Native people."[37] Fairchild's argument for the unique benefits afforded to hypermodern technologies by indigenous women exploited this currency to paint a new and appealing picture of both Indians and electronic culture as intimately joined rather than on opposite poles.

In Nick Montfort and Ian Bogost's immensely useful definition from *Racing the Beam,* a platform is "whatever the programmer takes for granted when developing, and whatever, from another side, the user is required to have working in order to use particular software."[38] The present essay is concerned with that "whatever," the material conditions that are usually invisible to the user and are necessary for digital media device creation. The existence of cheap female labor is absolutely taken for granted as a precondition of digital media's existence. . . . Innovation and development are impossible without access to hardware that can be produced flexibly, cheaply, and consistently, as it was on Navajo land from 1965 to 1975.

. . .

Montfort and Bogost, Wendy Chun, Alex Galloway, and Kathryn Hayles remind us that the digital *does* as well as *appears.*[39] . . . If the platform is defined as what is taken for granted, scholarship that "examines the relationship between platforms and creative expression" corrects the tendency to forget that digital media is "more than screen-deep," as Chun elegantly phrases it.[40]

Digital media has always had what Ted Friedman calls the "beige box problem": advertisers had a terrible time marketing a product that looked so dull: there was seemingly nothing to see.[41] In the 1997 "Intel Inside" campaign, . . . human figures fully covered in "bunny suits," or clean suits, danced and capered to catchy music. . . . However, if we look inside computing hardware, we will not see dancing bunny-suited clean room workers, happily making chips for free. Instead we see Asian women, Latinas, and Navajo women and other women of color. Looking inside digital culture means both looking back in time to the roots of the computing industry and the specific material production practices that positioned race and gender as commodities in electronics factories. This labor is temporally hidden, within a very early period of digital computing history, and hidden spatially. We must look to locales and bodies not commonly associated with these technologies, in out of the way places, to see how race operates as a key aspect of digital platform production.

Digital labor is usually hidden from users in closed factories in Asia, visible to us only as illegally recorded cell phone video on YouTube or through the efforts of investigative reporters who overcome significant barriers to access—again, nothing to see.[42] But as Nicholas Mirzoeff reminds us in *The Right to Look*, visual culture's political project enjoins us to look precisely at those objects, practices, and artifacts that either protest their own innocence or document subaltern experiences.[43] On the spectrum of digital labor, factory work soldering chips for iPhones, missiles, and servers is as close to the machine as one can get, as close to the means of digital production—the computer—as can be imagined. It is not creative labor, nor is it free. It is fascinating that, during a pivotal moment in early computing history, the industry's foremost electronics company represented it that way. This story of digital device manufacture on Indian land shows us how the discourse of women's indigenous cultural production has been used to explain the key role that women of color play within the integrated circuit of production.

. . .

The 1969 Fairchild Shiprock brochure does exactly that, by representing the labor of semiconductor manufacture as a "labor of love" or, more accurately, as agentive or creative race-labor rather than as alienated labor. Like weaving blankets, semiconductor production is posited as an intrinsic part of the Indian psyche, an expression of cultural essence imperiled, yet ultimately enabled, by the "modern" world.

Notes

My sincere thanks to Sara Lott, senior archives manager at the Computer History Museum in Mountain View, California, Jim Tutt, Christian Sandvig, Dan Scholler, Jonathan Sterne, Paul Edwards, and Iván Chaar-Lopez.

1. Donna Haraway, *Simians, Cyborgs, and Women: The Reinvention of Nature* (London: Routledge, 1991).
2. Aihwa Ong, *Flexible Citizenship: The Cultural Logics of Transnationality* (Durham, NC: Duke University Press, 1999), 64.
3. Tiziana Terranova, *Network Culture: Politics for the Information Age* (London: Pluto, 2004), vii, 184.
4. Wolfgang Ernst and Jussi Parikka, "Digital Memory and the Archive," *Electronic Mediations*, vol. 39 (Minneapolis: University of Minnesota Press, 2013), 265.
5. Ted Smith, David Allan Sonnenfeld, and David N. Pellow, *Challenging the Chip: Labor Rights and Environmental Justice in the Global Electronics Industry* (Philadelphia, PA: Temple University Press, 2006), 357; Jefferson Cowie, *Capital Moves: RCA's Seventy-Year Quest for Cheap Labor* (Ithaca, NY: Cornell University Press, 1999), 273; T. R. Reid, *The Chip: How Two Americans Invented the Microchip and Launched a Revolution* (New York: Simon and Schuster, 1984), 243; Karen Hossfeld, "Their Logic against Them: Contradictions in Sex, Race, and Class in Silicon Valley," in *Technicolor: Race, Technology, and Everyday Life*, Alondra Nelson and Thuy Linh Tu, eds. (New York: New York University Press, 2001), 34–63.
6. Semiconductors are the basis of all digital media devices, and as the Silicon Valley historian David Laws writes, "to some degree, almost every artifact based on semiconductor devices produced after the early 1960s depends in some way on technologies that were created by Fairchild or former Fairchild employees." Thus Fairchild ['s . . .] industrial history and labor practices set the tone for much of what was to become Silicon Valley. . . .
7. Cowie, *Capital Moves*, 273.
8. Kylie Jarrett, "The Relevance of Women's Work: Social Reproduction and Immaterial Labor in Digital Media," in *Television and New Media* 15, no. 1 (2014): 14–29.
9. See David A. Laws, "A Company of Legend: The Legacy of Fairchild Semiconductor," *IEEE Annals of the History of Computing* 32, no. 1 (2010): 60. Leslie Berlin, Lillian Hoddeson, and Christophe Lécuyer emphasize the centrality of Fairchild as

the first successful manufacturer of microchips and the leader in the field. See Michael Riordan and Lillian Hoddeson, *Crystal Fire: The Birth of the Information Age* (New York: Norton, 1997), 352; Leslie Berlin, *The Man behind the Microchip: Robert Noyce and the Invention of Silicon Valley* (Oxford: Oxford University Press, 2005), 402; Christophe Lécuyer, *Making Silicon Valley: Innovation and the Growth of High Tech, 1930–1970* (Cambridge, MA: MIT Press, 2006), 312.

10. Mike Cassidy, "What Went Wrong at Shiprock—The Fairchild Plant Promised a Better Future for the Navajos, but That Promise Was Never Fulfilled," *San Jose Mercury News,* May 7, 2000.

11. "Interview with Charlie Sporck," February 21, 2000, Los Altos Hills, CA. . . .

12. US Department of the Interior, Bureau of Indian Affairs, *Electronics Industry Expanding on Navajo Reservation,* September 2, 1966.

13. Cassidy, "What Went Wrong at Shiprock."

14. Fairchild Camera and Instrument, News Release, Shiprock, New Mexico, September 6, 1969.

15. Colleen M. O'Neill, *Working the Navajo Way: Labor and Culture in the Twentieth Century* (Lawrence: University Press of Kansas, 2005), 235.

16. Raymond Nakai, Fairchild Dedication (speech, Cline Library, Special Collections and Archives, Northern Arizona University, 1969).

17. Philip Joseph Deloria, *Indians in Unexpected Places* (Lawrence: University Press of Kansas, 2004), 178.

18. Guy B. Senese, *Self-Determination and the Social Education of Native Americans* (New York: Praeger, 1991), 218.

19. Peter Iverson, *The Navajo Nation* (Albuquerque: University of New Mexico Press, 1983), 273.

20. As Greg Harrison, a Fairchild Semiconductor employee, recalls, "sizable labor subsidies were available from the Bureau of Indian Affairs" and "the Native Americans on the reservations were badly in need of jobs and skills." "National Semiconductors Offshore Legacy Recalled as Company Readies China Facility," *Chip Scale Review: The International Magazine for Device and Wager-Level Test, Assembly, and Packaging Addressing High-Density Interconnection of Microelectrics IC'S,* May–June 2004.

21. Jim Tutt, personal communication, November 2011.

22. In a draft of a speech to be given at the 1970 Fairchild board of directors meeting, Nakai praised the board for taking a chance on the "Navajo workman" and wrote that he was "extremely pleased that your organization has decided to locate your machine tool division here at Shiprock employing additional Navajos, chiefly male. That you have made this decision enforces our belief that it is highly desirable to utilize the talents of the Navajo workman." "Remarks to Be Made at the Fairchild Board of Directors Meeting," Colorado Plateau Archives, Northern Arizona University, 1970.

23. "Industry Invades the Reservation," *Businessweek,* April 1970, my italics.

24. Senese, *Self-Determination,* 218.

25. For an excellent account of the relation between the space program and the nascent software industry, see David Mindell, *Digital Apollo: Human and Machine in Spaceflight* (Cambridge, MA: MIT Press, 2008), 359. . . .

26. *Shiprock Dedication Commemorative Brochure*, September 6, [1969], lot X5184.2009, folder 102725169, Computer History Museum, Mountain View, CA.

27. Richard L. Florida, *The Rise of the Creative Class: And How It's Transforming Work, Leisure, Community, and Everyday Life* (New York: Basic Books, 2004), 434.

28. Smith, Sonnenfeld, and Pellow, *Challenging the Chip*, 357.

29. A 1970 study of the correlation between birth defects and radiation, specifically from uranium mining among Shiprock Navajo workers, found that the "association between adverse pregnancy outcome and exposure to radiation were weak," but that "birth defects increased significantly when either parent worked in the Shiprock electronics assembly plant." Similar correlations were found at other assembly plants in California and elsewhere. L. M. Shields et al., "Navajo Birth Outcomes in the Shiprock Uranium Mining Area," *Health Physics* 63, no. 5 (1992): 542.

30. "Industry Invades the Reservation."

31. "Industry Invades the Reservation."

32. Shruti Rana, "Fulfilling Technology's Promise: Enforcing the Rights of Women Caught in the Global High Tech Underclass," *Berkeley Women's Law Journal* 15 (2000): 272.

33. Sadie Plant, *Zeroes + Ones: Digital Women + the New Technoculture* (New York: Doubleday, 1997), 305.

34. US Department of the Interior, Bureau of Indian Affairs, "Industries Turn to Indians for Precision Workers" [September 10, 1965].

35. Sporck attributes the rise of Silicon Valley industry and by extension computing culture to a successful resistance to unions—this attitude informed the Californian ideology, which depended on "sweat equity" or a radically entrepreneurial stance toward both labor and capital. . . . See Gina Neff, *Venture Labor: Work and the Burden of Risk in Innovative Industries* (Cambridge, MA: MIT Press, 2012), 195.

36. Tutt, personal communication.

37. Deloria, *Indians in Unexpected Places*, 4.

38. Nick Montfort and Ian Bogost, *Racing the Beam: The Atari Video Computer System* (Cambridge, MA: MIT Press, 2009), 180.

39. Sporck attributes the rise of Silicon Valley industry and by extension computing culture to a successful resistance to unions—this attitude informed the Californian ideology, which depended on "sweat equity" or a radically entrepreneurial stance toward both labor and capital. For a description of how this logic informed the technology industries in the late twentieth century, see Gina Neff, *Venture Labor: Work and the Burden of Risk in Innovative Industries* (Cambridge, MA: MIT Press, 2012), 195.

40. Tutt, pers. comm.

41. Deloria, *Indians in Unexpected Places*, 4.

42. Mike Daisey's popular and highly regarded one-man show, "The Agony and the Ecstasy of Steve Jobs," described horrible working conditions at Foxconn's Shenzhen plant . . . , where he claimed that Apple iPads were produced by workers whose hands were permanently disfigured by injuries and repetitive stress disorder. Though it turned out that some details of the account were fabricated, the *New York Times* went on to publish a multipart investigative feature story

about working conditions at Foxconn that revealed rampant overtime work and other labor practices that show the hidden stresses of consumer electronic manufacture. Charles Duhigg and David Barboza, "In China, Human Costs Are Built into an iPad, *New York Times*, January 25, 2012; Jack Linchuan Qiu, *Working-Class Network Society: Communication Technology and the Information Have-Less in Urban China* (Cambridge, MA: MIT Press, 2009), 303.

43. Nicholas Mirzoeff, *The Right to Look: A Counterhistory of Visuality* (Durham, NC: Duke University Press, 2011), 385.

CHAPTER 12

Inventing the Black Computer Professional

Kelcey Gibbons

From the late 1950s well through the 1980s, one could find quick, punchy stories about Black computer professionals in popular publications like *Ebony* magazine and the *Baltimore Afro-American.* These informative pieces were placed alongside advertisements for Tuxedo Club Pomade, editorials on the best cities for Negro employment, and even sensational tales of mixed-race love. For contemporary readers these stories were depictions of Black life and success, both imagined and actualized, present and future. The stories within these publications reveal significant entanglements between two revolutions: the civil rights movement and the so-called computer revolution. The same publications that covered Black life, with its highs and lows, were also sharing stories of burgeoning computer systems and the heroic experts who were building modern America.

This chapter focuses on one revolution in the light of the other. As the "computer revolution" gained momentum in the mid-twentieth century, another one with much older roots in the American experience was transforming the lives of those whose dreams were violently suppressed. By exploring what I call the "invention of the Black computer professional," we can understand that the computer revolution and civil rights movement were intertwined. At the forefront of modernization, the image of the Black computer professional framed the computer as a technology of opportunity and freedom for Black people. This professional was part of a vision of Black freedom that took shape in response to the long history of structural racism that forestalled Black futures, and it deserves to be more than a footnote in the history of computing.

Different conceptions of freedom have shaped imaginaries of the computer throughout its history.[1] In the histories of computing and community told by Fred Turner, for example, the freedom associated with computing was shaped by the privilege and power of the white American middle-class. In *From Counterculture to CyberCulture,* Turner explains how the computer was a part of a radical movement that would influence the social meaning of technology for decades. He proposes that modern computing culture took shape beginning in the 1960s, when the New Left and New Communalist, and later hackers and DIY hobbyists, wanted to escape and subvert the capitalist and bureaucratic social order that they believed restricted expression, identity, and information circulation in their predominantly white middle-class lives. They believed freedom was the ability for individuals to disengage from a manufactured social order and live independently, though not in isolation: freedom would be gained by detaching information from corporate, military, and State control, and making it possible for all to participate in a new democratic information economy.[2]

The Black computer professional, however, represented a quite different kind of "freedom," one shaped by what W. E. B. Du Bois, called the "problem of the 20th century": because they were not allowed to live as full citizens, organizing for "the betterment of the Negro race" became the mission of organizations like the National Urban League (NUL) and the Black press, who were working to bring Black people into the "program of America."[3] Part of their efforts included democratizing the labor and culture of technology; these communities fought for a future in which Black people could be both Black and American, Black and skilled, Black and professional, Black and technical, and Black and middle-class.[4] The freedom sought by the NUL and the Black press was not freedom *from* corporate America and the State but rather the opposite: recognition and inclusion within them.[5] They saw the Black computer professional as a means to that end.

There are different definitions of "freedom" that circulate in the history of computing, but both Black visions of freedom and Black people have largely been neglected in this history. Recent exceptions to this tradition include the work of Ruha Benjamin, Charlton McIlwain, and André Brock. Their work challenges the absence of Black people in computing history and the narrow focus on "great white men" like Stewart Brand, Steve Jobs, and William Shockley. Benjamin shows that, far from affording all of the promised "freedoms," the computer has been a device of engineered inequality that reproduces the same biases and forms of corporate and State control and surveillance that some believed it would eliminate. Instead of democratizing information and access, algorithm-based decision systems rely on encoded racial difference and extend economic, social, and legal inequality into a new age of suppressing Black freedom.[6]

Connecting networked, "online," activism of the past to online activism in our present, McIlwain describes the use of computer technology by a collective of Black technologists in the late twentieth century. Throughout the advent of networked computing, Black technologists developed their own online spaces for community building in the late 1980s and 1990s. Far from being excluded from the digital, they sought to wield its power for Black education, culture, expression, and freedom. However, by the early 2000s, the competitive market of the early internet stymied the success of different iterations of the "Black Net." McIlwain also examines the early correlation between computers and predatory policing systems in the 1960s and 1970s and argues that for Black people, the computer was a technology of oppression despite the language of inclusion and democratization that blessed its entry into mainstream culture.[7] Focusing on online discourse and internet infrastructure, André Brock writes about technologically mediated Blackness and the creation of Black cyberculture for joy, pleasure, connection, commentary, and leisure. Through discussions of platforms like Twitter, he shows how "everyday Blacks" are defining what freedom of expression means for them via information technologies. His discussion of Black Twitter highlights both the default whiteness of the internet and how inaccurate that whiteness is when thinking about creating identity and culture online.[8]

In this chapter, I explore an imagination of Black freedom that centered the computer prior to the networked contexts that Brock and McIlwain explore, and the planning, strategies, and organizing that followed from it. This vision of Black liberation was both embodied and abstract: it was a set of people and their stories and skills, images and photographs in magazines, and it was an idea around which training, professionalization, and community building was organized. I resituate the computer in American culture not as a technology of white freedom and Black oppression, but as a technology with personal stakes for Black communities long before personal computers. This is not an account of Black people in "white" spaces. The computer professionals in this chapter were individuals important for the communities that they lived in and moved through as neighbors, volunteers, and stories.

By focusing on the institutions and ideals that framed this Black community, I look at those who fashioned the Black computer professional into a symbol of Black freedom: members of the NUL and the "Black press." In tune with the needs of its audience, the Black press, known as the "fighting press," reveals a Black public, and the kinds of stories that resonated with them.[9] As the most influential information medium for Black people, the Black press amplified the voices of its audience while framing the computer as a tool for Black freedom by focusing on skill, education, and professionalization. Black media, the "press," and special interest publications worked to connect "all walks and levels of Negro life—businessmen, teachers, laundry workers, Pullman porters, waiters, and red caps; preachers, crapshooters, and social workers; jitterbugs, and Ph.D.'s" into a community of writers, doers, and readers.[10]

Focusing on these actors in this way, we are moving away from hardware and software concerns and the default whiteness of American technological mythology. We are examining a Black response to and incorporation of the computer in social and political culture, while refusing to allow very real historical disadvantage—and the resulting scarcity of sources—to remain a tool of erasure. While there is growing literature that explores how technology reproduces oppression, here I recover Black agency and imagination in the context of early computing. Of course, this could be read as a story of "failure"—"why aren't there more Black people in STEM?"—given these early efforts to create space for Black people in technological development. But if we focus on failure, we miss the power of what I describe as Hopeful Action and imaginations of alternative futures that this historical moment has to offer.

Prelude

In the spring of 1946, the Social Sciences Graduate School at Howard University hosted a conference entitled "The Post-War Outlook for Negroes in Small Business, The Engineering Professions, and The Technical Vocations." Conference attendants included representatives from DC public schools, the National Technical Association, the Senate's Small Business Committee, the US Office of Education, and the National Urban League. In addition to discussing expanding opportunities in industry and business, attendants also identified obstacles preventing Black people from participating in the new world of technology. One of the last to speak was Lewis Downing, dean of Howard's School of Engineering and Architecture, who said: "In the field of technology there lie ahead numerous opportunities for those Negroes who are properly prepared to enter that field. Still there are problem areas involved, and certain of these problem areas were very carefully enumerated to us at this conference. We have problems of tradition to overcome; counseling, training, placement, and employer-employee attitudes. We even have the Negro public to consider in the total picture—its attitude toward the manual arts and applied sciences."[11]

Downing and others foresaw a future demand for skilled labor in new technical fields; however, problems of "tradition" were obstacles in achieving democracy and true freedom. Despite the labor demands of the war effort, employer attitudes remained hardened against Black workers. The attitudes of many Black people, whose experiences of opportunity and citizenship were shaped by legacies of inequality and systemically restricted hope, also, argued conference members, needed to change. The publics referred to at the conference needed to know that the future would be different and to plan accordingly.

Reports from experts in engineering, education, and social work described the Black public as hesitant to invest in an America that constantly pulled investments from them, an America that produced and re-created Black social vulnerability in order to maintain white supremacy. For example, Black veterans who invested in the projects of America were high-risk targets for white violence, and between 1882 and 1968 approximately 4,742 Black people were killed by lynch mobs.[12] Considering the complex history of Black labor, where stability and opportunity for the worker ebbed and flowed, historian Joe Trotter explains, "limits on African-American access to labor unions, skilled jobs, and managerial positions reinforced their high concentration in general labor, domestic, and household service occupations."[13] Even gains made during the First World War were followed by troubling reversals that worsened gendered and racialized distinctions.[14]

There were anxieties around technical labor as a continuation of manual labor—was this another era in American technology when the value of Black bodies, not minds, would be used to build the future?[15] Education offered hope, but those who gathered the financial and social recourses needed to pursue formal and advanced education often faced difficulties finding and keeping well-paying work using their specialties; those who did succeed served Black communities as educators, lawyers, and doctors.[16] Traditions of injustice and structures of confined opportunity led to a practice of encouraging upwardly mobile Black youth to privilege solid middle-class jobs like teaching over engineering and to avoid dreaming outside of what was appropriate.[17]

Furthermore, even if attitudes could be changed, and more Black youth decided to pursue technical vocations, training was needed for them to become eligible applicants. This training had to be relevant, competitive, and relatively comprehensive.[18] Similarly, placing qualified applicants into jobs that paid and treated workers well, not discriminating against or underpaying them because of race, would be challenging without continued federal support, better relationships with industry, and an adjusted public image. Therefore, new vocational messages and relationships were needed to help youth find opportunities in the postwar era of technology.

Into the 1950s, social and professional organizations for Black people continued to challenge these prejudicial pasts as they funneled their resources into the fight for equality. A key organization was the National Urban League. In the mid-twentieth century, the NUL had 51 affiliate offices based in Black neighborhoods throughout 31 states. As federal pressure mounted on leading computing manufacturers to be more inclusive, the NUL's position was one of mediation and coordination. Its broad coalition of organizers positioned the NUL to mediate between the big picture of integration and equality and local enactments of those ideals. Between the national and local offices, the NUL focused on problems unique to individual Black communities and the nationally dispersed Black community.

What follows are three vignettes of "Black computer professionals." By looking at the work of the NUL and the kinds of stories published by the Black press, we can see both the *material* invention of the computer professional (via training and occupational opportunities) and its role as a *symbol* of modernity and progress for Black people. In line with the central thesis of this volume, the Black computer professional was both an embodied and an abstract entity. The first vignette looks at computer professionals in specific contexts of technological development and shows how the NUL looked for ways to *train* Black professionals by building institutional networks and relationships. Alongside this effort, the press looked for ways to *represent* Black professionals, bringing Black and American identities together for a new world of technology and freedom. The second vignette focuses on a national program launched by the NUL to connect Black people who had real or potential skills with jobs in computing. The third describes an NUL community initiative that used local social and educational institutions to create opportunities for students. Each vignette demonstrates Hopeful Action, and the kinds of relationships, networks, and strategies leveraged in its service, many of which were part of the fabric of "Black community."

The Southern Industry Project

In the summer of 1957, as a part of the NUL's Southern Industry Project, representatives from the Jacksonville League and the Atlanta regional offices surveyed Patrick Air Force Base in Cocoa Beach, Florida.[19] Radio Corporation of America (RCA) and Pan American Airways (PAA), in conjunction with the Air Force, had a presence on the base. They were developing technologies for missiles and spacecraft launching, tracking, and control.[20] NUL representatives reported that thirteen Black employees were working as stenographers and clerks and one as a mathematician. The local Jacksonville and Tampa branches were responsible for these job placements, as they had developed a relationship with on-base management. In the report that followed the visit, it was noted that RCA and PAA would need more physicists, mathematicians, engineers, secretaries, stenographers, and clerks. Encouraged by this information, the industrial field secretary for the NUL, Mahlon Puryear, organized a meeting between staff from HBCUs and department heads at RCA and PAA. He had three goals: to help the HBCUs to improve their curriculum, establish a relationship between the HBCUs and staff at the base, and put the HBCUs on recruitment lists for RCA and PAA.[21]

On January 15, 1957, representatives from five HBCUs arrived at the base. Having spent the night at the nearest HBCU, about 100 miles away, they discussed the kind of work RCA and PAA were doing, the current job openings for both companies, and what schools should stress in educational programs to meet the needs of these employers.[22] They also identified mechanical and administrative positions as a good point of entry for Black workers because of a growing need for such workers and the possibility of promotion into more advanced positions. The next phase of this meeting involved dividing attendees into two groups. One was composed of professors who specialized in mathematics and science, and the other was made up of professors of business administration, economics, and secretarial science.[23] The administrative manager for RCA outlined the work of the scientists at RCA for group one and explained how mathematicians with PhDs create formulas and how those formulas are then passed down to junior workers with BS degrees for programming and coding the computers RCA used. The RCA representatives recommended very specific kinds of courses for the HBCUs to prepare students for future employment. Group two focused on the benefits of working at the base, which included insurance, retirement, vacations, holidays, and overtime pay: perks that helped generate generational wealth and middle-class leisure.

Of course, the educators who visited the base had an advantage over those who had not. However, some aspects of high technological work could be found in Black newspapers, making the importance of the field and the valor of its professionals available to all readers. For example, in 1954, the *Chicago Defender*, a nationally distributed newspaper, published an article entitled "Bureau of Standards Sets Pace in More Than 1 Way." With generous photographs of Black technological experts, the text describes the work of "75 highly trained and highly skilled Negroes."

According to the story, machines bring order and consistency to the life of the consumer. Some calculate, verify, and "serve the world." The most useful machine for the work of the National Bureau of Standards (NBS) was one that was also important for "the layman," the Black reader, and citizen: "the thinking machine." Computers helped men order the world; some of these were people who looked like the readers. Even so, according to the article, race in the federal workplace was not *more* important than expertise; "Here there are neither Negroes nor any other different species. They are simply scientists contributing their abilities to the sum of the whole."[24] The article expressed a hope that skill-based meritocracy might create opportunities for Black people in this workplace, that it might dismantle the barriers of American racial hierarchy. Meritocratic accounts of skill-based inclusion are, in fact, often used to justify and perpetuate white dominance, but in this historical moment, Black organizers saw it as an opportunity for inclusion, if Black people could develop the skills.

A year later, 1955, in the *Baltimore Afro-American*, underneath the heading "Employees Prove Again That Democracy Works," the captioning image showed readers a mixed crowd of smiling employees at messy desks. This is what an integrated workplace with "a permanent staff of approximately 4,000 persons, a large percentage of whom are colored . . . holding responsible positions . . ." looked like. This particular story also pulls together two central ideas about freedom in the history of computing more broadly in this narrative of Black American labor. The article singles out an individual employee as a symbolic individual that readers could identify with: "Robert P. Stephens, 3742 Hayes St., NE, holds one of the important positions in the Bureau. He is a senior engineer in UNIVAC division. UNIVAC is an electronic automatic computer which 'frees the human mind for creative use by performing at rapid speeds almost unbelievable feats of computation, sorting and classifying' material both alphabetically and numerically."[25] Computing work was not just about freeing Black bodies from the undemocratic confines of the "color line."[26] The computer could free Black minds just as it was imagined to free white minds from tedious tasks, including computation, making room for creativity. The *Chicago Defender*'s 1954 article, and many other stories in the 1940s onward, has the theme of freedom for Black minds. Under the subheading "Cherish Freedom," explaining a benefit of working at NBS the article posits "freedom of thought and initiative" for employee retention.

Freedom of mind and creativity are not separate from the long-fought civil rights movement. At the 1942 "We Are Americans, Too" conference in Chicago, Dr. Lawrence Ervin argued that democracy "respects the personality of every individual" and it "directs him to respect himself and to make the best of his own natural gifts, to develop his own unique personality for the benefit of the whole."[27] The March on Washington Movement's push for skilled work and the NUL's networking for professional status was not only about better-paying work and recognition as a contributing member in American society. Their efforts also included the ability for Black people to have work that was fulfilling, important, and gave them room to express their innovative natures.

Under the heading "Publications Library Head," a story in *Ebony* magazine (fig. 12.1) emphasizing skill via promotion, middle-class status through leisure activities, and relationship with community reads:

PUBLICATIONS LIBRARY HEAD

Dorothy J. Boler, 25, publications librarian for International Business Machines (IBM), in Jackson, Miss., supervises the operation of the firm's data processing and periodicals library. She also assists computer salesmen and systems engineers in preparing IBM programs for both internal and customer use. Miss Boler, who holds a bachelor's degree in education from Tougaloo College, joined the firm three years ago as a secretary. A member of the Jackson Urban League and the United Negro College Fund, she is still single and enjoys traveling and playing tennis.

Figure 12.1. Dorothy J. Boler of IBM.

Dorothy J. Boler, 25, publications librarian for International Business Machines (IBM), in Jackson, Miss., supervises the operation of the firm's data processing and periodicals library. She also assists computer salesmen and systems engineers in preparing IBM programs for both internal and customer use. Miss Boler, who holds a bachelor's degree in education from Tougaloo College, joined the firm three years ago as a secretary. A member of the Jackson Urban League and the United Negro College fund, she is still single and enjoys traveling and playing tennis.[28]

In many stories through the 1950s and 1980s, the Black computer experts, be they programmer, engineer, technician, or secretary, had an artistic, creative, heroic mystique: "These earnest young people who play such highly important parts in making the nation run are the unsung unknown everyday heroes who want no fanfare about their work." The Black employees in these offices were examples of what was possible, and what the future would hold for African Americans. Giving voice to the sentiments of these professionals, the 1955 article ends with a quote from an unnamed employee: "It makes us feel that we are not Americans Too. We are Americans, period."[29]

The National Skills Bank

As the Black press was advertising images and narratives of Black success and progress, the NUL was going plant to plant, factory to factory, office to office working to make the Black professional possible. In 1963 the NUL established the National Skills Bank to create a database of people who had "actual or potential skills for sale." Trying to promote efficient and optimal job placement, the Skills Bank connected industries and companies looking to hire qualified Black people with those who could do the work, but it also went further. Utilizing the affiliate offices in NUL cities around the country, the Skills Bank established 56 skills bank locations in 31 states. These offices would identify:

- technically and vocationally trained men and women
- skilled and semi-skilled craftsmen
- clerical and related workers
- professional and semi-professional workers
- candidates for apprenticeship and related training programs
- candidates for on-the-job and other management-sponsored training programs.

Not just for Black people, the skills bank was for all minorities who wanted more than "token" employment: meaningful work and upward mobility.[30]

The success of the Skills Bank depended on its ability to have both applicants looking for new jobs and employers looking to hire Black people. To organize this effort, Skills Bank representatives would start by visiting their community League offices, whose staff were intimately aware of the best ways to locate and recruit local talent. Together they would conduct employment surveys to identify local prospects for job development and placement while working to build new and stronger relationships with local industry. Important tools for sharing the promises of the Skills Bank and motivating people to send in their resumes were NUL publications, newspapers, and community events. Black newspapers and some white newspapers in cities that had League offices provided an overview of the program as well as some stories of those who benefited from its efforts.

For example, an article published in 1964 in the *Pittsburgh Courier*, under the heading "Urban League's Skills Bank Aiding Thousands: Talented Applicants Have Been Placed in 5,000 Good Jobs," reads: "Computing Progress—Doris Wallace, accounting assistant at Wisconsin Telephone Co., Milwaukee is a programmer analyst working with IBM No. 1401 computers to develop improved accounting procedures. Miss Wallace, a top graduate of Talladega College, holds a master's degree from the University of Wisconsin."[31]

The Skills Bank was developed during a time when some industries were taking control over how they reached the equal employment opportunity threshold as the law required it.[32] As a part of the President's Committee on Equal Employment Opportunity (PCEEO), federal contractors, leaders in commerce and industry, and universities could voluntarily participate in the Plans for Progress program. Established the same year as the affirmative action mandate, Plans for Progress's voluntary compliance promised to encourage equal opportunity hiring practices within its own ranks.[33] As the NUL already was assisting federal contractors with the hiring of minorities, Plans for Progress companies relied on their NUL contacts for minority staffing and integration.

Representatives from RCA and IBM also served on the Skills Bank advisory board at the invitation of the NUL. Staff from the computer industry were also involved in training and outreach programs organized by the Skills Bank that were actively looking to place people in computer operator, programmer, technician, and engineer positions. In 1964, sponsored by the NUL in cooperation with IBM and funded by the Bureau of Apprenticeship and Training, the Customer Engineer Project (BAT-IBM) was a training program designed to prepare students for IBM computer maintenance jobs.[34] IBM provided training programs with the NUL, including secretarial training, computer operator, and programmer courses.

The NUL facilitated the ability for many Black people to move up the employment hierarchy by hosting conferences, vetting resumes, and processing applications for employment. It strove to be an organization that made Black workers competitive. It contributed to the creation of the Black computer professional by working to change employer attitudes and build relationships between HBCUs and technical industries; it consisted of counselors and placement specialists. It helped to integrate places of professional work and the opportunities for Black people to become professionals in the mainstream labor market. It also enlisted the help of other local community organizations in bring the dream of professionalization to Black youths.

Job Opportunities Clinic

In addition to making itself important for some of the big names in computing and electronic manufacturing, a key strength of the NUL's strategy was in different kinds of programming that connected Black betterment and social organizations to its work of making computer professionals.

For example, in 1964, the NUL's Southern Office and the Delta Sigma Theta Sorority, Knoxville Alumnae Chapter, hosted a Job Opportunities Clinic in Knoxville at Austin High School.[35] The joint goal of the clinic was to connect local students with jobs available in the community. Those who attended received a packet, the first page of which had a brief but loving note for the students:

> Dear Student,
> On the accompanying sheets you will find a list of some of the jobs that are open here in this community to qualified high-school graduates and to college students. We want you to know that many more job opportunities are opening to you daily notwithstanding race or creed. Your qualifications are your only limitations. Finish high-school and encourage your friends to do the same in order to qualify for better jobs! Sincerely yours, Delta Sigma Theta Sorority

Among the jobs available through the local League employment office were engineering and science technicians; operators for office machines and electronic computers; business machine servicemen; maintenance electronics; computer system operators; business machine operators; secretaries; and clerks. The Clinic also had occupational sessions where students could talk to a professional representative from different industries including IBM.[36] Other sororities and fraternal societies also worked with the NUL on a variety of vocational projects, including the Black Executive Exchange Program and the Scientists and Technicians of Tomorrow program.[37]

The Urban League ran a national youth incentive program called TST that organized science and technology career clubs in junior high and high schools nationally. A photo of a TST member was published in *Ebony* magazine (fig. 12.2). The caption reads, "High school student Ralph Temple learns about Honeywell's complex equipment during visit arranged by local Urban League's TST (Tomorrow's Scientists and Technicians) aid program for promising youths. Prince is a TST supporter."[38]

The NUL also recruited individual professionals who were members of Black communities, adding a second layer of responsibility. Not only were the professionals engaged in the work of integrating into the workplace, often working as perhaps the only Black employee, they were also community workers who participated in racial uplift by volunteering with and donating to organizations like the NUL. Some even let their image become an example of what was possible.

For those who attended the postwar meeting at Howard University, every Black professional was seen as an argument against inequality and a victory benefiting the race. Likewise, and contrary to the meritocratic myths that were used to keep so many Black people out of technical positions, being a Black professional was not just an individual position; it meant being creative and skilled on multiple fronts: in the office as a professional and in the community as a symbolic individual. It meant embodying the requisite skills and serving as an abstract idea for communities to organize around. Equality activists and professionals of the computer age had the opportunity and responsibility to make

Figure 12.2. Ralph Temple participates in the Urban League's "Tomorrow's Scientists and Technicians" program.

the computer matter in ways that shifted focus from what the machine can do to what the people who work with the machine are and can be for their communities.

Conclusion

Narratives of failure and success abound in the history of computing, but these are relational concepts, moving targets, porous, and open to alternate readings.[39] Given where we ended up—the underrepresentation of Black people in the computing industry and the tech industry's reproduction and maintenance of inequality and racism—what are we to make of the Black computer professional and the power of Hopeful

Action?[40] Was this just another account of "failure" for the footnotes of a whitewashed trajectory? I argue not. First, this is a story of an innovative process that is still in action today: professionalization messages around computer training and technological opportunity are ubiquitous on minority-serving community college campuses, social media, and in generational advice. Second, thinking of what professionals matter, the significance of computing in Black history cannot be limited to the contributions of a talented few but the many social innovators—the librarians, secretaries, technicians—whose presence in the American workplace challenged the sanity of segregation. Computers also played an important role in Black communities, representing both material possibility and the African American middle class, with all of the freedoms that they believed this kind of citizenship could afford. Computers, it seems, were a nexus around which multiple definitions of freedom took shape. Freedom, like failure, is a relational term.[41]

By keeping community in view, we have new ways of understanding innovation and success: the traditional locations of technological innovation, that have historically excluded Black people, are not the only places were technology is made, used, and gains its character.[42] The imaginations and actions of those whose worlds are obscured by historical blindness continue to shape cyber- and technoculture today. Viewing computing through the lens of freedom, and all that that meant for Black people from the 1940s through the 1980s, allows us to move beyond narratives of deficiencies or the digital divide—we can step back from the great white wall of American technological mythology and disrupt its power. Likewise, bringing the civil rights movement and the computer revolution (these simultaneous and overlapping moments of great change) together, we can understand that equality professionals, were co-creating new Black and American identities—with technology, and its symbolic and economic power, at the center.

Notes

1. Christopher Kelty, "Fog of Freedom," in *Media Technologies: Essays on Communication, Materiality, and Society,* ed. Tarleton Gillespie, Pablo J. Boczkowski, and Kirsten A. Foot (Cambridge, MA: MIT Press, 2014), 195–221.

2. Fred Turner, *From Counterculture to Cyberculture: Stewart Brand, the Whole Earth Network, and the Rise of Digital Utopianism* (Chicago, IL: University of Chicago Press, 2006).

3. W. E. B. Du Bois, *The Souls of Black Folk: Essays and Sketches* (Chicago: A. G. McClurg, 1903; New York: Johnson Reprint Corp., 1968); Herman Branson, "The Role of the Negro College in the Preparation of Technical Personnel for the War Effort," *Journal of Negro Education* 11, no. 3 (1942): 297–303, doi:10.2307/2292666.

4. Tamara Brown, Gregory Parks, and Clarenda Phillips, *African-American Fraternities and Sororities: The Legacy and the Vision*, 2nd ed. (Lexington: University Press of Kentucky, 2012), 75–100, 141–42, 213–30.

5. Robin Kelley, *Freedom Dreams: The Black Radical Imagination* (Boston, MA: Beacon Press, 2002).

6. Ruha Benjamin, *Race after Technology: Abolitionist Tools for the New Jim Code* (Medford, MA: Polity Press, 2019).

7. Charlton McIlwain, *Black Software: The Internet and Racial Justice, from the Afronet to Black Lives Matter* (New York: Oxford University Press 2019).

8. André Brock, *Distributed Blackness: African American Cybercultures* (New York: New York University Press, 2020).

9. Charlotte O'Kelly, "Black Newspapers and the Black Protest Movement: Their Historical Relationship, 1827–1945." *Phylon* 43, no. 1 (Spring 1982): 13.

10. Philip Foner and Ronald Lewis, *The Black Worker from the Founding of the CIO to the AFL-CIO Merger, 1936–1955* (Philadelphia, PA: Temple University Press, 2019), 251–52; O'Kelly, "Black Newspapers and the Black Protest Movement,"13.

11. Graduate School, With the Cooperation of the Howard University School of Engineering and Architecture, and a Special Community Advisory Committee, "The Howard University Studies in the Social Sciences: The Post-War Outlook for Negroes in Small Business, The Engineering Professions, and the Technical Vocations, Papers and Proceedings of The Ninth Annual Conference of The Division of The Social Sciences," *Graduate School Publications* 6 (April 9–11, 1946): 175.

12. Benjamin Bowser, *The Black Middle Class: Social Mobility—and Vulnerability* (Boulder, CO: Lynne Reiner, 2007), 46.

13. Joe Trotter, *Workers on Arrival: Black Labor in the Making of America* (Oakland: University of California Press, 2019), 90.

14. For an explanation of racialization, see Lisa Nakamura in this volume; Robert Zieger, *For Jobs and Freedom: Race and Labor in America Since 1865* (Lexington: University Press of Kentucky, 2007), 139–40.

15. Carroll Pursell, *A Hammer in Their Hands: A Documentary History of Technology and the African-American Experience* (Cambridge, MA: MIT Press, 2005), introduction.

16. For jobs analysis, see "I am Free, Black and 21: How Should I Earn a Living?" *The Crisis* (April 1933): 79–80.

17. Alphonse Heningburg, "The Future Is Yours," *Opportunity Journal of Negro Life* (Fall 1945): 181.

18. Graduate School, "The Post-War Outlook for Negroes in Small Business," 93.

19. The Southern Industry Project (1950–1957) was designed to examine the extent of Black employment in Southern industries and to cooperate with the heads of these industries to ensure the presence and success of democratic hiring and the promotion process. Initiated and concentrated in the South this project also examined industries located in the North.

20. *RCA Engineer* 24, no. 3 (October–November 1978), 27.

21. National Urban League Records, "Memorandum January 15th, 1952" and "A Visit to Guided Missile Plant Patrick Air Force Base Cocoa Florida," National Urban League, Part 1: Box II-D32, *For Development and Employment 1962–1963 Southern Industries Project.*

22. National Urban League Records, "Memorandum."

23. National Urban League Records, "Memorandum."

24. Ethel Paynk, "Bureau of Standards Sets Pace in More than 1 Way," *Chicago Defender,* March 27, 1954, 4.

25. "Inside U.S. Census Bureau: Employees Prove Again that Democracy Works," *Baltimore Afro-American,* December 10, 1955, 7.

26. W. E. B. Du Bois, *The Souls of Black Folk.*

27. Doctor Lawrence M. Ervin, Eastern Regional Director of the March on Washington Movement at the "We Are Americans, Too" Conference, held at The Metropolitan Community Church, Chicago, Illinois, June 30, 1943, and "Call To The March," *The Black Worker,* May 1941; Foner and Lewis, *The Black Worker,* 253, 262.

28. "Speaking of People," *Ebony,* August 1971, 7.

29. Paynk, "Bureau of Standards Sets Pace in More than 1 Way," 4; "Inside U.S. Census Bureau: Employees Prove Again that Democracy Works," 7.

30. National Urban League Records, "Six Months Activity Report National Skills Bank Project" and "The Economics of Equal Employment Opportunity by Brown University," National Urban League Part II: Box II-A45, *Program Department National Skills Bank Project, 1962–1965 Correspondence Concerning.*

31. "Urban League's Skills Bank Aiding Thousands: Talented Applicants Have Been Placed in 5,000 Good Jobs," *Pittsburgh Courier,* December 19, 1964, 15.

32. Executive Order 10925, signed by President Kennedy in March 1961.

33. "Kennedy's Equal Opportunity," *Jet,* April 12, 1962, 26–27; and "Taylor Still Aids Plans for Progress as a Consultant," *Ebony,* September 1967, 72; Zieger, *For Jobs and Freedom,* 153.

34. National Urban League Records, "Report of The Customer Engineer Project December 2nd, 1962," National Urban League Part II: Box II-A45, *Program Department: National Skills Bank 1962–65.*

35. The Delta Sigma Theta Sorority, one of the Divine Nine Black Greek letter organizations, originated because of racial exclusion from white Greek societies and made Black betterment one of its guiding principles. From its founding in 1913, Delta Sigma Theta has been involved in civil rights movements including the Women's Suffrage and the March on Washington.

36. National Urban League Records, "Delta Sigma Theta Sorority Knoxville Alumnae Chapter, Job Opportunities Clinic Worksheet 1964," National Urban League Part 2: Box D32, *For Development and Employment,* Southern Regional Office.

37. National Urban League Records, "Scientists and Technicians of Tomorrow Memorandum, San Diego Urban League June 1962," National Urban League Part 2: Box 1-D6, *Miscellaneous.*

38. "End of an Era," *Ebony,* December 1960, 76; "Tomorrow's Scientists and Technicians," *Journal of the National Medical Association* 50, no. 4 (July 1959): 285; National Urban League Records, "Memorandum," National Urban League Part 1: Box1-A52, *Tomorrow's Scientists and Technicians 1958 January–April;* "Memorandum," National Urban League Part 1: Box1-A67, *Vocational Services Office Memos and Reports 1960.*

39. See Mar Hicks and Lisa Nakamura in this volume.

40. Virginia Eubanks, *Automating Inequality: How High-Tech Tools Profile, Police, and Punish the Poor* (New York: St. Martin's Press, 2018); Simone Browne, *Dark Matters: On the Surveillance of Blackness* (Durham, NC: Duke University Press, 2015).

41. A theme also explored by Marc Aidinoff in this volume.

42. Rayvon Fouché, "Black Vernacular Technological Creativity," *American Quarterly* 58 (3): 639–61.

CHAPTER 13

The Baby and the Black Box

A History of Software, Sexism, and the Sound Barrier

Mar Hicks

In 1969, the year that the United States landed on the moon, a joint French-English project dubbed "Europe's Space Race" for its technological complexity and high political stakes, took off from Filton Airfield near Bristol. This plane, the Concorde, would not only break the sound barrier, it would reshape what was previously thought possible for international civilian air travel. By 1976 it was put into service as a passenger jet that could surpass Mach II, or twice the speed of sound.

This Anglo-French technological alliance had succeeded in creating the first, and to date only, supersonic passenger plane. Flying for British Airways and Air France, it could ferry passengers between New York and London, or New York and Paris, in under four hours—less than half the time of a normal jet. In an age before constantly online connection was common and a digital, wired world allowed employees of major corporations to collaborate across time zones and vast distances, the Concorde seemed to represent the vanguard of interconnectedness for business elites and the very wealthy.

Today, for people in the most privileged job classes, the importance of being physically present has receded; they can work together using the internet to collaborate daily or even cement multimillion-dollar deals. Due to the coronavirus pandemic, and the US and UK's mishandling of the crisis that caused death tolls to soar, interest in working from home has never been higher. Many tech corporations formalized the once informal measures that allowed their white-collar engineering staff to occasionally work from home. Meanwhile, teachers, nurses, waitresses, retail workers, and many others in traditionally feminized and service-oriented jobs have found themselves thrust into harm's way in order to stave off the complete breakdown of the US economy and health care system. At the same time, the majority male workforces in white-collar prestige positions for large multinational tech corporations have been granted the privilege to work from home for the foreseeable future, and perhaps indefinitely.[1]

The current situation represents a stark reversal from the mid-twentieth century, when office attendance was mandatory for most white-collar workers, face time was crucial for promotion, and provisions for working from home were usually nonexistent. During that period, the people clamoring for work-from-home arrangements were mostly women. Though working mothers of all social classes sought such accommodations, it was white-collar women in privileged fields, like computing, who got the furthest. In the 1960s, one such group of British women, after being discarded by employers who would not allow them to be both mothers and tech workers simultaneously, helped create a new model for working from home—while also helping design the fastest, most technologically advanced passenger plane in existence. These women programmed from their kitchen tables and living rooms, conducting meetings and code reviews by telephone, and in so doing gave us a preview of the future of tech work over half a century ago.

Modernizing a Stagnant Economy

As British leaders and citizens put the devastation of World War II behind them, the nation confronted an economically grim future. Through the 1950s, British austerity measures aimed at paying off wartime debts plunged many Britons into hardship. Food rationing persisted well into the 1950s, hurting the standard of living in peacetime. Rationing of fuel also meant that people were freezing in their own homes: during the winter of 1952, this led to thousands of fatalities when dirty-burning, lower-quality coal kept for the domestic market (so that higher quality coal could be sold abroad to pay down war debts) caused a deadly week-long smog that stopped all air and sea traffic in and out of London and led to the first clean air legislation in the world.[2]

Against this backdrop, the United Kingdom struggled to maintain some semblance of its former power. As technology and political might became increasingly intertwined during the nuclear arms race of the Cold War, the British government hastily set up a reactor in the north of England at Windscale to enrich uranium for bombs under the pretense that it was only a power plant. In their race to compete with the Soviets and to prove themselves to their American allies as nuclear equals, the UK Atomic Energy Authority literally cut corners on the project—shaving off the fins that allowed the uranium rods to dissipate heat properly—and caused a near catastrophe in 1957 that still ranks as one of the worst nuclear disasters in the world.

At the same time, British leaders increasingly began to understand that their economic future lay more in Europe than in the "special relationship" with the United States. Yet Britain tried to join the Common Market, or European Economic Community (EEC), in 1961 only to be vetoed by France, whose leadership doubted the United Kingdom would cooperate with—instead of steamroll—its European partners. In part as a result of this rejection, in 1962 the UK entered into a high-tech partnership with France as a show of cooperation and goodwill, which would become one of the most unique and ambitious international technological partnerships of the twentieth century.[3]

Aimed at showing off the nations' joint technological prowess through creating the world's first supersonic passenger jet, the project was half technological and half political. It was dubbed "Concorde" to reflect that it was meant to represent an alliance. Responsibilities for who would provision the equipment for each part of the airplane were carefully divided up along national lines—no small feat, given the interconnectedness of the plane's systems and its novel, untested design.[4]

Relations between the two nations became heated at more than one point in the years-long design and production process, with the United

Kingdom repeatedly fretting about cost overruns, while the French were determined to forge ahead. The British government seriously considered backing out in 1965 due to spiraling costs and setbacks with the design, but it did not dare because it was still seeking entry into the Common Market. British officials knew that pulling out of the Concorde project would create negative reverberations for their EEC application, proving French president de Gaulle's point that the UK could not really be trusted or relied upon to integrate itself with European goals.[5]

Portioning out the contracts to the companies that would build the different parts of the Concorde gave rise to a telling episode in the history of computing and labor that foregrounded gender, childcare, and early work-from-home jobs in cutting-edge, high technology. On the surface, this episode shows women's contributions to early computing, and is an important historical lesson on how gendered labor discrimination in high tech hurt national fortunes on both an economic and political level.[6] It shows how an explicitly feminist way of organizing a software company undid some of this sexist damage to allow the United Kingdom to advance technologically—and diplomatically—on the world stage. Below the surface, this story also highlights the contradictions inherent in technocratic, meritocratic ways of overcoming discrimination. The women in this instance were highly successful within a specific time frame and a specific situation of labor shortage that had been caused by Britain's attempts to push women technologists out of the workforce, yet their success could not undo the larger structural discrimination that it skillfully skirted.

This historical episode also connects the lived realities of many women's lives to the material needs of modernizing nations and offers insight into how telework, remote work, and "working from home" became an engine of the global economy—not just a preferable arrangement for workers. Today, "wfh" is a way to retain and foster talent, to save lives in a pandemic, and to bring technological collaborations to fruition that would have otherwise been impossible if the work were bounded by the limitations of physical presence and national borders.

Both then and now, relatively privileged workers are able to benefit from this system of labor hierarchy that allows some to work from home. This history shows how a group of women in the United Kingdom at the beginning of the computer age paradoxically succeeded in raising their own fortunes right as they found themselves being pushed out of their office-based tech jobs, by using a feminist business model that revolved around working from home.

Feminism in Technology

In 1962, a young woman named Stephanie Shirley (now Dame Shirley) found herself hitting the glass ceiling in her work in government, and then also in industry. Shirley was a child refugee of the Holocaust who had been brought to the United Kingdom as one of the 10,000 German Jewish children on the Kindertransport to escape genocide. Shirley often remarked that these grim beginnings deeply influenced her drive to succeed, saying that she felt she needed to make her life worth saving since so many others like her had died in Nazi extermination camps.[7]

As a technical worker and computer programmer, Shirley worked at Dollis Hill Research Station and later in industry with some of the brightest minds in UK technology. But despite her drive and ambition, she found herself passed over for promotion again and again. "The more I became recognized as a serious young woman who was aiming high—whose long-term aspirations went beyond a mere subservient role—the more violently I was resented and the more implacably I was kept in my place," Shirley recalled.[8]

After marrying at the age of 26, Shirley soon left her job, exasperated with the roadblocks put in the way of qualified women professionals, and had a baby. But she did not, like so many other professional women with family responsibilities, drop out of the workforce. Instead, with less than £100, Shirley set up her own software company in her home.

It was one of the first freelance programming companies in the world. She gambled on the fact that government and industry would not be able to write their own software in-house effectively and quickly enough and would need to outsource much of their programming work. Shirley knew firsthand how government and industry employers were pressuring women—who at this time made up the majority of programmers—out of their jobs. There was a growing programmer shortage as women left or were forced out, but Shirley realized these women were also a ready pool of labor.

Using her nickname "Steve," Shirley sidestepped clients' initial sexism by signing a "man's" name to her letters soliciting contacts and contracts, allowing her young company, then called Freelance Programmers Limited, and later called F International, to get off the ground. Shirley had stationery for the company made up in all lowercase letters, both because that was the style at the time, and also as an inside joke: "because we had no capital at all," she quipped.[9]

Shirley's company was unique because it was not just business as usual except with a woman in charge.[10] Shirley set her company up as an explicitly feminist business enterprise—one that would employ some of the many women like herself who had all of the technical skills to do this very much in-demand programming work but who were being shunted out of the labor force because they had gotten married or had children.[11] At the start, Shirley managed a remote team of women programmers while working out of her home—she would play a recording of typing sounds in the background when answering the phone in order to sound professional and cover any noises her young son might make.

Shirley's first major "help wanted" advertisement for her new company, published in the *Times* of London in 1964, noted that her company had "many opportunities for retired programmers (female) to work part-time at home." In this era, the term "retirement" was used to describe the situation women, who were as young as their twenties, found themselves in when they were all but required to leave the workforce upon getting married (fig. 13.1). Shirley's ad noted that the company offered "wonderful" opportunities but would be "hopeless for anti-feminists," clearly laying out the feminist mission and subtly alerting potential recruits that the company had a woman boss.[12]

Figure 13.1. A young woman named Anne Davis wears a punch tape dress at her "retirement party" as she leaves her job at ICL, a major British computer company, to get married.

ICL News, August 1970.

The Workplace of the Future Is the Home

In 1966, Shirley's four-year-old company—through a combination of networking and an excellent record of delivering on-time software—snagged one of the most prestigious programming projects in the nation: writing the software for the Concorde's black box flight recorder. The recorder would be used not simply in the event of a crash but during the whole design process of the plane to analyze test flights and help make adjustments to the plane's design. As such, it needed to be completed quickly, and work flawlessly.

The technological feat of the Concorde required bespoke equipment—everything from the downward sloping nose that moved out of the way to give pilots a better view of the ground as the plane landed at an uncharacteristically steep angle, to special paint to withstand the heat of flying for hours at over a thousand miles per hour. It required massive-yet-lightweight brakes, cabin radiation monitors, and windshields that could deflect a bird strike at over 500 mph, in addition to the innovative "delta wing" design to minimize drag and make the most of the massive thrust put out by its high-powered engines.

Its cockpit and cabin contained over 40,000 sensors and instruments whose readings would be funneled to the black box. Like the Space Race, the technologies designed for the Concorde trickled down into other areas. The heat-resistant rubber designed to keep its door seals intact under the friction of hours-long Mach 2 flights was adapted for use in premature infant units in hospitals because the high heat it could withstand made it possible for it to be easily sterilized.[13]

These technologies all had to be created in conversation and strict cooperation with the French side of the team, otherwise the plane would literally fall apart in the testing phase. A misunderstanding or measurement error could scrap months of work or even potentially destroy a test plane. To that end, the British and French teams set up control and command centers in Bristol, UK, and Toulouse, France, connected via a high-capacity telecommunications link, installed with the help of the British and French Post Offices. It carried communications of all kinds: from teleprinter and telephone messages to magnetic tape data transmissions, as well as facsimile transmissions of designs and drawings.[14]

Likewise, Steve Shirley's programming teams worked together linked by phone—an early requirement in her help-wanted listings was that the applicant needed to have access to a phone—but unlike the British team working on the Concorde with the French team via telelink, Shirley's workers remained in their homes, for the most part. They didn't work together in a physically centralized office space, as was the standard for white-collar professional jobs at the time. If what the main Concorde teams were doing with remote collaboration to accommodate an international partnership was unusual, Shirley's teams were going even a step beyond that in the service of accommodating women workers.

A rare image of Shirley and her staff in a machine room (fig. 13.2), using rented computer time to test and debug programs, shows the way programming in the 1960s had a completely different workflow from contemporary software development—one that was far less forgiving and required more linearity and less computer usage. One employee of Shirley's, whom Shirley recalled as being a smart programmer, was nonetheless let go for using far too much expensive computer time to test and debug her programs.[15] Writing bug-free code without much

Figure 13.2. Steve Shirley, Ann Moffatt, and co-worker Dee Shermer renting time on a mainframe, 1960s.

Photo courtesy of Ann Moffatt.

computer time to test and debug it was the mark of an excellent and careful programmer.

Baby Ex Machina

The woman who led the Concorde black box project was an experienced programmer and new mother named Ann Moffatt (née Leach). Thanks to Shirley's feminist management style, Moffatt did not get shunted out of her career like so many other women, who were encouraged or forced to quit upon marriage or having children.[16] Instead, Moffatt could work from home and balance the needs of her family with her career as a programmer.[17]

Moffatt was born in 1939, and perhaps fittingly one of her earliest memories was being put in the back of a plane to sleep while her father repaired spitfires on a British base during the war. The plane she was in was meant to stay in the hangar, but she woke up to find herself in the air. The horrified pilot then realized he had accidentally brought along a baby stowaway. Meanwhile, on the ground, her father was frantically looking for her. The plane turned around and "hastened down to the ground." She was only two years old.[18]

Growing up, she excelled at school, getting a scholarship to a prestigious grammar school. She recalls that her parents, however, didn't see the point: as a girl she was expected not to need much education, while her less academically gifted brother was encouraged throughout his schooling. Although her teachers argued for her to stay in school and then go on to university, her mother took her out of school at age 17, before she had earned the high school certificate needed to apply to college. She began working as a department store clerk, earning her school certificate at night school by passing exams in applied math, pure math, physics, and geography.

Moffatt then sought work at the UK government's Meteorological Office, in a role that was at that time usually given to men, explaining that she chose the job specifically because it offered her time off to study—her eventual goal was to get a university degree. "I was peculiar," recalled Moffatt, "because all the scientists were men . . . and so

I was looked on as this frivolous little girl who was going to uni and was clever at maths and so on, but was funny—with stiff petticoats, flouncing round the corridors and so on."[19]

From there, she went on to work at Kodak, not yet having a degree but having scored highly on the programmer aptitude test from IBM. She was offered a good job with a high salary in 1959 and began working in statistical analysis and later operations research. While at Kodak she was sent to a computer course at the British computer company Ferranti, where one of the teachers was Conway Berners-Lee, father of Tim Berners-Lee. Mary Lee Berners-Lee, Tim's mother, also worked at Ferranti as a programmer and was part of an effort to get equal pay for women there.[20]

Though Moffatt enjoyed her work immensely, at 25 she had a baby and her husband, who worked at Kodak as a chemist, did not want her to go back. But just before she had left Kodak, a co-worker had given her a newspaper cutting about a company that employed women programmers to work from home. As Moffatt explained, "The issue then was that they couldn't train enough programmers. There was a tremendous dearth . . . women programmers were leaving work and having babies and their skills were lost to the industry."[21]

After Steve Shirley started her own company, she began collecting more and more of these valuable workers, including, by 1965, Moffatt. For the first several years, Shirley managed her startup out of her home. But shortly before hiring Moffatt she rented a small office in London to take business meetings with customers, even though most staff worked remotely from their homes. Shirley interviewed Moffatt just after moving into the new office, and as Moffatt recalled, "Steve took me into an empty room and perched uncomfortably on the windowsill while I sat on a folding chair. I had dressed smartly in a 'business' suit for the interview. Dressed very casually, Steve smoked throughout the interview. I was mesmerised by Steve's open toed sandals and brightly painted toenails."[22]

Moffatt's first major assignment for Freelance Programmers was the Concorde project. Because of the complexity of this new aircraft and the rigors of testing and later maintaining it, its black box recordings needed to be analyzed after each flight, not just in the event of a crash.

This required analyzing massive amounts of data, because some sensors on the plane would record up to ten data points per second, and doing it quickly enough to keep the plane flying with minimum downtime. The data from each flight would be transferred from the black box to magnetic tape and then analyzed by computer.

Shirley's fledgling company was offered a whopping £20,000 bonus if they could get the project done in a short period of time: within the year. In the end, Moffatt supervised 11 people working on the project—all women programmers working from home, or occasionally coming together with their children in tow to work together in one person's house. The project budget would end up greatly expanding.[23]

The programming for the black box project was done in machine code, written in pencil on coding sheets, with the women using slide rules to do the needed engineering calculations. After that, the programming sheets would be picked up and thrown "in a mini [cooper]" and driven to a group of punchers for hire, and then driven back to Moffatt and her team to check over the accuracy of the punching. Finally, the cards would be taken by car to a computer center in Borehamwood, where the programs would be run on rented mainframe time. If all went well, the expected results would be sent back, and it was on to the next piece of the program. "If you were lucky," said Moffatt, "around midnight you could actually go in [to the computer center] yourself and, if you were capable of operating the computer . . . you could get closer and see how things worked."[24]

Moffatt had become the first manager of a team working entirely from home at Shirley's company, and it had turned out so well that from then on the company decided to let all of their project managers work exclusively from home. The company picked up more and more business and began to grow rapidly. By 1967, Freelance Programmers had several hundred workers.

In a photo from 1966 (fig. 13.3), Moffatt sits at her kitchen table, using a slide rule to do calculations while writing code for the Concorde. Soon after, Moffatt was promoted to technical director for the company, managing a team of what would eventually be roughly 300 home-based programmers, most of them women. "I mean I loved the work.

Figure 13.3. Ann Moffatt writes software for the Concorde as her baby daughter looks on. Moffatt's memoir, *The IT Girl*, details the story behind the photograph. Photo courtesy of Ann Moffatt.

I absolutely loved the work," recalled Moffatt. "I found yes, you could do it from home. Yes, you could even manage teams from home. You could be productive. The women were happy and productive and earning a lot of money. And we were getting lots of publicity from what we were doing."[25]

Faking IT until You Make IT

The scarcity of programming expertise, which was largely caused by the field's earlier feminization colliding with efforts in the 1960s to remove women programmers and replace them with management-aligned technocrats—men who would do both the management side and the technical side of the work—meant Shirley's company was able to tap into a deep need for outsourced programming expertise as a stopgap during this labor crisis.[26] "The company was growing. We were getting work from the government. The British government . . . were piling work on us. We were [also] getting work from big companies," recalled Moffatt.[27]

But despite this, clients were suspicious of the company's prices. Many clients believed that Freelance Programmers must be run like a "sweat shop" in order to produce the quality of software it did on such a tight schedule. In point of fact, this reflected the contrast between the amateurish men, who were pouring into the conventional office-based workforce's programming jobs in the 1960s with little experience, and the experienced team of highly productive, expert women programmers who were working from home. The women's treatment and camaraderie within the company was not apparent to the outside, however. Several times Moffatt opened the books to allow clients to "see the statistics for their own project and see how much each person had been paid. They then realised that . . . programmers were being paid well but were very productive." The productivity of onsite programmers employed by these companies was often half that of the Freelance Programmers' staff, who were working from home. Moffatt believed the ability to work from home was extremely beneficial to productivity; this has been borne out by later studies. At the same time, the amateurishness being introduced into the "regular" programming workforce during this time of labor shortage and masculinization also impacted relative quality and customers' expectations for time to completion. For all of these reasons, Freelance Programmers continually impressed customers with their speed and work quality: by the mid-1960s roughly 70% of their contracts were repeat business.[28]

Indeed, the need for this labor was so great, and the pace of work so intense, that Steve Shirley visited Moffatt only a few hours after the birth of Moffatt's second child in 1968, nominally to congratulate her, but primarily to persuade her to take over the technical management of the company. Within ten days Moffatt was visiting clients and managing a staff of 250 people.

While Moffatt was happy with her job, her husband was growing increasingly angry at her newfound work responsibilities. One day, in a rage, he drove her car to a dealer, sold it, and came back in a taxi, saying that now that she didn't have a car she wouldn't be able to keep her new management-level job. Furious, Moffatt went out and immediately purchased a new car, and within a few years they divorced. Moffatt went on to become the first woman on the British Computer Society council and continued to work with people at the highest levels of government and industry.[29]

The larger Concorde project continued into the 1970s, with the black box now finished. But despite their efficiency, Freelance Programmers had not been paid for their work. As the Concorde's development encountered delays and churned through ever more money—some estimate it ran over budget by well over a billion pounds—Shirley was unable to secure payment from the government for Freelance Programmers' relatively modest invoice. "I remember [the Concorde project] not really for the technology or the size, but on the financial side, because I didn't get paid for it," recalled Shirley. She had to personally visit the office of the project director, who contracted her company to do the work on several occasions, staying in the waiting room and making a general nuisance of herself because she needed the money for payroll. Eventually he sent word out through his secretary: "Tell Mrs. Shirley to come back tomorrow and her check will be ready."[30]

Concorde's Failure, Legacy, and Lessons

For all of its unrealized return on economic investment, the Concorde was a roaring technical success. It was a technological marvel that put the British back on the map after a string of internationally humiliating black eyes during the Cold War, from the Tanganyika Groundnut Scheme, to the Suez Crisis, to the Windscale disaster.[31] As nations once subjugated by the British fought for and gained their independence and British power receded, the United Kingdom floundered—struggling through multiple rounds of currency devaluation, wracked by strikes, adrift economically, and desperate to join the EEC. While the US and USSR were heading into space, the UK seemed decidedly earthbound.[32] The Concorde promised to change all that: to give the British a technological prestige project that would be the envy of the Western world and serve as an entrée into a stronger alliance with Europe.

Moffatt and her team's contributions to the Concorde helped bring the hope of a modern Britain, revitalized by technology and aligned with European industry, nearer to reality. The Concorde provided a much-needed shot in the arm, not only proving that futuristic supersonic passenger transport was possible but also that a joint venture between two nations so often at odds could be a technical success.

Yet, both the British and French knew that this was a blue-sky project that would run at an economic loss.[33] Despite taking hundreds of potential orders from global airlines, in the end only 14 planes were produced, 7 flying for British Airways and 7 for Air France on the London to New York and Paris to New York routes, respectively, at a cost many times the amount of a regular transatlantic jet ticket.[34] A by-product of a particular technological moment in the Cold War, the Concorde's popularity was undercut by cost and growing environmental concerns—its own history positioned it more as a political bargaining chip and a symbol of technological power than a viable product. When a crash led to all of the remaining Concordes being permanently grounded in 2003 after 34 years in service, no similar passenger plane would replace it.

Initially, the Concorde was also a political failure: in 1967 France blocked Britain's entry into the EEC for a third time, on the grounds that the British economy was underdeveloped. It was not until after French president Charles de Gaulle was no longer in office that the British were allowed into the Common Market in 1973, under the leadership of Conservative prime minister Edward Heath. The longer history of Britain in the EU, particularly post-Brexit, has tended to prove de Gaulle right: although the UK government wanted the economic benefits of EEC (and later EU) membership, the British remained politically unaligned with Europe, holding themselves at a remove while glorying in fantasies of reviving their past global empire.[35]

The Concorde story also shows how high technology does not indicate social progress: after entering the EEC, the UK soon fell afoul of European rules for the modernization of industry, because the UK refused to adequately implement equal pay for men and women. Britain was sanctioned twice by the EEC in the 1970s for having unequal pay and for passing equal pay legislation that was designed to meet the letter of EEC law but not actually implement equal pay. The British "equal pay" and national sex discrimination acts of the 1970s largely perpetuated and submerged the gendered economic inequality the EEC demanded member states fix.[36] This was because unequal pay had for so long been foundational to the functioning and prosperity of the UK economy, with computer corporations even measuring work in cheaper "girl hours" (instead of man hours) to create products at a lower cost.[37]

Ironically, the stellar work of Ann Moffatt's team on the Concorde was an extension of the legacy of unequal pay and unequal opportunity in the United Kingdom. The work of Moffatt and her team was part and parcel of this gendered—and sexist—environment in British industry and high tech. Moffatt, Shirley, and the other women who worked for Freelance Programmers cleverly subverted the sexism of the time by developing and participating in a model of feminist tech entrepreneurship that shifted the boundaries of the modern workplace and remade working conditions for the personal and collective gain of women workers.[38] Shirley's company allowed hundreds of women to continue careers that would have otherwise been cut short or interrupted, and Freelance Programmers produced around 70 women millionaires.

Shirley's company became an exception that proved the rule of Britain's sexist labor practices, cleverly taking advantage of the labor fallout from structural sexism in UK high tech. At the same time, Freelance Programmers' default work-from-home model raised productivity while lowering overhead, and it proved that having to come into an office, and get childcare that would allow one to do so, was not required to ensure the quality or quantity of the work women performed. It showed, rather, that traditional modes of office working were in place to maintain particular ideals of control and conformity for professional workers—ones which often boxed women out of the paid workforce.

Unfortunately, much like the Concorde, the technofeminism represented by the women who programmed its black box was a band-aid on a larger, older, technochauvinist system of inequality and power differences, which could not alone produce a widespread, structural fix.[39] Both Concorde and Freelance Programmers needed to build off of existing power structures to try to construct the future.[40] The success of Freelance Programmers notwithstanding, working from home or remotely has often been used to (re)create an exploitative and unfair economy of piecework. In the current economy, the digital piecework of the "gig" economy, as well as the tactic of using remote work to outsource internationally, disproportionately hurts women, people of color of all genders, and less affluent or privileged workers. These workers often form the silent infrastructure enriching the extractive technology corporations largely based in Silicon Valley.[41]

In addition, while the working conditions and women supervisors at Freelance Programmers were a godsend for many women, this model of employment still meant that working mothers had to do two jobs—even though it gave them the opportunity to do them simultaneously. One part of the legacy of the baby and the black box is that flexibility in working arrangements can be a double-edged sword, helping women stay in the labor market while also reaffirming larger structures that may keep them unequal players within it. To some, Moffatt's picture of her programming while watching her baby may represent how certain white women were able to flourish under conditions of scarcity and crisis by "leaning in" to their roles.[42]

As a result of this reconfiguration of work, Moffatt and her team became part of government and industry modernization in this period, and they were implicitly trusted to ascend the ranks of technocratic power, working from outside traditional workplace power structures while conserving their roles as caretakers within nuclear families.[43] Yet, many similarly qualified women like Moffatt could not use their skills after marriage. For every woman Steve Shirley employed there were hundreds—if not thousands—of women who would not get the chance to continue to fulfill their potential in the burgeoning high-tech economy. Eventually, this gendered labor drain would devastate the British computing industry, which sank into decline in the mid-1970s after several promising decades of innovation and growth.[44]

In the end, the most revolutionary part of Moffatt's work on the Concorde may have been the way it proved the viability of cutting-edge high-tech work from home at a very early stage in the digital era, presaging everything that was to come in terms of labor force reorganization in high tech. Shirley's company very early on gave the lie to the idea that in-office productivity was higher or that professional results required a physical presence in a traditional, centralized office.[45] As Moffatt notes in her memoir: "Steve's companies achieved high productivity and a reputation for top quality. There has now been a lot of research on why home-based workers achieve high productivity."[46] Shirley and her employees were the vanguard of an important new way of organizing work, one that has become apparent to most executives and corporations only much more recently. Indeed, many tech companies are now poised, given the changes wrought by the COVID-19 pandemic, to let programmers work from home permanently.

The benefits of this kind of labor flexibility bring major positive change not just to industries and national economies. Work arrangements geared to life patterns that are historically more common to women fundamentally change the material condition of many women's and children's lives in ways otherwise impossible under current economic systems. This is one reason why, unlike the Concorde, Moffatt's other project from 1966 is going strong well into the twenty-first century: her daughter Claire, the baby in the photograph, is now 57 years old.

Notes

1. Elizabeth Dwoskin, "Americans Might Never Come Back to the Office, and Twitter Is Leading the Charge," *Washington Post,* October 1, 2020; Rob McLean, "Facebook Will Let Employees Work from Home until July 2021," *CNN Business,* August 6, 2020; A. J. Horch, "How Major Companies Are Responding to Employee Needs in a Remote Work World that Has No End in Sight," *CNBC.com,* August 20, 2020; and Tom Warren, "Microsoft Is Letting More Employees Work from Home Permanently," *The Verge,* October 9, 2020.
2. Peter Thorsheim, *Inventing Pollution: Coal, Smoke, and Culture in Britain since 1800* (Athens: Ohio University Press, 2006).
3. Aeronautical Correspondent, "1,450 M.P.H. Airliner Pact Next Week," *Times* (*London*), November 24, 1962.
4. File AVIA 65/2004 (1960–1965), The UK National Archives, London. The British and French argued extensively about the spelling of the name "Concorde" with the

British preferring the English version of the word. Ultimately the French determined much of the shape of the project, including the name.

5. *Times* Correspondent, "Gen. De Gaulle's Blow to Brussels Talks," *Times (London)*, January 15, 1963.

6. Mar Hicks, *Programmed Inequality: How Britain Discarded Women Technologists and Lost Its Edge in Computing* (Cambridge, MA: MIT Press, 2017).

7. Mar Hicks, "Oral History of Dame Stephanie Shirley," conducted January 2018, held by the Computer History Museum in Mountainview, California (available at https://www.youtube.com/watch?v=TRlSaEZhFLg), and Stephanie Shirley, *Let It Go: The Story of the Entrepreneur Turned Ardent Philanthropist*, rev. ed. 2017 (London: Andrews UK, 2012).

8. Shirley, *Let It Go,* 58.

9. Shirley, 53.

10. For more and a comparison with Elsie Shutt's similar company in the United States, see Janet Abbate, *Recoding Gender* (Cambridge, MA: MIT Press, 2012), chapter 4.

11. Hicks, Oral History Interview with Dame Stephanie Shirley.

12. Freelance Programmers help wanted advertisement, *Times (London)*, June 26, 1964.

13. Arthur Reed, "From Hat Rack Brackets to Heavy Duty Brakes," *Times (London)*, November 28, 1972.

14. Reed, "From Hat Rack Brackets to Heavy Duty Brakes."

15. Author's interview with Dame Stephanie Shirley, January 2018.

16. For background, see Hicks, *Programmed Inequality,* chapter 3.

17. Abbate, *Recoding Gender,* chapter 4; Hicks, "Sexism Is a Feature, Not a Bug," in *Your Computer Is on Fire,* Thomas Mullaney, Benjamin Peters, Mar Hicks, and Kavita Philip, eds. (Cambridge, MA: MIT Press, 2021); Mar Hicks, "When Winning Is Losing: Why the Nation that Invented the Computer Lost Its Lead," *IEEE Computer* 51, no. 10 (2018).

18. Australian National Library, "Ann Moffatt interviewed by Sarah Rood in the History of ICT in Australia oral history project," ORAL TRC 6470/6 (nla.obj-220401631), recorded March 3 and May 30, 2014, in Lane Cove, New South Wales.

19. Australian National Library, "Ann Moffatt interviewed."

20. Australian National Library.

21. Ann Moffatt, *The IT Girl: 50 Years as a Woman Working in the Information Technology Industry* (London: Third Age Press, 2020).

22. Moffatt, *The IT Girl,* 105.

23. Shirley, *Let It Go,* and author's interview with Dame Shirley, January 2020.

24. Australian National Library, "Ann Moffatt interviewed."

25. Australian National Library.

26. Hicks, *Programmed Inequality.*

27. Australian National Library, "Ann Moffatt interviewed."

28. Moffatt, *The IT Girl,* 119–21.

29. Ironically, given the sexist banking practices of the time, Ann was not allowed to hold the mortgage to her own house, despite earning more than her ex-husband, and she had to ask her father to sign as guarantor on the loan. Moffatt, *The IT Girl.*

30. Author's interview with Dame Shirley, January 2020. Lord Arnold Weinstock, the director of GEC (General Electric Company plc), was responsible for bringing Freelance Programmers onto the Concorde project to do the programming for the black box.

31. The loss of many former colonized territories, including India; the multimillion-dollar boondoggle to grow nuts for cooking oil in Tanzania (then Tanganyika); the Suez Crisis, which revealed the UK's diminished imperial power; and the Windscale disaster of 1957, when the UK's rush to enrich uranium for nuclear weapons brought it to the brink of nuclear disaster, to name just a few.

32. On the political importance of technological prestige and women's work during the Cold War, see Margot Lee Shetterly, *Hidden Figures: The American Dream and the Untold Story of the Black Women Mathematicians Who Helped Win the Space Race* (New York: William Morrow, 2016).

33. CAB 168/161 (1964–1965), The UK National Archives, London.

34. By the time the Concorde was grounded in 2003 over safety concerns, tickets had reached well over $10,000 for a one-way trip.

35. On how the "soft" technological power of digital technologies has been used to extend imperial domination in the postcolonial period, see Halcyon Lawrence, "Siri Disciplines," and Kavita Philip, "The Internet Will Be Decolonized," in *Your Computer Is on Fire*.

36. Hicks, *Programmed Inequality*, chapter 4.

37. For instance, IBM UK measured its (feminized) manufacturing in cheaper "girl hours" in the 1960s. Hicks, *Programmed Inequality*, 21.

38. On the other hand, corporations used the ideal of the patriarchal family to structure workers' lives; see Corinna Schlombs, "Gender Is a Corporate Tool," in *Your Computer Is on Fire*.

39. Meredith Broussard coined the term "technochauvinism" to highlight the fact that "technological determinism" (the idea that technology determines or leads social change) ignored the racism, sexism, and white supremacy of sociopolitical systems that epitomized twentieth-century ideals of "futuristic" high tech and a more automated future. Broussard, *Artificial Unintelligence* (Cambridge, MA: MIT Press, 2018).

40. For examples of a similar dynamic, see Kelcey Gibbons, "Inventing the Black Computer Professional" in this volume; Lisa Nakamura, "Indigenous Circuits: Navajo Women and the Racialization of Early Electronic Manufacture" in this volume; Ruha Benjamin, *Race after Technology* (Cambridge, MA: Polity Press, 2019); Janet Abbate, "Coding Is Not Empowerment," in *Your Computer Is on Fire*; and Clyde W. Ford, *Think Black* (New York: Amistad, 2019).

41. See, for example, the work of activists Kristy Milland of Turker Nation (Kathryn Zyskowski and Kristy Milland, "A Crowded Future: Working against Abstraction on Turker Nation," *Catalyst: Feminism Theory Technoscience* 4, no. 2 (July 2018), and the large body of work revealing that tech companies exploit inequalities and extract labor value via "gig" work: Sarah Roberts, *Behind the Screen* (New Haven, CT: Yale University Press, 2019); Mary Gray and Siddharth Suri, *Ghost Work: How*

to Stop Silicon Valley from Building a New Global Underclass (Boston, MA: Houghton Mifflin, 2019); Safiya Noble, *Algorithms of Oppression* (New York: New York University Press, 2018); Ruha Benjamin, *Race after Technology*; Tressie McMillan Cottom, "The Hustle Economy," Veena Dubal, "Digital Piecework," Katrina Forrester and Moira Weigel, "Bodies on the Line," and Julia Ticona, "Essential and Untrusted" *Dissent* magazine, Fall 2020: Technology and the Crisis of Work.

42. Sheryl Sandberg, COO of Facebook, famously encouraged women to "lean in" and work harder within sexist systems to succeed, in *Lean In: Women, Work, and the Will to Lead* (New York: Alfred A. Knopf, 2013). The *Guardian*'s Zoe Williams described it as a "carefully inoffensive" book that perpetuated most of the sexist power structures that, on the surface, it claimed to critique. Williams, "Book Review: Lean In," *The Guardian*, March 13, 2013.

43. Not all women working at Shirley's company were operating within a nuclear family: Margaret Mears, who had initially been tapped to lead the Concorde Project but left to work in Australia at the Woomera military testing site, "had joined the company when she found she was pregnant. She was not married and had had her baby adopted at birth without even seeing him. The other woman manager, Pamela Woodman, was an unmarried mum who had decided to keep her baby." Shirley has confirmed that she also tried to hire—and help—women who were unmarried, single mothers, and also lesbians, and knew at least two trans employees who transitioned while at the company. Ann Moffatt, *The IT Girl*, 106–7, and author's interview with Dame Shirley, January 2020.

44. Hicks, *Programmed Inequality*, chapter 5.

45. Freelance Programmers kept data on their productivity to improve their workflow—much like current "agile" software development systems do. Ann wrote that this was key to the company's success, "especially the metrics used to estimate effort, time and cost and to track progress on the project." Other, more traditional companies asked Freelance Programmers if they could help them develop similar metrics to improve productivity. Ann Moffatt, *The IT Girl*, 120–21.

46. Moffatt, *The IT Girl*, 161.

CHAPTER 14

Computing Nanyang

Information Technology in a Developing Singapore, 1965–1985

Jiahui Chan and Hallam Stevens

On November 3, 1969, Ong Pang Boon, Singapore's minister for education, officially opened the Nantah Lee Kong Chian Computer Centre at Nanyang University. Although other computers had been operating in Singapore for several years, this was the first computer in the island nation that was to be used for training and education. "The artificial intelligence built into the computer is such that it has invaded every field of human endeavor," Minister Ong said—and taking advantage of this "revolution" meant having well-trained individuals to control computers and use them effectively.[1]

Singapore is now globally recognized as a high-tech city-state—from self-driving cars to "smart city" sensors, it is a place wired deep for the information technology age. The island's technological development owes much to the Singapore government's active cultivation of computing and communication technologies. In the late 1970s, the government recognized the need to develop information technology for Singapore. In 1980, the minister for trade and industry Goh Chok Tong appointed the Committee on National Computerisation to study Singapore's prospects. Its report recommended enhancing computer education in Singapore, the digitization of the civil service, and the establishment of a National Computer Board (NCB) to help oversee these efforts.[2] Due to its subsequent success in these activities, the NCB has received much of the credit for putting Singapore on a path toward digitization.

But the government's later role in developing IT has overshadowed the fact that, by the late 1970s, computer science was already an active discipline within Singapore, and a first generation of IT professionals had already been trained in computer use. Indeed, Singapore's computer pioneers were not the NCB but rather a small group of professors and technicians at Nantah Computer Centre and the Department of Computer Science at Nanyang University. Telling their story sets the history of information technology in Singapore in a different light. Most importantly, it relocates the initiative and agency in Singapore's modernization further away from the government and closer to other individuals and institutions. In particular, it suggests the importance of relationships between the educational and private sectors in fostering long-term technological development.

In the last decade there has been a concerted effort amongst historians of computing to decenter the West and offer a wider range of geographical and cultural narratives around information technology. In the domain of networking in particular, we now have excellent accounts of electronic networking in the Soviet Union, France, and Chile, amongst others.[3] Švelch has called attention to the role of gamers and hackers in developing computing in Eastern Europe, and in this volume he elaborates on these narratives by examining the cloning of computers in communist-era Czechoslovakia.[4] Likewise, Ekaterina Babintseva argues that we should not think of the development of Soviet versions of artificial intelligence as merely derivative of developments in the United States.[5] Within the United States, recent work by Joy Rankin has demonstrated the role of non-elite, non–Silicon Valley pioneers in the developing of computing, programming, and networking.[6]

Despite this turn away from Western, white, male narratives, there remain relatively few detailed English-language accounts of the development of computing in East Asia. Scholarship on Japan has focused on the development of robots and artificial intelligence.[7] For China, Jiri Hudeček has described Wu Wen-Tsun's efforts to create and utilize a uniquely Chinese version of mathematics and computation during the Cultural Revolution.[8] Thomas Mullaney's account of the Chinese typewriter looks toward, but does not yet outline, a more recent history of computing in China.[9] Elsewhere, histories of computing in Asia have either been focused on the role of governments and government agencies (e.g., the Ministry of International Trade and Industry in Japan), on the development of information and communication technology infrastructure, or on specific systems, machines, and companies.[10]

The historiography of computing in Singapore has followed this pattern. The vast majority of attention has been given to understanding the impact of government policies, government agencies, and leaders on the development of Singapore's information technology capacity.[11] This includes the government's role in attracting and fostering a strong microelectronics industry that played a significant role in Singapore's economic development. Fairchild semiconductor set up a factory in Singapore in 1969, followed by Hewlett-Packard in 1970, DEC in 1980, and Seagate in 1982. These facilities were centered on labor-intensive manufacturing and assembly processes.[12]

The story in this chapter, on the other hand, is about neither government policy nor electronics manufacturing. Rather, by examining how Singapore's first "public" computer was set up and used, we explore here the role played by academics, philanthropists, private companies, and white-collar Singaporeans in creating an IT industry in a newly independent and rapidly developing nation. This is a "people's history" of computing in Singapore and one that demonstrates that, as in the United States, a wide variety of people, institutions, and practices were involved in making and remaking computing.

The development of the Nanyang Computer Centre also offers a story about the diffusion of information technologies into a developing-world, postcolonial context. Although Nanyang University purchased its computer from an American multinational company (IBM), the university's aim was to utilize it in ways that would seed local capacity and talent. The scarcity of computers and individuals trained in their use necessitated that the Nanyang computer was put to work for a wide variety of purposes both for the university and for the wider community.

The First Computers in Singapore

Singapore's high level of technological development has been a product of its rapid transformation over the past 60 years. In 1956, the year Nanyang University was founded, Singapore remained a British Crown Colony, although one with growing powers of self-government. It took another ten years for Singapore to achieve full independence from both Great Britain and Malaysia. At its independence in 1965, Singapore faced a host of challenges—although Singapore had been a relatively wealthy trading port under the British, the economic future of such a tiny island nation remained uncertain. From the late 1960s, Prime Minister Lee Kuan Yew directed a process of aggressive "export-oriented industrialization," building up Singapore's industrial capacity through investment from multinational companies.

These conditions meant that the development of computing in Singapore necessarily followed a very different path than it did in the United States. There, massive calculating machines had been developed during World War II, largely to serve the interests of the military. During early years of the Cold War, massive investment in computers continued, with the Department of Defense providing money for both universities and corporations to develop and build computers. One of the main beneficiaries of such funding was International Business Machines (IBM). Although IBM had existed since 1911, the investment in electronic computers after World War II allowed it to rapidly expand its business. IBM sold its machines to the military and to the large companies such as railways, insurance companies, utilities companies, and steelworks.

In 1952, in the midst of this boom, IBM opened an office in Singapore, mostly selling office machinery such as typewriters and timing equipment (such as time clocks). In 1954, IBM supplied Singapore's first computing equipment to Overseas Assurance, the insurance company now known as Great Eastern General Insurance. Between 1954 and 1968, IBM also sold data processing and accounting machines to the Lee Rubber Company (1956), American International Assurance (1957), Public Life Insurance (1957), the Housing and Development Board (1961), the Central Provident Fund Board (1962), the Property

Tax Department (1962), the Ministry of Finance (1964), and Shell Eastern Petroleum (1968).[13] These computers were used for a range of specialized purposes, including inventory control and maintaining insurance and tax records. However, none of these machines were available for teaching or learning about computing or computers. They remained behind closed doors, dedicated to their particular tasks. Often those who used them were sent overseas for training; in other cases, foreigners were brought in to operate, repair, or update them.

Even by the late 1960s, the Singapore government's attitude toward computers remained ambivalent. One of the nation's most important challenges was employment. The automation offered by computers was perceived to be at least partially at odds with finding more jobs for Singaporeans.[14] This situation meant that by the end of the 1960s, although a handful of computers existed in Singapore, very little local expertise existed in how to build, maintain, or operate them.

Setting up a Computer Center

The inspiration for setting up some kind of institution for computer training and education belonged to Lu Yaw. In 1966, Lu had been appointed by the Singapore government to take up the position of deputy vice chancellor of Nanyang University with the mission of further developing the institution. One of Lu's ideas was to create a computer center for training Singaporeans in computer science and bringing computer research and expertise to Singapore.[15]

This would have been a bold step for any university in the 1960s, but Nantah's fragile position made it particularly risky. Nantah had been established with broad support from the Straits Chinese community (ethnic Chinese settled in Southeast Asia), but despite money and land coming from the local Chinese clan associations, the colonial government refused to grant permission for its establishment.[16] Undeterred, its supporters registered the university as a company and began to teach. Successive governments, through the period of decolonization, refused to recognize Nanyang degrees, questioning its standards of instruction.[17] This situation fomented student resentment toward the government, and by the early 1960s the university had become a hotbed of leftist activity. In 1964, many of these radical students were either arrested or expelled.[18] It was only in December 1967 that the Singapore government finally decided to recognize Nanyang degrees.[19]

Under such circumstances, creating a computer center was no easy task. For one thing, it would require a substantial sum of money. In 1966, the British computer company International Computers and Tabulators had set up a computer center in Singapore at a cost of 1.5 million Singapore dollars.[20] The university itself was recovering from a period of financial hardship and was not in a position to fund any machine by itself. Even if the university could come up with the cash, it would still need people to operate and maintain the machine, as well as to use it in their teaching and do research. Where would they come from?

These difficulties delayed the establishment of the computer center for two years. But eventually the solution came from Nantah's own alumni. Again, with the aim of further developing the university, Lu had traveled overseas to attempt to recruit alumni back to Singapore to teach at their alma mater. Several Nantah graduates had gone on to postgraduate studies at well-respected institutions in the United States and Europe; such graduates, he thought, could be used to reinvigorate Nantah's curriculum, not just in computer science, but in other fields too.

One of those who answered the call to return was Hsu Loke Soo. After graduating from Nantah, Hsu had ventured to the University of Rochester in New York state to complete a doctorate in physics. During his postgraduate studies, Hsu had been introduced to computers, using an IBM 7930 machine to perform calculations for his dissertation project. In particular, Hsu became familiar with the FORTRAN programming language commonly used for scientific and mathematical research in the 1960s.[21]

Hsu returned to Singapore in 1968 and began to revive the idea of creating a computer center. Working with Lu, he gathered together a group of returning Nantah alumni who had experience with using computers overseas. Along with the dean of mathematics, Teh Hoon Heng, the dean of science, Koh Lip Lin, and the head of finance, Chai Chong Yii, Lu and Hsu formed a committee to begin planning the computer center. At a time when very few Singaporeans were even aware of computers, this committee believed that a computer accessible for students and available for research would be important for Singapore's future.

But the problem of money remained. In 1967, Lu had unsuccessfully approached the Ford Foundation, a US-based philanthropic organization, for support for the computer center.[22] In the United States, many universities had developed close relationships with major computer companies such as IBM and Remington Rand. These corporations sponsored the installation of computers on campuses around the US in order to facilitate research and teaching (often resulting in developments that would benefit the companies themselves). Since Singapore had no native computer manufacturers, it could not rely on the same model.

Ultimately, the problem was solved by one of Singapore's own industrial giants. In 1968, Nantah's mathematics department had received funds from the Lee Foundation to sponsor the construction of classrooms. The Lee Foundation was the philanthropic arm of the massive Lee Rubber Company empire. As a philanthropist with a strong belief in education (Lee Kong Chian, the company's founder, had been a teacher) and an early investor in computing, the Lee Foundation seemed a good candidate for funding Nantah's computer center. Indeed, Lu's request to the Foundation met with immediate interest: Lee agreed to provide SGD $80,000 for a computer at Nantah.[23]

Using the Computer

The Nanyang University Lee Kong Chian Computer Centre was situated under the seating area in the basement of Nanyang Auditorium (fig. 14.1). The equipment consisted of an IBM 1130 computer with 16 kilobytes of memory and a 512-kilobyte hard drive, two punch card machines, a punch card reader, and a high-speed printer. These were spread out between a main computer room, two punch card rooms, and a library room. The Computer Centre also had several offices for its staff.

Running such a computer took a team of experts and technicians who could translate mathematical or scientific problems into punch

電腦中心
Computer Centre

Figure 14.1. Inside the Nanyang Computer Centre.
Nanyang Univ. Graduates Yearbook, 1975–76.

card code, who could actually do the punching, who loaded the cards into the machine, maintained the machine, and translated the cards back into numbers or mathematical symbols. The Nantah Computer Centre borrowed most of its staff from other parts of the university. Madam Chua Wu Ying, for instance, was a librarian who was recruited to the Centre and trained as a punch card operator. These activities and staff required continued funding that had to come from computer-use fees, businesses, and philanthropic organizations.

The Computer Centre had three immediate aims: to train Nanyang students in computing, to use the computer in scientific research, and to computerize the university's administrative systems. Together, these projects would all contribute to establishing computers and computing as a more widely known and accepted part of Singaporean society. The Nantah computer was envisaged not just as a tool for economic development (as the government would later promote it) but ultimately as a way to empower Singaporeans from different backgrounds and walks of life. Demystifying this new technology through education and usage would contribute to the anticolonial and anti-elitist missions of Nantah.

The teaching began with physics students. Beginning in 1969, physics undergraduates were offered an elective class in FORTRAN programming, taught by professors from the physics and mathematics departments.[24] Although these professors had not been formally trained in computing or computer science, their experience with computers allowed them to learn enough to teach Singapore's first computing courses. By 1973, the Computer Centre had its own independent teaching staff and was offering fifteen different courses to undergraduates across the university.

Nantah's computer was also used for research purposes by faculty members from a range of disciplines. Data processing, modeling, and the numerical solution of equations using computers was becoming an important part of research practice in many fields, from physics to linguistics. Access to and knowledge of a computer was critical for scholars to be able to remain on the cutting edge of research.[25] It was increasingly important for a university to have a strong computer staff that was equipped to facilitate the research work of its faculty members. One example of work taking place at this time was the processing of the large number of computations needed by the experimental Nuclear Research Group from the Physics Department.[26] Professor Hsu and the Computer Centre faculty staff also conducted research in Chinese character data mining by using information taken from the Chinese newspapers *Sin Chew Jit Poh* and *Nanyang Siang Pau* to identify two thousand commonly used Chinese characters. These characters were then coded into the punch card machine so that faculty members and students could input both Chinese and English characters.[27]

The faculty responsible for the Computer Centre, including Hsu and Dr. Tan Kok Phuang, also attempted to put the computer to work for the university administration. The Computer Centre assisted the university registrar in organization and in payroll calculations for the Finance Department. Later, the Computer Centre also computerized the university library by databasing library subscriptions and resources, and it assisted in the process of recording the inventory system of the entire Nanyang University to prepare for the merger of Nanyang University and the University of Singapore in 1980.[28]

Nanyang professors also established important connections with industry and with the local community. For example, the services of the Computer Centre were utilized by several local businesses.[29] These firms required services such as linear programming, data analysis, and payment and inventory control. The Computer Centre charged lower fees compared to other commercial computer service centers. These activities allowed the Computer Centre to build and maintain strong relationships with business, providing further training and employment opportunities for Nantah students and increasing the usage of computers in Singapore industries.[30]

The Computer Centre also offered its services to the Singapore government and its agencies. For example, it assisted with the computerization of the university entry system for the Ministry of Education. This involved database sorting, processing entry requirements, and allocating students into their respective university programs.[31] The Centre was later used as a model by the Ministry of Education to adjust its primary and secondary curriculum when it introduced computers at the lower educational levels.[32]

By the early 1970s, the Computer Centre had already established itself as an important locus of research and teaching work on the Nantah campus. Although the initial investment had been relatively modest, the computer was proving to be in high demand for a range of different types of work, including in the running of the university itself.

Training for the Future

Perhaps the most important legacy of the Computer Centre was the role it played in establishing a growing interest in and knowledge of computers amongst university students and the public. In comparison with more advanced economies, in the early 1970s Singaporeans had very little exposure to or awareness of computers or their potential. Creating such an awareness was one of the first steps necessary for expanding the role of computers in Singapore's society and economy.

The Computer Centre's inaugural courses taught, first, basic theoretical concepts in computing and, second, the FORTRAN programming language.[33] Apart from their immediate utility, the aim of these courses was to create interest in computers amongst students and to encourage them to be more receptive to using computers in their research work.[34] The courses were immediately popular. In the first year, the courses attracted over a hundred physics undergraduates taking computer courses as an elective.[35] Since the Computer Centre did not have funds to hire dedicated lecturers, the classes were taught by faculty from the Physics Department. Like Professor Hsu, these professors had prior experience and knowledge in computing from postgraduate training.

The Centre's staff also conducted classes for the public. For most Singaporeans, this was the first opportunity to even see, let alone use, a computer. In the early 1970s, there were only about 30 computers on the whole island, and their use was restricted to a few people inside the few companies that owned or leased them. Beginning in 1972, the Computer Centre ran basic but intensive four-week classes during university holidays. This effort proved immensely popular, attracting civil servants, teachers, engineers, senior executives, and accountants. Many of those who took the class saw it as a valuable qualification for advancing their careers. Within five years, the Computer Centre had trained hundreds of such individuals in the basics of computing. Importantly, the Computer Centre charged only a small fee for such classes—it was not attempting to make a profit, but rather to raise awareness and understanding of computing and computers in Singapore.

In addition to these pedagogical efforts, the Computer Centre also tried to promote and foster computing interests amongst a wider range of Nantah students by setting up the Nanyang Computer Society. The society, established on March 13, 1976, sought to assist in the advancement of students' knowledge of computer science and computing and foster a collaborative spirit among students in conducting computer-intensive work.[36] The society planned events, often working closely with industry (fig. 14.2). For example, the society organized a field trip to the computing center of Singapore Airlines to

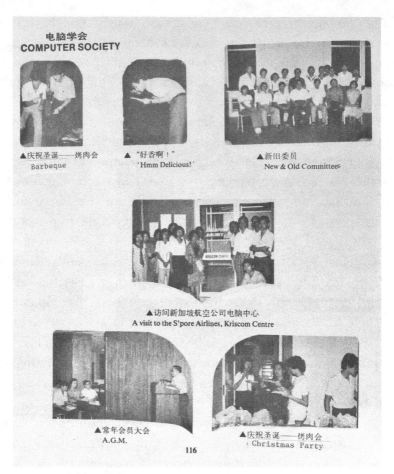

电脑学会
COMPUTER SOCIETY

▲庆祝圣诞——烤肉会
Barbeque

▲ "好香啊！"
'Hmm Delicious!'

▲新旧委员
New & Old Committees

▲访问新加坡航空公司电脑中心
A visit to the S'pore Airlines, Kriscom Centre

▲常年会员大会
A.G.M.

▲庆祝圣诞——烤肉会
Christmas Party

116

Figure 14.2. Activities of the Nanyang Computer Society.
Nanyang Univ. Graduates Yearbook, 1979–80.

expand the students' knowledge about the use of computers in business settings.[37] Such excursions demonstrated the connections between the academic study of computing and its industrial applications. Apart from field exposure, the society also held seminars on computers and computing and provided holiday classes on computers.[38]

Faculty and society members also worked together to publish articles for the Nanyang University Computer Society newsletter. Some of the topics included artificial intelligence, studies in programming languages like FORTRAN, and the latest news on computers and software.

These articles were designed to cultivate topics of discussion for students who were interested in computing. They also encouraged an exchange in opinions about the latest developments in the computing field.

Computing activities at Nanyang University were not restricted only to academic and educational work. Through classes for the public and through the Nanyang Computer Society, the Computer Centre began to have a broader influence, both on campus and beyond. The efforts of the Computer Centre contributed to growing awareness of computers and their influence in industry and public life. This promotion of computers came to have an important effect on the computer industry in Singapore.

Establishing Computer Science in Singapore

Computer science as a distinct field had emerged only in the late 1950s in the United States. In the 1960s, Singapore offered no formal education for computer training or computer science. The establishment of this discipline was to prove critical for the development of an information technology sector in Singapore. Hsu and other professors at the Computer Centre believed that computers would become increasingly important for Singapore as her financial, manufacturing, and service sector grew.

By the second half of the 1970s, Singapore had a growing need for graduates with high-level training in computing. By this time, the private sector was adopting a larger and larger number of computers. The number of computers in Singapore had grown from 32 in 1973 to 123 in 1976.[39] This was spurred in part by the introduction of minicomputers and microcomputers that were more affordable than large mainframes. These smaller and less powerful machines were capable of running the types of computer processes required in business and industrial settings, such as accounting, payroll, inventory control, and sales analysis.[40] If computers were to be widely used for these tasks, users would need specific training in hardware and software.

The technical and on-the-job training provided in many companies was sufficient for basic maintenance and operation of the machines.

Singapore had very qualified local engineers, without a degree in computer science, who had been trained by computer companies to operate the computer systems that the companies were selling.[41] But more complex computing work, such as programming and data processing, required a more sophisticated understanding of computers that was not provided by technical or industrial training. There were few locals who could handle this higher-level work. By the mid-1970s, industry needed a university-level program to produce local computer graduates with a deep knowledge of computer science.

The efforts to develop the computing scene by the Nantah Computer Centre played the key role in providing the island with local computer professionals. The Computer Centre had provided extensive computing training and service for the university and its members. However, as the need for computer expertise grew, it could not carry out in-depth research work and train students at a larger scale. A larger program and institutional base for computer science education was becoming necessary.

By 1975, computer courses had become sufficiently popular that the university decided to establish a full-fledged Department of Computer Science. By this point, the Computer Centre offered over twenty classes, organized into a minor for undergraduates. In addition to science and mathematics students, business students became particularly interested in learning how to use the machine, taking new classes in the COBOL programming language (Common Business-Oriented Language). Given this popularity and the growing need for well-qualified computer programmers, the government provided SG$926,000 for building a home for the new department.[42]

The Department of Computer Science at Nanyang University was set up under the Faculty of Science in 1975. Although the new department was administratively and physically separate from the Computer Centre, the two continued to work closely together. The department was headed by Hsu, and many of the first faculty moved over from the physics department into computer science. The department's primary mission was designing a curriculum for a computer science degree. This was offered as a three-year course, with students who performed well offered an additional year to complete an honors degree.[43] Having a department of Computer Science not only meant more computer classes for the students but also a greater cohort size. The department was the first and only institute in Singapore at the time to train local computer scientists and remained as such because computer systems were too expensive for most other educational institutions.[44]

Because Nantah was the only university to offer computer science in Singapore, the program was highly sought after by incoming students. Outstanding students from polytechnics and junior colleges across Singapore competed for places in the degree program. This helped the university to gain more students, but it also intensified rivalry with Singapore's other university, the University of Singapore.[45] The situation worsened, and Nanyang University was instructed by the government to close down the computer science department. In 1976, a government-commissioned study concluded that Singapore did not require a large number of computer science graduates at the time and that these graduates might not be able to get employed after graduation due to an oversupply.[46] However, the university continued to support the department and decided to keep the program. The department was eventually spared by a letter sent from the Prime Minister's Office to revoke the closure.[47]

Despite these growing pains, there was constant oversubscription for the computer classes conducted by the Computer Centre in the university.[48] In order to meet the demand, more computing classes were set up. The department recruited more faculty staff to teach the computer classes and increased the computing equipment in the university. The three-year degree in computer science continued to prove immensely popular, with many students competing for the limited spaces in the course.

After 1978, graduates from the Nanyang University Computer Science Department took up computing jobs in the public and private sectors, contributing significantly to the growing computerization of Singapore businesses and government in the 1980s. The department also played a role in influencing pre-university education. Responding to a request from the Curriculum Development Institute of Singapore, the department designed an intensive one-year diploma course to be offered to junior college teachers. Teachers graduating from this program were the first to teach computing at the pre-university level in Singapore.

The establishment of Singapore's first Department of Computer Science widened the influence of the Computer Centre. It significantly increased the number of individuals with training in computing and became a key source of graduates for Singapore's growing computer and information technology needs. Moreover, the department played a central role in the establishment of computer pedagogy at other educational levels.

Conclusions: The Legacy of the Nantah Computer Centre

In 1980, the Singapore government decided to merge its two universities, folding Nanyang University into the University of Singapore to form the National University of Singapore (NUS). Since the University of Singapore offered no courses in computer science, the Nantah Computer Centre and the Department of Computer Science were moved to the Kent Ridge (NUS) campus as part of the newly formed university. The Department of Computer Science at NUS became the Department of Information and Computer Science and finally the School of Computing. When Nanyang *Technological* University (NTU) was founded on the Nanyang University Campus in 1991, the Department of Computer Science remained at NUS. Computer science teaching at NTU was reestablished within the College of Engineering. Because of this peculiar history, the legacy of early computing on the Nanyang campus is almost completely forgotten.

This history is an important one primarily because it illustrates the wide variety of individual and institutional actors that contributed to Singapore's early technological development. Nanyang University's own mythology revolves around the idea of a university built by and for Chinese immigrants to Southeast Asia—a university built through donations from hawkers, cabaret dancers, and trishaw pullers.[49] This anti-elitist spirit seems also to have influenced the development of computing at Nantah. Lu, Hsu, and other pioneers at the Computer Centre did not see computing as an elite activity, to be restricted only to the highly educated portion of the population. Their efforts attempted to reach out to different segments of society. Their aim was not merely to teach computing to university students but also to bring the power and versatility of computers to the attention of as many people as possible. During the 1970s, the predominant attitude toward computers in Singapore was one of caution and fear. Many believed that the increasing automation that computers offered would end up putting them out of a job. The Nantah Computer Centre did much to dispel such worries. By offering training to not only university students but also businesspeople, professionals, civil service employees, teachers, and ordinary citizens, the Computer Centre contributed much to the normalization of computers within Singapore society. "Spreading the message" of computers laid an important foundation for the next steps of computerization of Singapore in the 1980s. Although personal computers were nonexistent when the Nantah Computer Center opened (the Apple I arrived only in 1976), the Nantah pioneers anticipated that computers would become widely used in many parts of society.

The story of the Nanyang Computer Centre shows that the emergence and diffusion of computing into Singapore was not purely a top-down, government-driven initiative. In fact, in some instances the government actively opposed the expansion of computer education at Nantah, further galvanizing supporters of these efforts within the university. Establishing a base of computing talent was the work of educators, philanthropists, students, and professionals who saw the potential of computers for both personal and national economic advancement. Like Hicks's story of Concorde's "baby and the black box," this account shows how a diverse range of actors, operating across a range of unexpected locations—and sometimes even working against government and industry norms—contributed to "computerization" in its various forms.[50]

This account also challenges the focus on machines and hardware that is common in the history of computing. The emergence of Nantah as a center of computing demonstrates that having a computer itself was hardly enough to create national computing capacity. Rather, this capacity had to be built by fostering local expertise and practice through pedagogical and institution-building activities. Much of this work was performed by Nantah graduates who returned from the West (especially the United States), bringing with them their embodied computing expertise. These returnees were motivated by a desire to contribute to the mission of a young, anticolonial, Chinese-language university in Southeast Asia. This reverse diaspora foregrounds the role of ethnic identity, nationalism, and alumni networks—rather than military or corporate actors—in expanding electronic computing beyond the West.

As Joy Rankin has suggested in a very different context, such "people's histories" of computing are vital for showing that computing culture did not emerge only from Silicon Valley and personal computers.[51] Sites such as the Nanyang Computer Centre highlight the role of a diverse range of actors and networks in building an alternative vision of computers and computing. This was not just a vision of modernization and economic development in a postcolonial nation; it also brought computers and computing education to as wide a range of Singaporeans as possible.

Notes

1. "Minister on the Need to Master Computers," *Straits Times,* November 4, 1969. "Nantah" (or "Nan Da") is the abbreviated Chinese form of "Nanyang Daxue" (Nanyang University).
2. Committee on National Computerisation, "Report on the Committee on National Computerisation," October 1980.
3. Benjamin Peters, *How Not to Network a Nation: The Uneasy History of the Soviet Internet* (Cambridge, MA: MIT Press, 2016); Julien Mailland and Kevin Driscoll, *Minitel: Welcome to the Internet* (Cambridge, MA: MIT Press, 2017); Eden Medina, *Cybernetic Revolutionaries: Technology and Politics in Allende's Chile* (Cambridge, MA: MIT Press, 2011).
4. Jaroslav Švelch, *Gaming the Iron Curtain: How Teenagers and Amateurs in Communist Czechoslovakia Claimed the Medium of Computer Games* (Cambridge, MA: MIT Press, 2018). See also Švelch, this volume.
5. Babintseva, this volume.
6. Joy Rankin, *A People's History of Computing in the United States* (Cambridge, MA: Harvard University Press, 2018).
7. See, for example, Yulia Frumer, "Cognition and Emotion in Japanese Humanoid Robotics," *History and Technology* 34, no. 2 (2018): 157–83; and Colin Garvey, "Artificial

Intelligence and Japan's Fifth Generation: The Information Society, Neoliberalism, and Alternative Modernities," *Pacific Historical Review* 88, no. 4 (2019): 619–58.

8. Jiri Hudeček, *Reviving Ancient Chinese Mathematics: Mathematics, History, and Politics in the Work of Wu Wen-Tsun* (New York: Routledge, 2014).

9. Thomas Mullaney, *The Chinese Typewriter* (Cambridge, MA: MIT Press, 2018).

10. For example, Tessa Morris-Suzuki, *The Technological Transformation of Japan: From the Seventeenth to the Twenty-First Century* (Cambridge: Cambridge University Press, 1994); Chigusa Kita and Hyungsub Choi, eds., "History of Computing in East Asia," *IEEE Annals of the History of Computing* (April–June special issue, 2016): 8–10.

11. Gurbaxani et al., "Government as the Driving Force toward the Information Society: National Computer Policy in Singapore," *The Information Society* 7 (1990): 155–85.

12. Liu Fook Thim, *The Fairchild Singapore Plant 1967–1989: The Story of a Pioneer Semiconductor Assembly and Test Factory and Its Former Employees* (Singapore: Liu Fook Thim, 2016).

13. "Singapore Chronology," available at https://www-03.ibm.com/ibm/history/exhibits /asia/Singapore_ch1.html.

14. C. K. Lee, "The New Breed that Moves with the Times" *Straits Times,* December 14, 1958; "Computer Use May Not Make Sense," *New Nation,* September 12, 1973, 2.

15. Lu Yaw, interviewed by Lu Chi Sen, April 28, 1994, accession number 001599, Development of Education in Singapore (Prewar–1965) in the Education in Singapore Collection (Part 1: English), Archives Online at the National Archives of Singapore, transcribed by Roger Khong on September 28, 1995, Reel 26 of 26.

16. B. K. Ji and G. Q. Cui, eds., *Nanyang da xue li shi tu pian ji* [A pictorial history of Nantah] (Singapore: Times Media for the Chinese Heritage Centre, 2000), 23–24, 43.

17. Nanyang University Commission, *Report of the Nanyang University Commission, 1959* (Singapore: Printed at the Govt. Print. Off., 1959), 29.

18. "Action at Nanyang" *Straits Times,* June 29, 1964, 8.

19. "Nantah Degrees: Students Are Pleased," *Straits Times,* December 27, 1967, 4.

20. "ICT to Set Up Centre Soon" *Straits Times,* April 27, 1966, 12.

21. Hsu Loke Soo, interviewed by Lye Soo Choon, December 24, 2012, Accession number 003784, Education in Singapore Collection (Part 2: Chinese), Archives Online at the National Archives of Singapore, translated by Chan Jiahui, January 18, 2016, Reel 6 of 10.

22. Teh Hoon Heng, *Zhenfenxin jiang nanda gushi* [Teh Hoon Heng tells the story of Nantah] (2011): 122–23.

23. Teh Hoon Heng, *Zhenfenxin jiang nanda gushi,* 62–63.

24. Nanyang University, *Nanyang daxue gaikuang* [Nanyang University calendar 1969–1970] (Singapore: Nanyang University, 1970), 143–44.

25. Li Te, *Tong xulesi fujiaoshou tan nanda diannao zhongxin* [Interview with Prof. Hsu Loke Soo about Nantah Computer Centre], *Nanyang Siang Pau,* November 7, 1969, 3.

26. Nanyang University, *Research Programs in Singapore: Proceedings and Papers of a Seminar Held under the Auspices of Nanyang University from 6 to 7 August 1969* (Singapore: Nanyang University, 1970), 63.

27. Loke Soo, Hsu, email message to Chan Jiahui, February 22, 2016.

28. Li Te, interview with Prof. Hsu Loke Soo, 3. The Head of the Computer Centre at the time, Dr. Leong Kuo-Sing, worked with Dr. Tan Kok Phuang to process the inventory system. Tan Kok Phuang interviewed by Chan Jiahui.

29. "Centre's Not for Profit, Says Nantah," *New Nation,* May 21, 1973, 3.

30. "Centre's Not for Profit, Says Nantah."

31. *Nanda diannao zhongxin kuozhan gongcheng wancheng tianjia sanzhong yiqi* [Nantah Computer Centre expansion plans completed, three new machines added], *Nanyang Siang Pau,* October 15, 1974.

32. *Nanda diannao zhongxin kuozhan gongcheng wancheng tianjia sanzhong yiqi.*

33. Hsu Loke Soo, Education in Singapore Collection, Reel 7 of 10, Nanyang University, *Nanyang daxue gaikuang* [Nanyang University calendar 1969–1970] (Singapore: Nanyang University, 1970), 143–44.

34. Li Te, interview with Prof. Hsu Loke Soo, 3.

35. Hsu Loke Soo, Education in Singapore Collection, Reel 7 of 10.

36. *Nanda diannao xuehui chengwei fading tuanti—diyijie lishihui yi chengli* [Nanyang University Computer Society formalized—The first Board of Advisors was set up], *Sin Chew Jit Poh,* June 19, 1976, 9.

37. Nanyang University, *Nanyang University Graduates Year Book 1979/1980* (Singapore: Nanyang University, 1980), 116.

38. *Nanda diannao xuehui niding huodong jihua* [Nanyang University Computer Society finalized future plans for events], *Nanyang Siang Pau,* August 12, 1979, 35.

39. Ministry of Science and Technology Singapore, *Survey of Computers in Singapore, 1976* (Singapore: Ministry of Science and Technology, 1976), 6.

40. *Nanda zhanghaotang boshi suo chouhua diannao zhuanti yanjiang jieshu—Yanjiangzhe zhichu weilai diannao yongtu riguang xuyao tidao zhefangmianzhi jiaoyu* [Nantah's Dr. Thio Hoe Tong points out computers will be more widely used and talked about education in the field of computing], *Nanyang Siang Pau,* February 19, 1975, 9.

41. "More than 200 Firms Using Computers," *New Nation,* May 25, 1972, 15.

42. Tan Kok Phuang, interviewed by Chan Jiahui, February 22, 2016.

43. Nanyang University, *Nanyang daxue gaikuang* [Nanyang University calendar 1975/76] (Singapore: Nanyang University, 1976), 104–5.

44. Chuang Saw Choon, interviewed by Chan Jiahui, February 22, 2016.

45. Thio Hoe Tong, email interview by Chan Jiahui, January 28, 2016.

46. Teh Hoon Heng, *Zhenfenxin jiang nanda gushi* [Dr. Teh Hoon Heng tells the story of Nantah] (2011), 64–65.

47. Thio Hoe Tong, email interview.

48. Thio Hoe Tong, email interview.

49. Nanyang University, *A Brief Sketch of Nanyang University* (Singapore: Lam Yeong Press, 1958), 2. See also, Tan Kok Chiang, *My Nantah Story: The Rise and Demise of the People's University* (Singapore: Ethos Books, 2017).

50. Hicks, this volume.

51. Rankin, *A People's History of Computing in the United States.*

CHAPTER 15

Engineering the Lay Mind

Lev Landa's Algo-Heuristic Theory and Artificial Intelligence

Ekaterina Babintseva

In his 1973 study of human problem-solving, Soviet psychologist Lev Landa criticized the American approach to formalizing human cognitive functions. Landa particularly disagreed with Allen Newell and Herbert Simon, who gained acclaim among artificial intelligence (AI) practitioners for their work on the Logic Theory Machine (LTM), a computer program that solved mathematical proofs and, ostensibly, imitated the way humans solve problems. An educational psychologist, Landa believed that the LTM hardly captured the mind's actual work. His skepticism stemmed from his 1950s studies of students solving geometry proofs and whether their thinking methods could be analyzed in terms of logical structures. In the 1960s, his work attracted the attention of the Council on Cybernetics—a powerful Soviet scientific institution—and the Soviet state. Both thought that Landa's research could offer a methodology for cultivating Soviet citizens' technoscientific creativity, considered by the Soviet state as a paramount cognitive quality in the age of computerization. Along with other cognitive psychologists and educators of the 1960s and the 1970s, Landa regarded the ability to solve problems as the quintessence of creative thinking. Since problem-solving is central to scientific research and engineering, Landa and his colleagues viewed creative and scientific thinking as equivalent. His general theory of human mental operations, which both described how the mind solves problems and served as a methodology for the teaching of efficient problem-solving techniques, sought to make accessible the nebulous process of creative thinking to any mind—and, potentially, even to a machine.

This chapter examines the origins of Landa's theory of human thinking, named the Algo-Heuristic Theory (AHT), and its contributions to AI research. It seeks to decenter the US-dominated historiography of AI by recognizing the influence of Soviet scientific culture and, thus, giving a more complete picture of AI from the 1960s to the 1980s. I make three interconnected claims. First, I argue that the unique Soviet institutional setting gave rise to an approach to human cognition previously unregistered in historiography. Historians Stephanie Dick and Hunter Heyck have shown that twentieth-century cognitive science in the West was dominated by the idea that human reasoning essentially works as logical deduction.[1] However, the history of Landa's AHT explicates a very different perspective on human thinking. The major premise of Landa's approach was that deductive logic could account only for the solution of simple problems, being incompatible with tasks that require more complex mental processes. The unique disciplinary status of Soviet psychology played a key role in shaping this premise. Until 1972, Soviet psychology was primarily a pedagogical discipline, with the Academy of Pedagogical Sciences (APN) being the umbrella institution for all research in the field. Landa studied how actual people solve problems. In contrast, his RAND counterparts, Simon and Newell, were more interested in computer algorithms than the mind's working, leaving unquestioned the assumption that mathematical reasoning is the standard of human thinking.[2]

Second, recognizing the importance of local contexts in the history of computing, this essay claims that Landa's research offered an alternative meaning for "algorithm" that emerged at the intersection of educational psychology and cybernetics. Historian of computing Ksenia Tatarchenko has shown that, unlike in the United States, where the term "algorithm" applied exclusively to computer-executed instructions, Soviet programmers also viewed algorithm-writing as a way to discipline the mind.[3] Working at a time when the USSR's concern about the automation of its production was at its highest, Landa suggested that algorithms could optimize not only the functioning of Soviet plants and factories but also the working of the human mind. In their chapter for this edited volume, Liesbeth De Mol and Maarten Bullynck demonstrate how managerial practices and industrialization gave rise to the notion of "programming," which, in the nineteenth century, was understood as the process of ordering an activity. This case study shows that twentieth-century concerns about efficiency and automation also shaped the notion of an algorithm as a mind-ordering procedure.[4] The desired mind, in Landa's theory, was the problem-solving and, hence, creative mind.

Third, in line with emerging postcolonial scholarship, I call attention to the importance of seemingly peripheral and mundane contexts for AI. The history of AI is largely yet to be written, but where academic histories exist, they tend to focus on a very small community of researchers in elite academic and military research institutions such as the RAND Corporation.[5] However, the AI of those niche American communities represents only a small fraction of research at the intersection of psychology and computing. In contrast, this case study explores the development of AI at public education institutions.[6] While in American AI research, "learning" was more of an abstract idea, devoid of substantial empirical findings, in the Soviet Union, AI practitioners painstakingly worked to understand the *actual* working of the learning mind. Hence, it should not be surprising that Landa's ideas traveled across the Iron Curtain and became instrumental in the expert systems approach to AI in Western countries. Echoing Landa's criticism of Simon and Newell, expert systems developers insisted on the need to concentrate on actual human cognitive functions. In his essay for this volume, Jaroslav Švelch argues that Eastern bloc computer developers creatively adapted Western technology to their local contexts. Landa's AHT was also far from a mere adaptation of American AI to the Soviet context: while initially it made use of American research, later it was circulated back to the West as an independent theory.

This chapter is divided into four parts. Part one examines the cybernetic origins of Landa's algorithmic studies of thinking. The second part explains how Landa's Algo-Heuristic Theory synthesized Soviet and American approaches to AI. The third section follows Landa through his 1976 immigration to the United States, where he founded his own company and used AHT to train private companies' and government organizations' personnel. As section four explains, Landa's work was exactly what American AI practitioners needed to model the domain-specific knowledge of human experts. In the 1980s and 1990s, some of them turned to AHT to model the instructor's expertise in special educational computer programs. The history of AHT thus reveals how mundane sites—secondary education and corporate training— were simultaneously the sources of ideas about AI and the places where AI research found its application.

Cybernetic Psychology: Lev Landa's Approach to Programmed Instruction

Landa's work on the formalization of problem-solving developed as a result of the application of cybernetic methods to Soviet psychology. In 1959, the Academy of Sciences of the USSR, the center of Soviet civilian research, was joined by a new powerful research unit named the Council on Cybernetics. Envisioned by mathematician Aleksandr Liapunov and founded by admiral and professor of engineering Aksel Berg, the council sought to introduce the quantitative methods of cybernetics to all Soviet fields of natural, human, and exact sciences.[7] Berg was especially keen on introducing cybernetic methods into Soviet psychology, thinking that those two could help attract more government funding and raise the overall prestige of psychology in the USSR. In 1959, Soviet psychology was a pedagogical discipline. Moscow psychologists worked at a lower-rank institution named the Academy of Pedagogical Sciences, where they concentrated on developing school curricula. Cybernetics, Berg believed, could help psychology evolve from a marginal field into a discipline that would be a part of Soviet Big Science research.[8]

Berg's first attempt to *cybernetize* psychology was marked by the opening of the council's Section on Psychology and Cybernetics in 1962.

He invited Landa to supervise research on programmed instruction, which Berg considered a promising field within psychology. In the Soviet Union, programmed instruction or *programmirovannoe obuchenie* was a subfield at the intersection of cybernetics, psychology, and education that sought to analyze human learning in terms of logical structures and develop algorithms that would order how humans acquire new information and solve problems. The subfield's emphasis on algorithms correlated with the major principle of Soviet cybernetics, which was the algorithmization of a vast array of scientific disciplines that previously had little to do with mathematics and logic.

Landa's interest in problem-solving dated back to his graduate studies at the Institute of Psychology of Moscow's APN, where, in 1955, he defended his dissertation "The Psychology of Reasoning Methods Formation (Based on How Seventh and Eighth Grade Students Solve Geometry Proofs)." Two years later, he was appointed a research fellow at the APN's Institute of Theory and History of Pedagogy, where he worked to identify the algorithmic patterns of solving geometry problems.[9]

In the 1960s, he extrapolated his finding to understand how humans solve all kinds of problems, not necessarily geometry proofs. By that time, Western cognitive psychologists contended that there are general mental mechanisms that are responsible for how humans solve all sorts of tasks.[10] In the Soviet Union, general problem-solving only started gaining prominence among Soviet scientists. Working at the forefront of this research field, Landa sought to create algorithmic methods that would enable students to solve any kind of problem. Landa's definition of an algorithm came from the research of mathematician Liapunov: "By algorithm is usually meant a precise, generally comprehensible prescription for carrying out a defined (in each particular case) sequence of elementary operations (from some system of such operations) in order to solve any problem belonging to a certain class (or type)."[11] Applying this definition to pedagogy, Landa suggested that students should be taught with prescriptions that would walk them in small, incremental steps toward a problem's solution. To build such prescriptions, one needed to discern all the mental steps taken to solve the problem. Once these steps are visible, researchers could write teaching programs called "instructional algorithms" to direct students' thinking to correct solutions and proper understanding of the material.

Berg believed Landa's research had immediate relevance to Soviet computing and automation. He was especially enthusiastic about applying Landa's work to the design of special teaching computers that would interact with students without any human interference.[12] By following Landa's algorithms, Berg thought, computers could gain full control of how students master general problem-solving abilities. As the state viewed the efficient learning mind as a crucial economic resource, the cybernetics-inflected studies, such as Landa's, proposed computers as the best instruments to cultivate this resource.[13] Additionally, Berg recognized that Landa's work on instructional algorithms could contribute to the council's major objective, the development of computer algorithms that would imitate human thinking.

The Algo-Heuristic Theory and the Soviet Approach to AI

In the late 1960s, working at his own Laboratory of Programmed Instruction (LPO) at the Academy of Pedagogical Sciences, Landa became concerned with a formal description of creative thinking, the kind of thinking that leads to the production of new knowledge and the solution of complex problems. For Landa, "complex problems" were intractable to simple logical algorithms. The latter relied on the discrete means of expressing human thinking, but, as Landa learned from his studies, we do not think in discrete steps when dealing with a complex task. Landa chose to draw on Soviet and Western research in AI to make the mechanism of creative thinking visible. The Soviet state generously supported his studies, regarding creativity as the driving force of the Soviet scientific-technical revolution and computerization.[14]

Landa was not alone in criticizing existing logical algorithms for their rigidity and, therefore, inapplicability to complex problem-solving. For example, in their collaborative work, Soviet mathematician Dmitriĭ Pospelov, psychologist Veniamin Pushkin, and philosopher Vadim Sadovskiĭ explained how well-known and widely used algorithms for the solution of linear algebraic equations could not imitate the creative thinking that humans engage when they solve complex problems.[15] Such algorithms, they explained, tried out random solution variants until they stumbled upon the most appropriate one. While they were appropriate for simple problems, such as solving logic theorems or a crossword puzzle, they were useless in solving more complex tasks for midcentury computers neither had the memory or operational capacity to try out the vast number of possible problem solutions.[16]

In the United States, mathematician Allen Newell and psychologist Herbert Simon established the field of heuristic programming to decrease the number of problem solutions tested by a computer. In Greek, the adjective "heuristic" means "serving to discover." Heuristic programming, thus, became an approach to AI that improved the efficiency of a system by equipping it with the human-like ability to make logical shortcuts in its reasoning and limit the number of potential solutions.[17] In 1958, Newell and Simon wrote the first information processing program named The Logic Theorist (LT), which, they argued, imitated the ways the human mind solves complex problems.[18] The two scientists assumed that solving a problem is akin to finding one's way through a maze. To solve a problem, humans reduce the number of corridors in the maze, test the decreased number of alternatives, and then choose the most appropriate one. Therefore, the heuristics designed by Simon and Newell helped the system reduce the number of possible paths out of the problem maze.

Reacting to Newell and Simon's research almost a decade after their first publications on LT, Soviet psychologist Pushkin disagreed with the maze premise. A doctoral graduate of Moscow State University, he became the director of the Laboratory of Heuristics in 1967. One of the first Soviet laboratories working on the questions of formal description of human thinking, the lab belonged to the Academy of Pedagogical Sciences. Earlier in his research career, Pushkin had studied how dispatchers and operators of airports and railway stations solve complex problems they encounter in their work. He argued that their primary task—the optimization of transportation flow—was a perfect example of a complex problem, as there was no ready set of instructions to enable its solution.[19] Pushkin defined problem-solving as a characteristic of productive thinking, an exclusively human cognitive process responsible for developing new patterns of mental activity and allowing an individual to orient in new circumstances.[20] Opposing Newell and Simon's idea of a ready-made maze, Pushkin proposed that instead of testing possible solutions, the mind constructs "a dynamic information-processing model of the external world" to establish the relations between the elements of a problem.[21] Based on his research on how chess players move their eyes during a match, he concluded that when the mind models the problem, it also constructs the paths to the solution.

Landa synthesized Soviet and American approaches to problem-solving to develop a classification system of all types of possible problems and related cognitive procedures.[22] Maintaining that neither American nor Soviet conceptions of problem-solving were universal, he suggested that while in some cases humans try out different solution variants, in other cases, the mind itself has to construct the paths to the solution.[23] From Newell and Simon, Landa borrowed the idea of a *problem space*. When people solve problems, proposed Newell and Simon, they search in a problem space that consists of the initial state of the problem, the goal state, and all possible states in between. They named such states *nodes*, and the actions people take to proceed from one node to another—*operators*. To elaborate the criteria for his classification system, Landa used terminology similar to Simon and Newell's.[24] In his English-language publications, he used the word *field* to describe a notion corresponding to Simon and Newell's *problem space*. *Object* stood for what American cognitivists named *node*, and the terms *transformation rule* and *action* were similar to Simon and Newell's *operator*.[25]

Drawing on Newell and Simon's terminology, Landa formulated four general types of problems: algorithmic, semi-algorithmic, semi-heuristic, and heuristic. To be solved, each of these problems required a corresponding prescription. Algorithmic problems had a known set of transformation rules that led to the problem solution. Finding the common denominator of two numbers is an example of an algorithmic problem. According to Landa, to solve such a problem, one needed to follow an algorithmic prescription that fully determined the solver's thinking.

In contrast, when solving semi-algorithmic problems, one knew only what *may* be done to have the problem solved. When solving such problems, one has to choose the solution through trial and error, but one does not have a strict prescription regarding what transformations should be made in a problem set. Therefore, solving such second-type problems required semi-algorithmic prescriptions. According to Landa, everyone who followed such prescriptions solved the semi-algorithmic problem correctly, even though their solutions were not identical.[26]

The third type was semi-heuristic problems. One does not know what transformations should be performed to get from the initial to the goal state, nor does one have a set of options to choose from. Instead, one has to search for a solution outside of the problem set, which means that one first has to identify a new conceptual field which may contain the solution: "It is clear that problems of the third type cannot be solved by choosing, since they have an undefined field of choice (alternatives from which to choose are not given and are unknown before the problem-solving process begins, although they may be stored in memory)."[27] Landa continues: "The difficulty posed by these problems consists not in knowing which alternatives to choose and test, but in determining the field in which a solution is to be found and in making the transition into this field."[28] When solving problems of the third type, humans use semi-heuristic rules rather than algorithms. The following instructions to edit a text is an example of a semi-heuristic prescription and semi-heuristic problem: "When in two neighboring sentences the same words are repeated, and this produces an impression of monotony, then replace one of them by its synonym. If no synonym comes to mind, consult a dictionary of synonyms."[29] All students following semi-heuristic prescriptions solve a problem, however, not all of them solve the problem correctly, and not all of them arrive at the same solution.

While people apply their prior knowledge to find a solution to semi-heuristic problems, the solvers of the fourth type of problems have to discover this knowledge "by an active process of cognition, yielding new information."[30] Landa classified such problems as heuristic ones. They are solved with heuristic prescriptions that help discover "a property of some object which was so far unknown" or "a principally new course of actions or a new procedure."[31] The following are examples of heuristic prescriptions: "examine the object from various vantage points," "recall a similar problem," "try to apply some other method if the previous one was not successful." Landa explained that "such instructions specify the process of solving a problem to a still lower degree than semi-heuristic ones and do not guarantee the solution of a problem."[32]

This typology of problems and their corresponding cognitive procedures received the name of the Algo-Heuristic Theory. The algorithmic and heuristic prescriptions of the AHT provided a model of "general methods of thinking" and, at the same time, offered the theory and methodology of teaching and learning such methods.[33] On the one hand, Landa's conception of algorithmic and semi-algorithmic problems and mental procedures was premised on Newell and Simon's theory that one is provided with a field of solution variants to choose from when solving a problem. On the other hand, Landa's idea of heuristic and semi-heuristic problems and cognitive processes was based on Pushkin's approach, which maintained that there are no ready problem variants to choose from and, instead, the mind itself has to construct possible problem solutions.

According to Landa's theory, heuristic thinking procedures fully represented creativity, while algorithmic cognitive processes were most remote from human creative thinking. He explained that algorithmic procedures were most determined by instructions, while the solution of heuristic problems was a less regulated process:

> The creative process occurs in those situations in which the solution is not directly determined either by instructions or by past experience. Of course, a creative problem may be solved only on the basis of experience, but creative processes are characterized by the fact that in them experience must be retrieved, that a specific non-algorithmic search through experience (i.e. knowledge stored in memory) must take place, as a result of

which experience yields objects (images, concepts, actions) which would not have been recalled in and of themselves (automatically, associatively).[34]

Two elements were essential to creativity, according to the AHT: self-organization and independence. Once a prescription began to determine one's thinking procedures, "independence and self-organization disappear" from the problem-solving process.[35] Landa believed that Newell and Simon's heuristic programs left no space for independence and self-organization. Therefore, they could only imitate the cognitive procedures responsible for the solution of noncreative tasks. Landa even called American heuristic programs "incomplete algorithms" because they highly determined the process of problem-solving.[36]

Landa believed that Hungarian-born mathematician George Pólya developed the most adequate heuristic and semi-heuristic rules. A Stanford professor, Pólya identified two kinds of reasoning employed in mathematical reasoning: demonstrative and plausible. The former "has rigid standards, codified, and clarified by logic (formal or demonstrative)." Many think that mathematicians rely solely on demonstrative reasoning. However, in much of his work, including his famous *Mathematics and Plausible Reasoning* and *How to Solve It*, Pólya aimed at dismantling that myth.[37] He showed that, contrary to the popular assumption, before they make a proof, mathematicians often have to "guess a mathematical theorem." Pólya posited that the "hazardous, controversial, and provisional" plausible reasoning was at the heart of creative thinking. His work inspired Newell and Simon's heuristic programming. During his undergraduate years at Stanford, Simon, ostensibly, even took all Pólya's courses.[38] However, in the 1980s, Newell would admit that the heuristic rules used by AI practitioners, for the most part, were quite different from Pólya's prescriptions.[39] Landa's vision of heuristics was more in line with Pólya's theory than the work of his American counterparts.

Pólya developed a set of heuristic rules that could introduce students to the art of guessing and, hence, creative thinking. Landa provides one of his rules: "If you are unable to solve a problem, try to remember some similar problem the solution to which you know."[40] This rule,

Landa explains, is rather unspecific, but "it elicits search for similar problems" and, therefore, may guide the problem-solver to a solution.

Based on Pólya's heuristics and his own experimental work with geometry students, Landa developed the following list of heuristic rules:

1. Begin solving a problem by looking to see what is given and what is to be proved; separate the two.

2. Draw the most direct and obvious conclusions from the given information. For example, if the problem says "given an isosceles triangle," ask yourself the following questions: "what properties of isosceles triangles do we know?" . . . We emphasized that this rule had to be followed in order to connect the givens of the problem with that which was to be proved.

3. Now proceed to that which is to be proved, and ask yourself the question: "Which attributes are sufficient in order to prove that the given figure to be proved is such-and-such?"

4. Think over, to yourself, all the sufficient attributes of the figure to be proved that you know; compare some or even each of them with the given information and with the diagram and then decide which of them seems to be best for proving what you need to prove. Then search among the sufficient attributes for the chosen attribute; then search among the sufficient attributes the next attribute; and so on. If one attribute does not work, or seems not to be promising, try another one.

5. Isolate from the diagram the figures and elements between which you must prove some relationship (for example, equality of angles or segments), include these elements into various relations with other possible elements, and keep asking yourself: "What are these elements and what else could they be at the same time?" Try to answer this question by looking for all possible relationships between the elements in question. We emphasized that this is what is meant by "seeing" or "looking at" a diagram. In showing the subjects a systematic approach to examining a diagram based on consecutively correlating each element with other elements and, thus, regarding it from all

points of view, we at the same time showed them how to "rethink" or "reconsider" or "reconceptualize" it.

6. If the diagram contains none of the figures or elements which are necessary in order to use the attributes which you thought of above, then construct the ones you need.

7. If the question: "What else could these elements be?" leads you to see certain elements as elements of a geometrical figure which is not in the given diagram, but the properties of which could be useful in your proof, then construct this figure.

8. While you are constructing these figures, draw all possible conclusions from them—use all of the attributes which result from these constructions.

9. Remember what was given in the conditions of the problem, and, in case of difficulty, check to see if you have forgotten something.

10. Since difficulties can also arise as a result of not following the rules, when you are having difficulties, refer again to the rules and check to see if you have forgotten to apply one of them.[41]

Landa's AHT sought to resolve the apparent tension between control and creativity. He posited that following his heuristic rules, ultimately, students internalized them, meaning that they maintained independence and self-organization, the key attributes of creative thinking in Landa's theory. In other words, as opposed to algorithms that fully determine one's actions, heuristic rules were the means of "implicit [learning] control."[42] Thus, in Landa's mind, rule-boundedness and creativity were not opposed to each other. The former could cultivate the scientific creativity of problem-solving.

While Landa had written his algorithmic and semi-algorithmic prescriptions in logical notation back in the 1960s, he never formalized his heuristic and semi-heuristic prescriptions. On the one hand, at that time, Landa did not have the logical means for the formalization of his prescriptions. On the other hand, he emphasized that his prescriptions were designed first of all for humans, and, therefore, they did not even need to be as elementary as computer algorithms.[43] It seems that he understood that the translation of human processes into computer operations is always alienating. For example, Stephanie Dick has shown that Newell and Simon, who wanted to re-create the practice of problem-solving in the machine, had to change it into something new to accommodate the affordances of the computer.[44] By keeping the human in the loop, Landa avoided the alienating traps of full formalization.

Expert Training in the United States

Landa's work on the AHT commenced in the USSR. However, his monograph, *Instructional Regulation and Control: Cybernetics, Algorithmizing, and Heuristics in Education,* which offers a full-fledged description of the AHT, never came out in the Soviet Union. Instead, it was published in English in the United States in 1976, the same year Landa fled the Soviet Union and settled down in New York City.[45] A year earlier, Landa's son Boris, a Soviet dissident and one of the founders of the Moscow chapter of Amnesty International, had followed his American wife to the United States.[46] Boris's immigration cast a deep shadow on Landa's reputation as a trustworthy Soviet citizen and scientist. The APN lowered his professorship position to a mere research fellow. As a persona non grata in Soviet scholarly communities, he had no choice but to follow his son.

This geographic transition marked a remarkable career shift in Landa's life. His AHT found a practical application on US soil. A former state scientist, in New York, Landa became a business consultant. He founded his own company, Landamatics International, which consisted of "himself, a part-time secretary, and an office in a run-down building in Queens."[47] The company developed training materials for the employees of private businesses and government organizations. Landamatics' exclusive offer was that its training materials could improve companies' productivity by 75 to 90 percent.[48] At different times, Landa used AHT to develop employee training materials for American social services offices, banks, insurance companies, and industries.[49]

Drawing on AHT, Landa developed algorithmic and heuristic prescriptions for white- and blue-collar workers. Each prescription was unique for each task. Landa explained that to perform manual or intellectual tasks, employees have to make a large number of mental operations. Yet, he claimed, only 10% to 20% of workers are capable of taking up correct mental operations and solving a problem quickly and correctly. Even those 10% to 20% can hardly explain what mental operations allowed them to solve a problem. Others waste time and, therefore, corporate resources looking for, sometimes unsuccessfully, a proper approach to solving a problem. The AHT, on the other hand, could turn a novice into an expert performer in only a few weeks.[50]

The AHT, thus, analyzed and explained the unobservable, nonconscious, and intuitive mental processes that "underlie expert performance, learning, and decision making."[51] For instance, when working for the Iowa Department of Social Services, Landa developed a flowchart that made visible the cognitive processes that stood behind clerks' decisions whether a client was eligible for benefits (fig. 15.1.). The AHT also offered techniques to break down those processes into "relatively elementary component operations"; develop detailed descriptions of expert cognitive operations; and write down algorithmic and heuristic prescriptions that would lead to a problem solution.[52]

From Instructional Design to Knowledge Engineering

During his years as a Soviet scientist and an American business consultant, Landa developed the following definition of intelligence: "Intelligence is a hierarchically organized system of well mastered general methods of thinking, with different methods having different degrees of generality, which provide for the ability to effectively solve problems that have never been encountered in a person's experience."[53] The AHT demystified and made visible general methods of thinking and, therefore, made it possible to develop intelligence in students. In other words, as Landa claimed in one of his interviews, the AHT allowed teachers to "create intelligence."[54]

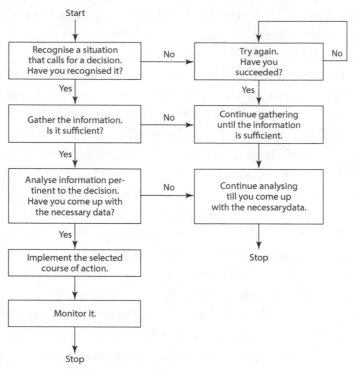

Figure 15.1. A flowchart developed by Landa for the Iowa Department of Social Services.

Lev N. Landa, "The Creation of Expert Performers without Years of Conventional Experience: The Landamatic Method," *Journal of Management Development* 6, no. 4 (April 1987): 47, https://doi.org/10.1108/eb051652.

This definition of intelligence had several implications. First, it put Landa in the same camp with those American psychologists who considered intelligence to be a teachable, rather than an inborn, skill.[55] Second, and more importantly, Landa's definition was quite similar to the one employed in the design of expert systems, special computer programs that work with a specific database and employ heuristic and logical rules to assist humans in decision-making.

In the 1960s, AI practitioners Ed Feigenbaum, Bruce Buchanan, Randall Davis, and Eric Horvitz suggested that the task of creating general intelligence was unattainable.[56] Instead, they proposed replicating domain-specific thinking. This idea was at the heart of the expert system approach, which sought to replicate experts' inductive thinking. Landa's work had many similarities to the mid-twentieth-century expert system named LHASA, whose developer E. J. Corey was deeply interested in the questions of pedagogy, mainly in making synthetic planning teachable.[57] Just like Landa, both in his pedagogical and LHASA work, he sought to identify "the elemental units" of chemical thinking.[58]

Landa's contemporaries recognized the overlaps between the AHT and expert systems design.[59] After his relocation to the United States, Landa chose to refer to his work as *instructional design,* emphasizing that regular teachers and corporate training personnel lacked the qualifications required to analyze cognitive procedures.[60] American expert systems developers also recognized how difficult it is to analyze experts' thinking. For instance, Edward Feigenbaum suggested that "the principal bottleneck in the development of expert systems" was *knowledge engineering.*[61] This term underscored that expert systems were the product of the joint work of computer scientists and experts. In this ensemble, the former observed and uncovered "by the careful, painstaking analysis" how experts solve problems to replicate their thinking computationally.[62]

When the editor of the journal *Educational Technology* interviewed Landa, he asked him what he thought of the overlaps between his instructional design and knowledge engineering. Landa's answer was that, essentially, he and AI practitioners working on expert systems were doing the same thing, as they all worked to elicit and make visible expert thinking. Yet, Landa underscored that while expert systems served as a "[mental] crutch" to its users, helping them solve a problem but not teaching them how to do it independently of a computer program, the AHT actually sought to develop students' independent capacity for problem-solving.[63] Allowing for some degree of independence was important, for, in Landa's theory, creative problem-solving was bound to independence.

While Landa thought of expert systems as technology that de-skills humans, in the course of the 1980s and the 1990s, educational psychologists began considering the idea of employing the AHT in the design of special educational expert systems. For instance, some scholars suggested using Landa's work in the development of expert systems for accounting training.[64] And at the University of Quebec, educational psychologists Nicole Lebrun and Serge Berthelot drew on Landa's research to build an educational expert system to teach human sciences.[65]

Conclusion

The case of Lev Landa's Algo-Heuristic Theory reveals a little-known site of research and development in artificial intelligence. In the twentieth century, Soviet educational cognitive psychology developed instruments and techniques for the analysis of problem-solving that were quite similar to those employed by AI practitioners. Far from being confined to the military-industrial complex, the analysis of human cognition happened in pedagogical institutions and training workshops. The unique position of Soviet psychology as a pedagogical discipline shaped Landa's keen interest in understanding the very essence of how humans solve problems and think creatively. In contrast, American AI practitioners of that time had little interest in understanding the actual mechanisms of human thinking. Landa's institutional affiliation also led him to the notion of an algorithm as a prescription that guides human—rather than computer—intelligence.

The existing academic accounts of twentieth-century AI have predominantly focused on American research conducted in high-secret military and elite academic institutions. This chapter, however, demonstrates that AI theories were both developed and applied in much more prosaic spheres of twentieth-century life, such as secondary education and corporate training in the Soviet Union and the United States. Thus, the story of Landa's AHT is the story of how ideas about AI seeped into the mundane domains of life, shaping the ways lay minds were learning and thinking in the twentieth century.

Notes

1. Stephanie Dick, "Of Models and Machines: Implementing Bounded Rationality," *Isis* 106, no. 3 (2015): 623–34; Hunter Heyck, "Defining the Computer: Herbert

Simon and the Bureaucratic Mind—Part 1," *IEEE Annals of the History of Computing* 30, no. 2 (2008): 42–51.

2. Alison Adam, *Artificial Knowing: Gender and the Thinking Machine,* 1st ed. (London and New York: Routledge, 1998), 35.

3. Ksenia Tatarchenko, "Thinking Algorithmically: From Cold War Computer Science to the Socialist Information Culture," *Historical Studies in the Natural Sciences* 49, no. 2 (2019): 194–225.

4. For the history of automation of white-collar work, see Jon Agar, *The Government Machine: A Revolutionary History of the Computer* (Cambridge, MA: MIT Press, 2003).

5. Jamie Cohen-Cole, *The Open Mind: Cold War Politics and the Sciences of Human Nature,* reprint ed. (Chicago, IL: University of Chicago Press, 2016); Hunter Crowther-Heyck, *Herbert A. Simon: The Bounds of Reason in Modern America* (Baltimore, MD: Johns Hopkins University Press, 2005); Stephanie Dick, "After Math: (Re)Configuring Minds, Proof, and Computing in the Postwar United States" (Cambridge, MA: Harvard University Press), accessed August 30, 2018, https://dash.harvard.edu/handle/1/14226096.

6. Joy Lisi Rankin argues for the importance of educational institutions in computing research and development in *A People's History of Computing in the United States* (Cambridge, MA: Harvard University Press, 2018).

7. On the cybernetization of Soviet science, see Slava Gerovitch, *From Newspeak to Cyberspeak: A History of Soviet Cybernetics* (Cambridge, MA: MIT Press, 2004).

8. Ekaterina Babintseva, "'Overtake and Surpass': Soviet Algorithmic Thinking as a Reinvention of Western Theories during the Cold War," *Cold War Social Science* (2021): 45–72.

9. "Lev Naumovich Landa," *Voprosy Psikhologii* [The issues of psychology], 1999.

10. Cohen-Cole, *The Open Mind,* chap. "Scientists as the Model of Human Nature."

11. Lev Nakhmanovich Landa, *Algorithmization in Learning and Instruction* (Englewood Cliffs, NJ: Educational Technology Publications, 1974), 11, quoted in E. Babintseva, "'Overtake and Surpass,'" 53.

12. A. I. Berg, "Kibernetika i Pedagogika [Cybernetics and pedagogy]," August 20, 1964, f. 1810. d. 27, ARAN.

13. Babintseva, "'Overtake and Surpass,'" 53.

14. Ekaterina Igorevna Babintseva, "Cyberdreams of the Information Age: Learning with Machines in the Cold War United States and the Soviet Union" (PhD diss., University of Pennsylvania, 2020), chap. "Creativity for the Information Age."

15. Dmitriĭ Aleksandrovich Pospelov, Veniamin Noevich Pushkin, and Vadim Nikolaevich Sadovskiĭ, "Ėvristicheskoe Programmirovanie i Ėvristika Kak Nauka [Heuristic programming and heuristics as a science]," *Voprosy Filosofii,* no. 7 (1967).

16. Pospelov, Pushkin, and Sadovskiĭ, "Ėvristicheskoe Programmirovanie," 45–46.

17. James R. Slagle, *Artificial Intelligence: The Heuristic Programming Approach* (New York: McGraw-Hill, 1971), 3.

18. A. Newell, J. C. Shaw, and H. A. Simon, "Elements of a Theory of Human Problem Solving," *Psychological Review* 65, no. 3 (1958): 151–66; Dick, "After Math"; Crowther-Heyck, *Herbert A. Simon.*

19. Veniamin Noevich Pushkin, *Operativnoe Myshlenie v Bol'shikh Sistemakh* [Operational thinking in large systems] (Moskva: Ėnergiía, 1965), 21–23.

20. Veniamin Noevich Pushkin, *Ėvristika i Kibernetika* [Heuristics and cybernetics] (Moskva: Znanie, 1965), 43.

21. V. N. Pushkin, "Heuristic Aspects of the 'Man-Large System' Problem," *IFAC Proceedings* 2, no. 4 (1968): 718.

22. For the description of the experiment, see L. N. Landa, "Some Problems in Algorithmization and Heuristics in Instruction," *Instructional Science* 4, no. 2 (July 1975): 104.

23. Lev Nakhmanovich Landa, *Instructional Regulation and Control: Cybernetics, Algorithmization, and Heuristics in Education* (Englewood Cliffs, NJ: Educational Technology, 1976), 118.

24. In many of his publications, Landa admitted that Simon, Newell, and Shaw's work on the formalization of geometrical proofs was "analogous" to his 1960s research. For example, see L. N. Landa, "Teoretichekie Problemy Algoritmizaťsii i Ėvristiki v Obuchenii [Theoretical questions of algorithmization and heuristics in teaching]," *Voprosy Psikhologii* [The issues of psychology] 4 (1975): 64.

25. I have not found any of Landa's publications in Russian where he uses these three terms. One of his Russian articles, however, discusses *oblast* of problem which, if translated directly, means "field," and, judging from the context, is similar to Simon and Newell's problem space. Lev Nakhmanovich Landa, "Priem, Metod, Algoritm [Technique, method, algorithm]," *Voprosy Psikhologii* [The issues of psychology] 4 (1973): 71–83.

26. Landa, *Instructional Regulation and Control*, 147.

27. Landa, *Instructional Regulation and Control*, 140.

28. Landa, *Instructional Regulation and Control*, 140.

29. Landa, *Instructional Regulation and Control*, 147.

30. Landa, *Instructional Regulation and Control*, 140.

31. Landa, *Instructional Regulation and Control*, 140.

32. Landa, *Instructional Regulation and Control*, 147.

33. Lev N. Landa, "Landamatics Instructional Design Theory and Methodology for Teaching General Methods of Thinking" (Annual Meeting of the American Educational Research Association, San Diego, CA, 1998), 3.

34. Landa, *Instructional Regulation and Control*, 163.

35. Landa, *Instructional Regulation and Control*, 166.

36. Landa, *Instructional Regulation and Control*, 111.

37. George Pólya, *Mathematics and Plausible Reasoning: Patterns of Plausible Inference* (Princeton, NJ: Princeton University Press, 1954); Pólya, George, *How to Solve It: A New Aspect of Mathematical Method* (Princeton, NJ: Princeton University Press, 1945).

38. Dick, "Of Models and Machines," 627.

39. Allen Newell, *The Heuristic of George Polya and Its Relation to Artificial Intelligence* (Pittsburgh, PA: Department of Computer Science, Carnegie Mellon University, 1981), 46.

40. Landa, *Instructional Regulation and Control*, 167.

41. Landa, *Instructional Regulation and Control*, 256.

42. L. N. Landa, "Teoretichekie Problemy Algoritmizat͡sii i Ėvristiki v Obuchenii," 72.

43. Lev N. Landa, "The Creation of Expert Performers without Years of Conventional Experience: The Landamatic Method," *Journal of Management Development* 6, no. 4 (April 1987): 47.

44. Dick, "Of Models and Machines."

45. Landa, *Instructional Regulation and Control;* "Lev Naumovich Landa."

46. "Lev Naumovich Landa."

47. "Lev Naumovich Landa."

48. Otis Port, "Lev Landa's Worker Miracles," *Businessweek* 3284, September 21, 1992, 72.

49. Port, "Lev Landa's Worker Miracles." That was far from the only case of the exchange of cybernetics-inflected methods of management and governance between the Western and the Eastern blocs. For an analysis of the Soviet and American co-production of the cybernetic instrumentation of control, see Egle Rindzeviciute, *The Power of Systems: How Policy Sciences Opened Up the Cold War World* (Ithaca, NY: Cornell University Press, 2016).

50. Landa, "The Creation of Expert Performers," 42.

51. Landa, "The Creation of Expert Performers," 43.

52. Landa, "The Creation of Expert Performers," 43.

53. "The Improvement of Instruction, Learning, and Performance Potential of 'Landamatic Theory' for Teachers, Instructional Designers, and Materials Producers: An Interview with Lev N. Landa (Part Two)," *Educational Technology* 22, no. 11 (1982): 14.

54. "Improvement of Instruction," 14. On notions on intelligence embedded in American expert systems, see Stephanie A. Dick, "Coded Conduct: Making MACSYMA Users and the Automation of Mathematics," *BJHS Themes* 5 (2020): 205–24.

55. Arthur Whimbey, *Intelligence Can Be Taught* (New York: Bantam Books, 1976); Robert J. Sternberg, "How Can We Teach Intelligence?," *Educational Leadership* 42, no. 1 (1984): 38–48; David Perkins, *Outsmarting IQ: The Emerging Science of Learnable Intelligence* (Simon and Schuster, 1995), cited in Lev N. Landa, "Landamatics Instructional Design Theory and Methodology for Teaching General Methods of Thinking."

56. David C. Brock, "Learning from Artificial Intelligence's Previous Awakenings: The History of Expert Systems," *AI Magazine* 39, no. 3 (2018): 7.

57. Evan Hepler-Smith, "'A Way of Thinking Backwards': Computing and Method in Synthetic Organic Chemistry," *Historical Studies in the Natural Sciences* 48, no. 3 (2018): 304.

58. Hepler-Smith, "'A Way of Thinking Backwards,'" 304.

59. James C. Taylor and Noel R. Thomas, "A Knowledge-Engineering Approach to Accounting Education," *Accounting Education* 3, no. 3 (1994): 237–48.

60. "Improvement of Instruction," 9.

61. Edward Feigenbaum and Avron Barr, *The Handbook of Artificial Intelligence,* vol. 2 (Stanford: Heuristech Press, 1981), 84.

62. E. Feigenbaum, *The Art of Artificial Intelligence: Themes and Case Studies of Knowledge Engineering* (Stanford, CA: Computer Science Department, Stanford University, 1977): 4.

63. "Landamatics Ten Years Later: An Interview with Lev N. Landa," *Educational Technology* 33, no. 6 (1993): 11.

64. Taylor and Thomas, "A Knowledge-Engineering Approach to Accounting Education."

65. Nicole Lebrun and Serge Berthelot, "Utilisation d'un système expert pour l'apprentissage de concepts de nature heuristique en sciences humaines au primaire," *Revue des sciences de l'éducation* 19, no. 3 (1993): 463–82.

CHAPTER 16

The Measure of Meaning

Automatic Speech Recognition and the Human-Computer
Imagination

Xiaochang Li

In an unpublished report titled "The Truly Sage System or Toward a
Man-Machine System for Thinking" dated August 1957, psychologist
and computer researcher J. C. R. Licklider laid out a novel vision of
computers not as high-speed calculating instruments but dynamic information systems integrated with both user actions and communication networks.[1] Licklider would go on to become the founding director of ARPA's Information Processing Techniques Office in 1962. There,
he redirected the course of computing research and development in
pursuit of his vision of an interactive information processing and communication technology, funneling a massive defense research budget
into the development of key components for computer interaction, including graphical displays, time-sharing, and distributed networks.[2]

The realization of the computer as an interactive system in close
partnership with human endeavor, Licklider explained in 1957, required
a refashioning of the boundaries between humans and computers. "We
should study the possibility of making the machine encroach upon the
human domain—of making it learn, recognize patterns, get at the essence of a complex idea," he explained, advocating for the inclusion of
artificial intelligence research in interactive computing.[3] "In fact," he
went on to add, "the problem of coupling between the man and the machine would be greatly simplified if the machine could be developed

in part in man's image."[4] That 1957 essay also included a detailed explanation for the design and use of a system that was feasible in the near term using technology already available at the time, with one notable exception. Flagging the page with a footnote, Licklider acknowledged a single function in his hypothetical design that remained "clearly beyond present capacity": it could be operated by speaking to the computer.[5]

This chapter considers the history of automatic speech recognition research in the United States and its curious role in the imagining and implementation of computing as a communication medium. Defined by researchers in the 1950s as an effort to map the "meaning-measurement relation" between the abstractions of language and the material properties of the voice,[6] speech recognition was enlisted as a stage for competing theories in artificial intelligence and human-computer interaction. As such, it became not only a site where a particular understanding of speech and language was being formalized but one in which the boundaries of human and machine perception were being imagined, defined, and negotiated through communication technology.

Licklider's 1957 proposal marked a critical intersection, as speech recognition expanded from acoustics and telecommunications engineering into computer science, appealing to a potent fantasy of frictionless interaction between humans and machines. At the same time, computers themselves were being imagined toward new aims, as growing commercial interests helped recast the computer from a calculating instrument for automating rote clerical tasks into an information system used in support of knowledge production and decision-making, consolidating its technical and managerial potential.[7] Speech recognition thus emerged as a computational problem in a dual sense. As a problem of computational knowledge, it spanned questions of pattern recognition, machine perception, and language understanding, becoming a benchmarking task for investigating the limits of computational reason and the automatic extraction of useful information from data. As a problem of computational practice, speech recognition was envisioned as a means of interacting with computers themselves, shifting its application from the transmission of communication *between* machines to communication *with* machines.

The chapter outlines three episodes in the history of speech recognition in the United States in the twentieth century. Each traces the emergence of a characteristic approach to modeling the "meaning-measurement relation" as speech recognition was defined first as the simulation of physiological and perceptual processes in the body in signal processing, then as the representation of expert knowledge and linguistic intuition in artificial intelligence, and finally as the detection of statistical patterns in speech and text corpora in the lead-up to data-driven machine learning. Part one follows the search for stable, universal patterns of speech in the early twentieth century that merged principles from physiology and experimental phonetics with the economic interests of the telecommunications industry. What began as mechanized traces of bodily movements evolved into electronic templates of auditory perception by the 1950s, as telephone engineers sought to mathematically replicate the sensory-motor properties of speech. Picking up from there, part two tracks speech recognition research through its transition from signal processing into computing in the 1960s, as it was caught up between the conceptual turbulence of defense-funded artificial intelligence research and new visions of interactive computing. Speech recognition in this period took the form of AI expert systems, which extended the simulation of the "measurement-meaning relation" from the physiology of the vocal apparatus and auditory nervous system to include ever-growing layers of linguistic expertise and contextual knowledge as representations of human reasoning. The final section then doubles back to trace the parallel development of the so-called statistical approach led by the IBM Continuous Speech Recognition group in the 1970s, which abandoned the search for measurement-meaning relations altogether. In a radical departure, speech recognition was turned away from the simulation of human speech and reasoning processes and toward data-intensive, statistical pattern recognition, replacing the speaking body with the text corpus. Through these, the history of speech recognition offers a story of how the parameters of computational knowledge was tethered to a problem of embodied communication, and how speaking bodies and recognizing machines were mutually defined in their wake.

Vocal Gestures and Artificial Ears

Speech recognition began close to the body, growing out of experimental phonetics in the late nineteenth and early twentieth century as researchers turned instruments from physiology toward the investigation of language. Language was brought into the laboratory as "a source of communication" through the physiology and acoustics of speech and made available to visual and mathematical analysis with the aid of graphical inscription and sound recording technologies.[8] A scattering of partially realized devices for automatic speech recognition emerged as efforts to attach the study of speech physiology to principles of sound transmission, proposing to uncover standard forms for speech sounds that could be encoded as a substitute for stenographic shorthand or as a means to operate a typewriter.[9]

These devices were recognition systems in theory only, with components for identifying the traces of the voice that were proposed but never implemented. Instead, their inventors distinguished their devices from mere inscription by claiming to have identified a "compact system of natural characters" that directly captured the "true nature" of speech in contrast to the abstract representations of the written alphabet.[10] Quite notably, the reliable production of these "natural characters" was achieved only through the use of highly affected forms of speech, such as whispering, requiring deliberate modifications to the process of speaking.[11] In other words, what made these physiological traces more "natural" was the reduction of acoustic variation rather than the absence of alteration, thus equating the "true nature" of speech with uniformity of measurement.[12]

The idea that the linguistic identity of speech sounds could be reduced to a set of uniform and stable measurements made the prospect of speech recognition appealing to communication engineering. Formal research programs for speech recognition were established within industrial telecommunications laboratories following World War II, resulting in the first wave of operational recognition systems in the 1950s.[13] Most influential among these was the work of AT&T Bell Laboratories, where speech recognition was pursued as a part of longstanding economic interests in devising more efficient methods of speech transmission by studying the capacities of the body.[14] Speech recognition grew from ongoing research to isolate the essential "information-bearing elements" within the acoustic signal that made speech intelligible to the human ear and consequently to reduce the amount of bandwidth required to transmit speech by removing extraneous or "redundant" acoustic information from the signal.[15]

AT&T's overwhelming influence over the first decade of speech recognition research established a transmission-centered view of speech based on acoustic-phonetic principles, which presumed a relatively stable correspondence between measurable acoustic features and phonetic categories of language. The process of speech recognition was thus narrowed to a task of acoustic classification, of sorting segments of the acoustic signal into the appropriate phonetic groupings. This premise, along with AT&T's overwhelming influence over the field, was secured in large part through the broad adoption of the sound spectrograph, an instrument developed by Bell Laboratories as an aid in cryptanalysis during World War II.[16] Considered a breakthrough instrument for visualizing speech, the spectrograph consolidated a diverse range of speech analysis frameworks, from language education to electronics manufacturing, around the investigation of the acoustic signal.[17] One of the spectrograph's principal designers, Director of Transmission Research Ralph K. Potter, also saw its potential use in automatic speech recognition, proposing a program to develop "speech operated devices" based on the spectrograph in 1946.[18]

Crucially, the sound spectrograph further anchored the design of speech recognition by machine in the simulation of human processes, though it relocated the bodily referent from the physiological traces of the voice to the perceptual capacities of the ear. The spectrograph recorded areas of acoustic energy concentration along the frequency spectrum and, unlike the previous instruments such as the oscilloscope, was able to show the changing intensities across distinct "bands" of component frequencies rather than as a single waveform. Drawing a link to earlier physiological models of speech analysis, Bell Laboratories engineer Gordon Peterson likened the spectrographic images of frequency patterns to "vocal gestures [that] seem to carry most of the essential significance of speech sounds."[19] However, the tracking of "vocal gestures" was only the first step to recognition, as it had "long been recognized . . . that not all of the details present in the sound spectrogram are essential to the interpretation of speech sounds."[20] Potter himself began working to devise a new version of the spectrograph that could perform as an "artificial ear." This new "aural spectrograph" was envisioned by Potter to replicate the ear as an acoustic analyzer, transforming the absolute frequency values from raw spectrographic data into patterns that purportedly visualized acoustic information as perceived by the ear's nervous system (fig. 16.1).[21]

Potter's "artificial ear" became the basis of the Automatic Digit Recognizer (nicknamed "Audrey") at Bell Laboratories, which identified the spoken digits 0–9 for a single speaker. First demonstrated in 1952, the Audrey became one of the earliest and most influential speech recognition systems, introducing a "template-matching" approach that became the dominant method for speech recognition into the 1960s.[22] As envisioned in the "artificial ear," this method presented acoustic classification as a proxy for recognition, transforming spectrographic data from speech input into an acoustic pattern, or template, which was then compared to a set of reference templates stored in the memory circuits of the computer to determine the nearest match. The process for transfiguring the direct measurement of "vocal gestures" of the body into the acoustic templates of the artificial ear required a sequence of mathematical transformations that Potter, along with J. C. Steinberg, Gordon Peterson, and Harold Barney, believed to closely simulate the "normalizing transformations" carried out within the inner ear.[23] Though it is not possible to go into technical detail here, these templates were constructed to reduce the precision and granularity of spectrographic data in an effort to accommodate the acoustic variability of the speech signal. Early

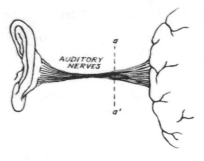

FIG.1 SCHEMATIC OF EAR AND AUDITORY NERVES TO BRAIN

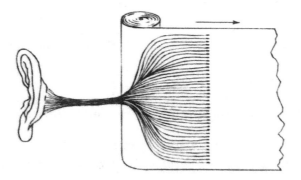

FIG. 2 HYPOTHETICAL CASE IN WHICH NERVE FIBERS ARE
SEPARATED, EQUIPPED WITH RECORDING TIPS AND
ARRANGED ACROSS A MOVING BAND OF RECORDING
PAPER IN THE ORDER OF THEIR TERMINATION ALONG
THE BASILAR MEMBRANE.

Figure 16.1. Illustrations of the "Aural Spectrograph."
Ralph K. Potter, "The Aural Spectrograph—Case 37874-13," Technical Memoranda, AT&T Bell
Laboratories, January 28, 1947. Courtesy of AT&T Archives and History Center.

speech recognition systems like the Audrey thus presented a complicated tension between abstraction and embodiment, enlisting methods
of abstraction in an effort to more closely simulate the body.

Template-matching systems, however, were not the only method of
speech recognition in the 1950s. A handful of systems were developed
that sought to reduce the amount of storage and processing required
to maintain reference templates of every word or phonetic element
contained within the recognizer's vocabulary of accepted utterances.
These relied on a "rules-based" assessment that classified speech utterances using phonetic rules regarding acoustic properties in speech

rather than through comparisons to individual reference patterns. These systems, even when based on the same underlying acoustic-phonetic principles of template-matching devices, anticipated two defining features in the AI expert systems that would take over from acoustics-centered recognition by the 1970s. First, the process for determining acoustic rules was far less standardized than those for generating templates, and often determined in an ad hoc manner through the judgment of any phoneticians that research teams had access to.[24] And second, rules-based models offered a more generic format than acoustic templates, which were designed specifically around the display and calculation of spectrographic data. As a result, they offered a more flexible modeling approach for the incorporation of non-acoustic factors in speech recognition.

Computational Symbiosis and Engineered Knowledge

A new force began to uproot speech recognition from its foundations in physiology and acoustics in the 1960s in the form of digital computing, with wide-ranging implications for the mutual definition of humans and machines. The shift in focus from signal compression to computer interaction transformed the relationship between humans and machines that speech recognition was imagined to manage. Earlier tasks centered around efforts to optimize transmission, such as voice-controlled crossbar switches for telephony, were replaced by an interest in using speech input for command-and-control, in aid of integrating computers into knowledge production and decision-making as an active participant. Speech recognition was refashioned under the pressure of changing institutional aims and technological circumstances, bringing new approaches for mapping linguistic meaning to acoustic measurement.

By the end of the 1950s, speech researchers were coming to realize that the acoustic characteristics of speech alone were not sufficient for recognition. Correspondences between acoustic features and phonetic units were neither as unique nor as stable as previously believed, at least once vocabularies expanded beyond a few words. Researchers acknowledged that the speech signal may simply carry far too much of the messiness of the body to be resolved by acoustic analysis alone, however elaborate the mathematical compensations enlisted to the task, and began to seek out ways to incorporate language information and contextual knowledge alongside acoustic measurements. Growing interest in computing galvanized this expansion beyond acoustic classification by simultaneously offering faster, more flexible tools for research and experimentation as well as new sites of application.[25]

Computing gave speech recognition a glamorous use-case, one that unhooked its value from the economic interests of the telecommunications industry and tied its purpose—and funding—to matters of artificial intelligence, interaction design, and national defense.[26] As highlighted in this chapter's opening reference to Licklider's hypothetical "man-machine thinking system" in 1957, advocates for speech recognition in computing research considered speech input capabilities to be a pivotal step in tapping into the full potential of computers, a position that Licklider elaborated in his landmark 1960 article, "Man-Computer Symbiosis."[27] Opinions regarding the utility of speech recognition, however, were far from uniform, and arguments over its feasibility and desirability staged a larger debate about what it was that computers were not only capable of but also for whom they were intended. That is, debates surrounding speech recognition were effectively debates around both the purpose of the computer and the identity of its user.

In 1961, a panel entitled simply "What Computers Should be Doing," was held at MIT as part of its centennial celebration lecture series, "Management and the Computer of the Future." The panel featured a lecture by J. R. Pierce, director of communications research at Bell Laboratories, with Vannevar Bush acting as moderator, Claude Shannon and Walter Rosenblith as discussants, and an additional host of notable figures in computing, artificial intelligence, and machine sensing, including Marvin Minsky, J. C. R. Licklider, John McCarthy, Hubert Dreyfus, and Paul W. Abrahams contributing from the audience. Pierce's lecture focused on what he saw as a failure to confine computing research to problems that were better suited to computers than to humans. "The fact that a general-purpose computer can do almost anything does not mean that computers do all things equally well," he argued, repeating as a refrain throughout his presentation, "Machines are not people."[28] Speech recognition, Pierce maintained, along with related tasks such as optical character recognition and machine translation, while "a favorite pastime of computer experts," was particularly emblematic of this lack of restraint:

> No one can argue that computers cannot read words; they read words off punched cards, perforated tape, and magnetic tape every millisecond (or less). Selfridge has even made a computer read hand-sent Morse Code. . . . What computers do not do very well . . . is to read varieties of type faces in various orientations, or to read handwritten script, or to recognize many words when spoken in a variety of voices. . . . In contrast, my bank's accounting machinery has no trouble with the numbers printed in magnetic ink on my checks, though to me they are strange and obscure. . . . If a machine is to be forced to read records . . . it should not have to thumb through dog-eared pages.[29]

The debate Pierce defined, in other words, centered around the type of material deemed appropriate for computer processing. Speech, handwriting, and dog-eared pages were formats produced directly by the body, inconsistent in execution and capricious in purpose, compared to those that had been constrained to the uniform and explicit standards of the machine in the manner of morse code and magnetic print. At question was thus not the suitability of computers to pattern recognition in general but rather, as Pierce clarified during the panel's general discussion, to a specific class of patterns that "are very important to us [humans]: the spoken word, the written word, and so forth."[30] At stake, in other words, was a question of the role of computers in communication.

Speech and language input, as communication, thus transformed the problem of pattern recognition into a different class of activity. In his remarks following Pierce's presentation, Claude Shannon characterized these "usually difficult and cumbersome" operations as problems of "judgement, insight, and the like."[31] The particular challenge of mapping acoustic measurements to linguistic meaning, burdened with both the variability of speech sounds and the ambiguity of language, encompassed several "central problems in the field of artificial intelligence," making it a compelling site for AI research "beyond purely pragmatic goals of producing a usable tool in man-machine communication."[32] Speech recognition was adopted as an ideal benchmarking task for "expert systems" approaches based in "knowledge-engineering," which rose to prominence in AI research beginning in the 1960s.[33] This approach sought to explicitly represent expert knowledge of the relevant physical, linguistic, cognitive, and contextual processes involved in speech perception and understanding.

In 1971, the United States launched its first major government-sponsored speech recognition project, the ARPA Speech Understanding Research (SUR) program, operated under the Information Processing Techniques Office (IPTO). The project, which funded research across nine academic and industrial laboratories and produced four working systems after the initial period of five years, explicitly operated under a program mandate of artificial intelligence research and prominently featured the application of expert systems approaches.[34] In comparison to the acoustic templates of the 1950s, expert systems dramatically expanded both the scope of human faculties being modeled and the corresponding complexity of the formal procedures required to model them, growing from reference patterns of sensory-motor functions to elaborate logical architectures that sought to represent, integrate, and prioritize a myriad of physical, physiological, psychological, and contextual knowledge sources. What remained intact, however, was an unquestioned dedication to the direct and explicit replication of human speech and reasoning processes.

Allen Newell, as chair of the SUR project steering committee, in fact listed the direct emulation of human speech understanding as the *primary* "scientific payoff" of successful speech recognition.[35] Beyond its functional role as an exemplar of computational reasoning, speech recognition was of particular interest to AI researchers as an essential means of communication with computers. As Newell further explained, "To keep the channel to computers closed to speech, is to cast computers into . . . a device to be used only deliberately and in constrained ways. To permit speech with computers is to open them up to a wider participation in human affairs."[36] Speech recognition, in its communicative capacity, was framed once again as an essential object of study in relation to digital computers, imagined to be integral for their broad adoption as general information systems rather than specialized instruments.

Noise in the Channel

In January of 1972, just a few months after the SUR project began, IBM launched its Continuous Speech Recognition (CSR) group at the Watson Research Center in New York. There, a team of researchers would spearhead efforts to pry speech recognition away from the study of human sensory and reasoning processes and toward the production of massive quantities of computer-readable speech and text data. Rather than integrating increasingly complex layers of linguistic and contextual knowledge in the effort to mimic human speech perception and understanding, the team at IBM abandoned knowledge representations in favor of brute-force pattern recognition trained on speech and text data corpora. In a drastic departure from previous approaches, the IBM CSR group were not particularly concerned with emulating the production, perception, or interpretation of speech. Fred Jelinek, the group's director and an information theorist by training, framed their approach as a conceptual shift away from the fixation on human processes, explaining, "We thought it was wrong to ask a machine to emulate people. Rather than exhaustively studying how people listen to and understand speech, we wanted to find the natural way for the machine to do it."[37]

Though frequently referenced as the "statistical approach," what truly defined the work of the IBM CSR group was less the inclusion of statistical techniques than it was the rejection of the field's founding premise: that machine recognition depended on the replication of human faculties. On a purely technical level, IBM's approach was, in fact, based on the same information-theoretic concepts that undergirded the spectrograph and acoustics-centered speech recognition at

Bell Laboratories. What had changed, however, was the status of the speaker. The body was no longer the source of the speech signal, whose physiological and cognitive processes could be analyzed to uncover the linguistic identity of speech sounds based on a presumed correspondence between acoustic and phonetic features. Instead, the IBM model treated both the acoustic measurements and linguistic content of the speech signal as effectively equivalent by moving the body from the position as the source of the signal and absorbing it into the transmission channel itself.

Drawing from information theory, the CSR group reformulated speech recognition as a "noisy channel" problem. As described in signal processing, the noisy channel problem assumes the existence of a message that is distorted as it passes through a communication channel that contains "noise," or irrelevant variations in the signal that obscure the original message. The task at hand then becomes the process of decoding the distorted output to retrieve the original message from the added noise.

In the previously established approaches using template-matching and expert systems alike, the speaker's body served as a source of input that transmits speech into the recognizer, and the speaker's sensory and interpretive faculties must be effectively reverse-engineered and replicated as the acoustic and language models. The noisy channel approach, in contrast, absorbed both speech and speaker into the transmission channel. As shown in the diagram from the IBM group's watershed 1983 article (fig. 16.2), the speaker is no longer an external source of input, but rather paired with the acoustic processor as a source of noise interference *within* the acoustic channel, from which a prior input must then be retrieved.[38]

As IBM CSR researchers Lalit Bahl, Fred Jelinek, and Robert Mercer explained, "The speaker and acoustic processor are combined into an *acoustic channel*, the speaker transforming the text into a speech waveform and the acoustic processor acting as a data transducer and compressor."[39] The noisy channel approach effectively sutured the speaker and the acoustic processor, making speech part of the "noise" that distorts a message as it passes from sender to receiver. As Jelinek

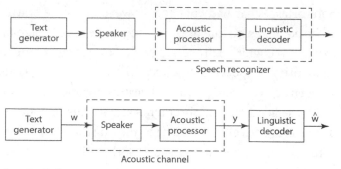

Figure 16.2. Block diagrams comparing the standard view CSR (*top*) with the "noisy channel" model (*bottom*). Note that the "text generator" in these diagrams represents the use of an experimental control.

Lalit R. Bahl, Fred Jelinek, and Robert Mercer, 1983, "A Maximum Likelihood Approach to Continuous Speech Recognition," *IEEE Transactions on Pattern Analysis and Machine Intelligence* 5 (2): 179–90.

elaborated in his 1997 textbook *Statistical Methods for Speech Recognition:* "The total process we are modeling involves the way the speaker *pronounces* the words, the ambience (room noise, reverberation, etc.), the microphone placement and characteristics, and the acoustic processing performed by the front end."[40] Speech and the speaker were together reduced to another component of the transmission process, part of the "total process" of noise interference. As such, they became categorically indistinguishable from not only other sources of acoustic interference like room noise and reverberation but also the material and spatial properties of the recording apparatus and environment, and even the effects of digital compression performed by an acoustic processor.

Speech itself therefore ceased to be of primary concern: "From the point of view of your speech recognizer, all you're interested in are the words; everything else is simply noise."[41] Speech in the body, rather than being the underlying, external object of investigation to be perceived and simulated by the mechanical system, was reduced instead to a component of the data process, one that degrades rather than transmits information. The problem of speech recognition, in short, became one of recognition *despite*, rather than *of*, the distinct qualities of human speech. Speech recognition was in this way conceptually unanchored from the body and reformulated as a general problem of computational pattern recognition. Rather than determining any specific properties of the "meaning-measurement" relation, the noisy channel model consolidated acoustic features and language-use alike as statistical regularities detected in large quantities of speech and text data, and then simply calculated the likelihood of co-occurrence between one set of measurements and another. The transformative abstraction was not a new mathematical codification of the body and the processes of perception, interpretation, and understanding by computational means. It was, instead, the generalization of what it meant to know by computation, in which acoustic measurement and linguistic meaning alike were merely surface patterns in data that could be modeled in the absence of perception, interpretation, and understanding altogether.

In place of Licklider's vision of human-computer collaboration, IBM's approach of data-intensive statistical modeling reframed speech recognition as a purely computational process that was distinct from, it not outright antithetical to, human perceptual faculties and linguistic expertise. This approach overtook AI expert systems by the end of the 1980s, leading to the successful commercialization of speech recognition software and becoming standard across natural language processing.[42] The transformation in speech recognition would also play a critical role in popularizing both the conceptual framework and key technical methods that gave rise to what historian of science Matthew Jones refers to as the "instrumentalist culture of prediction" in machine learning and data science that has come to dominate computational analysis in the past two decades.[43]

Speech recognition thus offers a site to examine the making of computational knowledge as it was burnished against the peculiar problem of communication in the body. The challenge of making the irrepressible variation of the voice and the supple ambiguity of language amenable to the rigors of algorithmic procedure resulted in a complex and ever-shifting epistemic choreography between abstraction and embodiment. The history of speech recognition research and the shifting approach to modeling the meaning-measure relation highlights the ways in which the numerical abstraction of computation and the embodied performance of communication were mutually informing, and how the understanding of what computation is and does was shaped around the tasks that became thinkable and desirable for computers to do.

Notes

1. J. C. R. Licklider, "The Truly Sage System or Toward a Man-Machine System for Thinking," NAS-ARDC Special Study (August 20, 1957): 1–2. Special thanks to Dr. Daniel Nemenyi for sharing a scanned copy of this document amid archive closures.
2. Martin Campbell-Kelly et al., *Computer: A History of the Information Machine*, 3rd ed. (Boulder, CO: Westview Press, 2014), 208. See also M. Mitchell Waldrop, *The Dream Machine: J. C. R. Licklider and the Revolution that Made Computing Personal* (New York: Penguin Books, 2002), 176–77, and M. Mitchell Waldrop, "The Origins of Personal Computing," *Scientific American*, December 1, 2001, 85–91.
3. Licklider, "The Truly Sage System," 9.
4. Licklider, 12.
5. Licklider, 2, footnote.
6. E. E. David Jr., "Voice-Actuated Machines: Problems and Possibilities," *Bell Laboratories Record*, August 1957, 282.
7. Thomas Haigh, "Inventing Information Systems: The Systems Men and the Computer, 1950–1968," *Business History Review* 75, no. 1 (2001): 15–61.
8. Robert Brain, *The Pulse of Modernism: Physiological Aesthetics in Fin-de-Siècle Europe* (Seattle: University of Washington Press, 2015), 69.
9. E.g., W. H. Barlow, "The Logograph," *Journal of the Society of Telegraph Engineers* 7, no. 21 (1878): 65–68, and John B. Flowers, "The True Nature of Speech: With Application to a Voice-Operated Phonographic Alphabet Writing Machine," *Proceedings of the American Institute of Electrical Engineers* 35, no. 2 (February 1916): 183–201. For a detailed analysis of these devices in the context of speech recognition, see Xiaochang Li and Mara Mills, "Vocal Features: From Voice Identification to Speech Recognition by Machine," *Technology and Culture* 60, no. 2 (2019): S129–60.
10. Flowers, "The True Nature of Speech," 183.
11. Flowers, 186–89.
12. Efforts to quantify typical and universal features in speech for the purpose of operating *machines* were embedded within the broader context of efforts to quantify human difference, particularly along lines of race, gender, and disability. Physiological instruments were closely related to the tools of anthropometry and helped extend the classification of bodies from the measurement of features to that of functions and behaviors. See Fatimah Tobing Rony, *The Third Eye: Race,*

Cinema, and Ethnographic Spectacle (Durham, NC: Duke University Press, 1996); Lisa Cartwright, "'Experiments of Destruction': Cinematic Inscriptions of Physiology," *Representations* no. 40 (1992): 135. A consequence of this logic can be seen in the example of racialized surveillance using facial recognition in Cierra Robson's chapter in this volume.

13. K. H. Davis, R. Biddulph, and S. Balashek, "Automatic Recognition of Spoken Digits," *Journal of the Acoustical Society of America* 24, no. 6 (November 1952): 637–42, and H. Olson and H. Belar, "Phonetic Typewriter," *IRE Transactions on Audio* AU-5, no. 4 (July 1957): 90–95.

14. K. H. Davis, "The Cathode Ray Sound Spectroscope," *Bell Laboratories Record*, June 1950, 263. Both Mara Mills and Jonathan Sterne have written extensively about research programs at AT&T that studied speech production and hearing in an effort to reduce the speech signal to only what was essential for perception. See Mills, "Deaf Jam from Inscription to Reproduction to Information," *Social Text* 28, no. 102 (March 20, 2010): 35–58, and Sterne, *MP3: The Meaning of a Format* (Durham, NC: Duke University Press, 2012).

15. Gordon E. Peterson, "The Information-Bearing Elements of Speech," *Journal of the Acoustical Society of America* 24, no. 6 (November 1, 1952): 629–37.

16. According to Allen Newell, who advised the launch of the Speech Understanding Research program at (D)ARPA, "only since the sound spectrograph may we date the present era [in speech recognition]." See Allen Newell, "A Tutorial on Speech Understanding Systems," in *Speech Recognition: Invited Papers Presented at the 1974 IEEE Symposium*, ed. D. R. Reddy (New York: Academic Press, 1975), 4. For the use of the spectrograph in cryptanalysis, see C. H. G. Gray and Bell Telephone Laboratories, Incorporated, "Spectrographs for Field Decoding Work," Final Report, Project 13.3–86 (New York, NY: Communications Division, National Defense Research Committee of the Office of Scientific Research and Development, May 30, 1944).

17. R. C. Mathes and J. C. Steinberg, "Notes on Speech Analysis Conference" (AT&T Bell Laboratories, December 7, 1949), Case 38869–1, AT&T Archives and History Center.

18. R. K. Potter, "Speech Operated Devices—Development Program" (AT&T Bell Laboratories, January 31, 1946), File 38869–1, AT&T Archives and History Center. Potter had originally imagined the spectrograph as an aid to deaf education. See Ralph K. Potter, "Visible Speech," *Bell Laboratories Record*, January 1946, 7.

19. Gordon E. Peterson, "Vocal Gestures," *Bell Laboratories Record*, November 1951: 500–501. Peterson developed the acoustic specification of vowel sounds that was used as the basis of the influential "Audrey" speech recognizer at Bell Labs.

20. Peterson, "Vocal Gestures," 500.

21. R. K. Potter to R. C. Mathes and J. C. Steinberg, "Notes Concerning the Selective Voice Control Problems," October 21, 1947, File 38869–1, AT&T Archives and History Center.

22. Though the Audrey is preceded by Jean Dreyfus-Graf's "sonograph" in 1950, the Audrey is widely depicted within speech recognition field as the first significant system in terms of impact. See, for example, Nilo Lindgren, "Machine Recognition of Human Language Part I—Automatic Speech Recognition," *IEEE Spectrum* 2,

no. 4 (March1965): 117–18. The template-matching approach is also commonly referred to more generally as "pattern-matching," though I have chosen to avoid that term here in order to avoid confusion with statistical pattern recognition methods in machine learning.

23. R. K. Potter and J. C. Steinberg, "Toward the Specification of Speech," *Journal of the Acoustical Society of America* 22, no. 6 (1950): 812. This determination was based on a series of speech and listening experiments conducted by Potter, Steinberg, and others between 1946 and 1951. See Li and Mills, "Vocal Features."

24. Lawrence Rabiner and Biing-Hwang Juang, *Fundamentals of Speech Recognition* (Englewood Cliffs, NJ: Prentice-Hall, 1993), 50, and Alex Waibel and Kai-Fu Lee, "Knowledge-Based Approaches," in *Readings in Speech Recognition,* ed. Alexander Waibel and Kai-Fu Lee (San Mateo, CA: Morgan Kaufmann, 1990), 197. Similar concerns over expert judgment were also raised while knowledge-based systems were still heavily in favor. For instance, in a 1976 overview of the field, Raj Reddy noted that the "most elusive factor" that led to inaccurate labeling in speech recognition was the lack of "objective phonetic transcription" due to variations in the "subjective judgements of phoneticians." See R. Reddy, "Speech Recognition by Machine: A Review," *Proceedings of the IEEE* 64, no. 4 (1976): 519.

25. Edward E. David, Jr., "Computer-Catalyzed Speech Research," in *Proceedings of the Fourth International Congress on Acoustics,* Copenhagen, 1962.

26. B. Beek, E. Neuberg, and D. Hodge, "An Assessment of the Technology of Automatic Speech Recognition for Military Applications," *IEEE Transactions on Acoustics, Speech, and Signal Processing* 25, no. 4 (1977): 310–22. There was both a major influx of defense funding in the 1970s for speech recognition specifically, as well as in many related subfields relevant to military applications, such as voice identification/authentication, language identification, and audio surveillance.

27. J. C. R. Licklider, "Man-Computer Symbiosis," *IRE Transactions on Human Factors in Electronics* HFE-1, no. 1 (1960): 4–11. See also Wayne A. Lea, "The Value of Speech Recognition Systems," in *Trends in Speech Recognition,* ed. Wayne A. Lea (Englewood Cliffs, NJ: Prentice-Hall, 1980), 3.

28. J. R. Pierce et al., "What Computers Should Be Doing," in *Management and the Computer of the Future,* ed. Martin Greenberger (Cambridge, MA: MIT Press, 1962), 300.

29. Pierce et al., "What Computers Should Be Doing," 297–98.

30. Pierce et al., 318.

31. Pierce et al., 308.

32. G. M. White, "Speech Recognition: A Tutorial Overview," *Computer* 9, no. 5 (1976): 41.

33. On the epistemic and practical underpinnings of knowledge-engineering, see Ekaterina Babintseva, "Engineering the Lay Mind: Lev Landa's Algo-Heuristic Theory and Artificial Intelligence," in this volume.

34. Wayne A. Lea and June E. Shoup, "Review of the ARPA SUR Project and Survey of Current Technology in Speech Understanding," Office of Naval Research, January 16, 1979: 11, and Lawrence Roberts, "Expanding AI Research and Founding ARPANET," in *Expert Systems and Artificial Intelligence: Applications and Management,* ed. Thomas C. Bartee (H. W. Sams, 1988), 230.

35. Newell, "A Tutorial on Speech Understanding Systems," 48.

36. Newell, 51.

37. Frederick Jelinek quoted in Peter Hillyer, "Talking to Terminals . . . ," *THINK*, 1987, IBM Corporate Archives.

38. Lalit R. Bahl, Fred Jelinek, and Robert Mercer, "A Maximum Likelihood Approach to Continuous Speech Recognition," *IEEE Transactions on Pattern Analysis and Machine Intelligence* PAMI-5, no. 2 (1983): 179–90.

39. Bahl, Jelinek, and Mercer, "A Maximum Likelihood Approach," 179, italics in the original.

40. Frederick Jelinek, *Statistical Methods for Speech Recognition* (Cambridge, MA: MIT Press, 1997), 7, italics in the original.

41. Roberto Pieraccini, *The Voice in the Machine: Building Computers that Understand Speech* (Cambridge, MA: MIT Press, 2012), 110.

42. On the role of statistical approaches in the success of commercial speech recognition software, see Nils J. Nilsson, *The Quest for Artificial Intelligence* (New York: Cambridge University Press, 2009), 281.

43. Matthew L. Jones, "How We Became Instrumentalists (Again): Data Positivism since World War II," *Historical Studies in the Natural Sciences* 48, no. 5 (2018): 673. It is not possible to go into detail here about the wide-ranging impact of statistical speech recognition on natural language processing and machine learning in general. As a brief example, in a 2011 article published in *IEEE Intelligent Systems,* three research leads at Google depict the company's approach to NLP and other applications of machine learning "at web scale" in terms of lessons from statistical speech recognition and machine translation. See Alon Halevy, Peter Norvig, and Fernando Pereira, "The Unreasonable Effectiveness of Data," *IEEE Intelligent Systems* 24, no. 2 (2009): 8–12.

CHAPTER 17

Broken Mirrors

Surveillance in Oakland as Both Reflection and Refraction
of California's Carceral State

Cierra Robson

Paul Beatty's satirical novel, *The Sellout,* follows an African American
narrator's account of the events that led to his own landmark Supreme
Court case, *Me v. United States,* where he argues for the reinstitution of
legal segregation. The narrator has grown up with a single father, a controversial sociologist, in the "agrarian ghetto" of Dickens near modern Los Angeles. Here, the narrator's father subjects him to "Liberation Psychology" studies, racially charged experiments aimed at proving
that the Black psyche is more humane than its white counterpart. Dickens is crime ridden, drug infested, and otherwise dangerous. After his
father is killed by the police, the state gives the narrator a 2-million-dollar
settlement which he uses to continue to grow and sell watermelons and
artisanal marijuana. One day, the city of Dickens mysteriously dis-
appears from maps of California, instead engulfed by a sprawling Los
Angeles. The narrator determines that the poverty and violence of Dickens was the cause of its removal from maps. Announcing that he plans
to "bring back the city of Dickens," by any means necessary, the narrator takes to literally painting the bounds of the city. The entire community unexpectedly joins in, blocking off streets and highways to paint
a thick white line illustrating the border of the former Dickens. Despite
this community interest, these efforts are fruitless.

One seemingly unrelated day, the narrator presents his work at a
career fair at a racially diverse middle school where the children are so

rambunctious that they are unable to learn. Somewhat jokingly, he suggests that the principal segregate the school by race. Frustrated and looking for any answer, the principal allows it. Much to their surprise, it actually works: the children become calm, engaged, and more open to learning than they had previously been. After witnessing the effects of this segregation campaign, the narrator determines that the best way to bring Dickens back is to reinstitute segregation and modern slavery, as these are the only structures that might make the city more orderly. Continuing his city-establishment project, the narrator begins segregating buses and agrees to a mentally ill neighbor's request that the narrator be his slave master. These racially charged practices of city establishment land him in front of the Supreme Court.[1]

Materializing the themes Beatty saw fit to satirize in *The Sellout*, this chapter explores the mapping technologies of the Domain Awareness Center—a surveillance system that targets communities of color in Oakland, California. I open this chapter with a description of Beatty's novel to raise questions about the connection between race, place, maps, and state violence. In *The Sellout*, the mere practices of painting city lines and of demanding to be included in maps are radical acts. At the same time, these lines are imbued with racial meaning. This fictional account of community engagement reveals the power imbued in the lines drawn on paper. In Oakland, too, community engagement in conversations about digital mapping technologies also bring to light the relationship between cartography, race, and power.

In 2008, the Oakland City Council approved over two million dollars in funding to create the Domain Awareness Center, a citywide surveillance center designed for counterterrorism. Colloquially termed the DAC today, the project is a 10.9-million-dollar city-wide construction that combines streams of surveillance footage from cameras throughout the city with facial recognition technologies, license plate readers, and crime hotspot maps. The project was meant to help first responders appropriately react to terrorist threats, medical needs, and environmental emergencies. The DAC continues to spark heated debates between the residents of Oakland, who believe that the surveillance center will be used to empower the Oakland Police Department—already known for its long history of abuse—and the city officials, who understand the DAC as an opportunity for "social equity."[2]

Though the DAC's construction officially began in 2013, its origin traces back to at least 2008, when Ahsan Baig—the manager of information systems for the city—and Renee Domingo—of the Office of Emergency Management Services—first pitched a 2-million-dollar federal funding proposal to the Oakland City Council.[3] At the same time, cities around the country pitched similar centers in response to a new pool of federal funding that the Department of Homeland Security (DHS) created to curb terrorism after the September 11 terrorist attacks. Between 2001 and 2008, over 52 surveillance centers were created nationwide. Cities, eager to tap into this funding, were enthusiastic to create their own surveillance centers.[4] Though the federal government has rhetorically positioned these centers as a national security measure, protesters have critiqued these surveillance centers for their ability to bolster already strong police departments. These claims were well founded: in 2007, for example, reports found that surveillance centers were being used more for anticrime and other all-hazard functions more than they were being used for antiterrorism.[5]

In Oakland, a city that had never experienced a terrorist attack, skepticism about the true function of the DAC was widespread. Beyond an egregious history of violent and racially charged policing in the city, residents were especially concerned that the cameras used in the system were hyperconcentrated in only the most racially diverse locations. Further, the proposed contractor, the Science Application International Corporation (SAIC), regularly partnered with the Department of Defense to create military-grade weapons. The company's website describes its mission to "keep our military, intelligence, federal, local and commercial customers—as well as our citizens—connected, better informed, productive, and secure."[6] That "our citizens" comes only after an em dash is telling. Positioned grammatically as an afterthought, the well-being of the residents comes second to the interests of government partners.

Concerned that this partnership might facilitate a militarization of the Oakland Police Department, residents protested the DAC, calling for its complete destruction. Instead, the City Council unanimously approved the funding with only one question asked, more in jest than in earnest: "We won't be looking into people's living rooms, will we?"[7] At each subsequent City Council meeting, the group of protesters grew, ultimately forming the foundations of a grassroots antisurveillance organization, Oakland Privacy. The protests reached a boiling point in 2014, when over five hundred residents filled the City Hall to speak out. Instead of dismantling the DAC, the City Council offered protesters a Resident's Council on Privacy. Today, the surveillance center's construction is ongoing, despite continued protest.

For the Oakland protesters, the DAC was not a method of terrorist identification. Instead, the surveillance center was an amplification of local policing tactics. Every trip to the grocery store, walk to school, run in the park was recorded and could be used by the police to build a case against any Oakland resident in ways that were previously impossible. The surveillance center, then, reinforced existing social relations between the police and Oakland residents. More, the DAC represented what Foucault termed "discipline," a method of controlling behavior of residents under the threat of constant supervision.[8]

But beyond constant surveillance, the technology was both massifying and individualizing, simultaneously stripping and assigning individual identity. This, Fanon describes as being returned "spread-eagled": the act of dissecting one's body into smaller and smaller pieces, drawing meaning from each piece, yet still failing to contextualize it.[9] Much like Li's exploration of the abstraction of the human body from speech processing software detailed in "The Measure of Meaning" in this volume, to be returned spread-eagled is to have "fact" read from one's body without having one's personhood considered. In rendering individual bodies as nodes on crime hotspot maps, blobs on grainy surveillance footage, and latitude and longitude coordinates, the DAC dissected each Oakland resident, returning a set of threats to be responded to rather than people to be understood.[10]

At this point, one might be tempted to ask, *but what is so bad about a map?* How do simple depictions of geospatial data become not only political but also racializing under a capitalist structure? How, if at all, do maps fit into the existing theoretical frameworks that link technology and power? Beatty's novel begins to answer these questions. The removal of Dickens from state maps reveals that maps are imbued with state authority and that they often reinforce state violence: rather than rehabilitate an underresourced Dickens, the state literally erased it from its vision.[11]

Though Beatty explores what it means to have those lines taken away, in this chapter, I investigate what it means to have those lines superimposed. I ask, what power do maps carry? I consider how maps, particularly those used by the DAC, come to be laden with racial logics, and how those logics represent an extension *and* transmutation of histories of racism. Importantly, I argue, the technical elements of the DAC are what make the extension and transmutation of these histories possible. It is through the crime hotspots, the facial recognition technologies, and the license plate readers of the DAC that power manifests. These technical functions digitize, reinforce, and reconstitute what sociologist Barbara Reskin calls "the discrimination system," or the set of interdependent, mutually reinforcing racial disparities across domains of social life like housing, education, work, policing, and health.[12]

This chapter argues that Oakland's DAC is both a reflection and a refraction of one particular historic discrimination system: the dialectical relationship between residential segregation and racialized over-policing. In exploring this case, I extend existing historiography of computing to engage further with histories of spatial racism, racial capitalism, and the prison-industrial complex.

In the first part of this chapter, I situate the story of the DAC within the longer history of residential segregation and the California prison boom. I show how the surveillance center reflects this discrimination system by surveilling only the neighborhoods still impacted by the practices of redlining. In the second part of the chapter, I examine how the DAC also refracts this discrimination system: the DAC enabled both governments and companies to extract profit from this system in ways previously unimaginable. In part three, I explore how existing theories of technology and power help explain the historical continuity of the DAC, but inadequately explain the ways that capital makes this discrimination system profitable. Like Nichols does in "Patenting Automation of Race and Ethnicity Classifications" in this volume, I theorize how racially neutral language obscures how companies profit from existing inequalities. I show how the DAC—at once advertised as a counterterrorism project aimed at protecting the people—is built to protect some at the expense of others. As Aidinoff's description of a "Jeffersonian internet" in this volume shows, politics shape the narratives we tell ourselves about technology.

The Reflection: No Innocent Map

The DAC, practically speaking, is a series of detailed maps. Integrating predictive policing technology with real-time footage, the technology combines predictions of where crimes are likely to occur, surveillance footage, and geospatial maps of the area in order to make the Oakland Police and Fire Departments more efficient. Combined with these maps, the DAC integrates facial recognition software and license plate readers into the system.[13] Figure 17.1 shows where the DAC cameras are placed throughout the city of Oakland.

Produced more than eighty years earlier, figure 17.2 was created in 1937 by federal agencies. It depicts the popular practice of redlining. Named for the practice of literally drawing a red line on maps, home brokers used maps like this to delineate who could buy homes and where along racial lines from the 1930s to the 1960s.

Home brokers became the foot soldiers of racial segregation, using these maps to create, maintain, and reinforce racial segregation in cities. The legend in the top right corner of the map describes that the green areas were marked for the "First Grade," the blue for "second grade," the yellow for "third grade," and the red for "fourth grade." More than a simple grouping of the land, these grades were intimately associated with the material benefits of an area. The green sections had the best resources, the red had the worst. While the maps make no explicit reference to race, it was well known that brokers would sell homes in the first grade only to native-born white homebuyers while

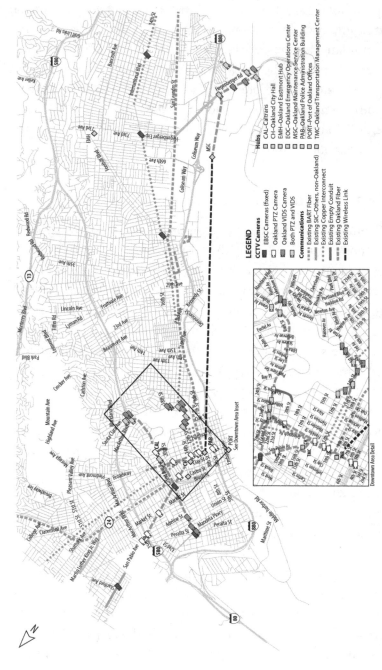

Figure 17.1. Placement of Domain Awareness Center (DAC) cameras in the city of Oakland.

Figure 17.2. 1937 federal redlining map. The darkest areas on the map represent the blue (second grade) areas; the green areas (first grade) are the second-darkest; red areas (fourth grade), located mostly in the southwest, are the third-darkest areas; and yellow areas (third grade) are the lightest. Colors in the original.

they relegated Black residents to fourth grade. Even though redlining was outlawed under the Fair Housing Act of 1968, its effects ripple into today as these logics, even if unspoken, have permeated the housing market.[14] Today, the Lake Merritt Area—a grade 4 red area on the redlining map—houses one of the most diverse groups of residents in the United States.

To our modern sensibilities, maps like this seem flawed and outdated at best, and severely racist at worst. Yet at the time, the government widely recognized the practice of redlining as a legally enforceable positive social force. A tool of state-sanctioned racism, residential segregation maps were used to hoard resources for native born white residents while barring nonwhite residents from these resources. This map is only one example of the ways that graphical depictions of bodies in space are reflective of existing power structures.

Beyond the generalized idea that maps hold power, a comparison of these two maps reveals that the area densest with DAC cameras is the Lake Merritt area, previously highlighted in the redlining maps as one of the few areas Black people were allowed to live in Oakland in the 1930s. Looking at the maps in figures 17.1 and 17.2 side by side, one can see that, more than just a graphical depiction of where things are in space, the map of the DAC cameras reasserts the lasting influence of the redlining. A clear example of digital redlining, the DAC construction entrenches a longstanding tradition of state-sanctioned racism.[15] In tracing this area, both the Oakland City Council and SAIC rely on the assumption that the most diverse areas are also the most dangerous and need the most policing.[16]

This assumption is far from new. After redlining was officially outlawed in the 1960s, the prison population in California skyrocketed to over 450% of state capacity.[17] By the turn of the millennium, the California prison population surpassed that of all US states, despite decreases in crime.[18] Governor Schwarzenegger declared a state of emergency after several Supreme Court cases deemed that the overcrowded prisons were a form of cruel and unusual punishment. By 2008, the federal government mandated that the State of California reduce its prison population to only 137.5% of capacity—a process which would require releasing up to 46,000 prisoners.[19] Thus, the state embarked on a project of "realignment."

Beginning this project, the state released many nonviolent offenders and—importantly—shifted the burden of supervision responsibilities away from the state itself and toward individual counties. Statewide, residents and state officials alike worried that decarceration would increase crime throughout the state. These statewide shifts correspond exactly to the moment at which Oakland's citywide surveillance center was created. After realignment, surveillance technologies likely emerged as a means by which local governments sought to manage individuals released from prisons and quell fears of a potential increase in crime.

Because state funds for such projects were scarce, local governments looked to an ever-increasing pool of federal funding earmarked for counterterrorism surveillance centers. While these funds were specifically earmarked for terror prevention, city officials nationwide began using the centers for crime prevention. In Oakland, antiterrorism is never mentioned during the preliminary meetings or reports about the DAC.[20] Crime deterrence had always been the plan: in a memo detailing the DAC's purpose, the chief of police noted that the system was meant to "lead to the rapid *identification* of those responsible for *crimes* committed in view of a camera, *deter* those who (without the presence of a camera) might seize an opportunity to *prey* on others, [and] support the *prosecution* of *criminals* whose activity is captured."[21] More than looking to prosecute criminals, the DAC was meant to enact social control by regulating the behavior of Oakland residents.

Despite protests, the city's commitment to the project was blatant. While protest against the DAC was largely peaceful, Councilwoman Lynette Gibson McElhaney objected to resident rebuttals, stating, "What has concerned me about protests in Oakland is how destructive they have been."[22] In the same meeting that the DAC was approved, the City Council also approved a restriction on protest rights. The measure banned certain "objects" from all protests, like hammers and spray paint. Any violation of this new ordinance resulted in jail time. It is not a coincidence that the very system which was built to surveil residents deemed criminal was passed on the same day that the definition of criminality expanded to include the means by which residents could thwart the surveillance system.

Oakland is not the only place where technologies like the DAC exist. Worldwide, digital maps buttressed with algorithmic tools continue to proliferate as a new iteration of carceral logics. They aim not only to catch existing criminals but to socialize the population away from crime and even legal protest under the threat of constant surveillance. Ignoring historic practices of racialization and criminalization, city officials deploy the most resources in places deemed the most threatening, even when that reputation has little to do with actual rates of crime.[23] Advertised as objective and innocent, these maps hold the power to convict, indict, and mark as criminal.

Maps, a form of what Bourdieu terms informational capital, have always been central to state control. Like the census and birth certificates, maps have long been used to make residents legible to governments interested in maintaining order.[24] In rendering bodies visible in space, maps operate as a form of biopower, described by Foucault as the state's use of population-level statistics to "make live and let die."[25] Cartography, like any science, has always been a political tool. What we choose to draw, what we choose to neglect, and how we choose to label carry political weight. Though seemingly objective, these features of maps make social logics like racism seem objective. In this way, maps simultaneously deny that social rules exist under the façade of objectivity, while they legitimize this social order. Viewed in this way, maps—even in their simplest form—are a form of surveillance, "spatial panopticon[s]" that come to embody state power.[26]

As the maps of redlining show, state racism is also imbued through cartographic practices.[27] The DAC digitized the interlocking relationship between residential segregation and criminalization. In so doing, it reveals that racist logics continue to both justify and be justified by cartographic representations.

A Radical Cartographic Practice

Just as cartographic practices have been used to track bodies, police space, and reinforce racist ideology, so too might the map be used for liberatory ends. Race theorist Ruth Wilson Gilmore's political-economic evaluation of the California prison boom is a prime example. Trained as a geographer herself, Gilmore creates and explores a new map that explains the deep networks of land, labor, and capital that connect prison to society. Practicing a "radical cartography," Gilmore's map of the California prison boom is fundamentally deconstructionist, and it underscores how false understandings of prisons reinforced by maps reproduce racism under the veneer of objective science.[28] Gilmore's map radically dispels the political myths found in the maps against which she writes.

Similarly, surveillance theorist Simone Browne deconstructs the plan (read: map) of the Brooks slave ship in her discussion of the origins of racialized surveillance. She draws upon the measurements of the space allotted to each slave—"two feet, seven inches"—to discuss the "trauma of the Middle Passage as multiply experienced and survived."[29] She asks, "What does it mean that I now look to this plan, but not from the elevated and seemingly detached manner as it was first intended to be looked upon?"[30] In her deconstruction of a seemingly objective map, Browne is able to launch a theory of surveillance attentive to race, difference, and the multidirectional act of looking.

In their radical readings of maps, these authors provide their own maps to disrupt deeply held beliefs that come to be taken as fact. These beliefs, indoctrinated into the American sociopolitical consciousness through the illusion of cartographic objectivity, shape our institutions. These authors launch a powerful critique of otherwise accepted systematic acts of racism and state violence.

On the ground, the protesters in Oakland also practiced this radical cartographic practice. At one of the City Council meetings, a young Black woman approached the podium to express her concerns about the DAC. Given one minute of speaking time, she introduced herself first as an Oakland resident. She then discussed how her grandparents,

100 and 94 years old, moved from Arkansas to Oakland 60 years ago for a better life and economic opportunity. They saw Oakland as a beacon of hope where they could make a life for their great-great grandchildren. She argued that her family has contributed to the community for over half a century and continues to do so. She concluded by declaring that her family deserves to "live in a peaceful community, not a police state with surveillance."[31] Though short, her comments implicitly reference the South of the 1940s, where economic opportunity for Black people was largely unavailable; they invoke the Great Migration of Black Southerners to the north and west of the country; and they acknowledge that surveillance provides safety for some at the expense of others.

Following from Gilmore, Browne, and the protesters in Oakland, the final section of this chapter practices a cartographic deconstruction of the DAC. As I have outlined the case thus far, one can speculate how the DAC in Oakland might be the next logical step of the California prison boom and state-sanctioned segregation. Yet, memos from the chief of police reveal that despite these links to the prison boom and practices of redlining, surveillance in Oakland was meant to be something *more* than policing alone: the DAC in Oakland was meant to be *profitable*. In providing a new map of Oakland's DAC—one that reveals one of many political-economic feedback loops—I aim to understand the DAC as something different from what has been theorized before.

The Refraction: Duopolies of Power

Systems like the DAC hold immense political and economic incentives all over the county. Microsoft, for example, actively competed for the DAC contract, boasting that it had created a similar system in New York. Correspondence between the company and city officials revealed that the city of New York would reap 30% of all profits that Microsoft made from any sale of its surveillance center technology to other cities. Mayor Bloomberg of New York announced the plan with excitement, noting that the City of New York could "even make a few bucks!"[32] Yet, this deal was not simply a matter of economics: an in-flow of cash favored the mayor's reelection. In a city with over 116 billion dollars' worth of debt, any mayor who makes a deal to regularly infuse millions of dollars into the economy is very likely to be supported.[33] This economic incentive makes it incredibly difficult for residents to launch a political critique: how does a resident of New York choose between their privacy and the well-being of their city?

Mayor Bloomberg understood that money and politics go hand in hand; his comments reveal the deep political and economic networks that underpin the construction of a surveillance technology like those in New York and Oakland. In just a few words, Bloomberg's comments reveal how surveillance is more than just a police tactic: these centers are also used to produce more wealth for the state *and* for multi-billion-dollar companies at the expense of the residents. These networks operate in ways that are distinct from the prison-industrial complex.

Though Oakland did not choose Microsoft as the contractor for the Oakland DAC project, the mission of the selected contractor, SAIC, highlights the state's interest in linking practices of security to the economy. Self-advertised as a government partner, SAIC defined its mission to be to *"defend against attacks and protect the economy."*[34] In many respects, Oakland's DAC adheres to the theory of surveillance capitalism, a new form of capitalism that relies on clicks, user data, and statistics "as free raw material."[35] For the DAC, this "raw material" came in the form of location data, crime data, and video footage. Surveillance capitalism holds that such an economic order displays "radical indifference"—it cares nothing about the people it collects data from, only that it collects data. Yet, Oakland—one of the most diverse cities by population—was the only city in California to deploy such a system: though the state's major metropolitan cities have increased surveillance slightly, the DAC was by far the largest construction up until that point. In fact, shortly after the construction of the DAC, nearby San Francisco passed a citywide ban on similar surveillance technologies.

That Oakland was the only city in California with a surveillance center of this magnitude contradicts the claim of radical indifference.[36] One asks why, if companies do not care whose data they are collecting, they would be so interested in collecting and aggregating the geospatial data about Oakland's residents, one of the state's most diverse populations.[37] Some may suggest that this is because Oakland is at the most risk for terrorist attacks; yet, data indicates that there is little statistical evidence to justify spending state funds on counterterrorism in the United States writ large, let alone in Oakland.[38] News sources fail to produce evidence that Oakland is or has historically been at risk for terrorist attacks. So, if not counterterrorism, what explains this increase in discriminate surveillance despite the theory that it is in the best interest of companies to act otherwise?

This question is in part answered by another. The theory of surveillance capitalism implies that companies and state actors are diametrically opposed. This theory holds that the answer to surveillance capitalism is rule of law. But this presumption takes for granted the ways in which the state itself is implicated in the production of these technologies.

The City of Oakland had always sought to benefit financially from the DAC. In a preliminary report detailing the "fiscal impact" of the construction, the city justifies the cost of the DAC—over 11.3 million dollars in installment costs and 1.5 million dollars in recurring costs—by arguing that the DAC could be an opportunity through which "the City may experience economic growth through business development," which may "bring customers who may contribute to the [local] economy."[39] It was widely recognized by the protesters that these "customers" would likely be the newest residents of the city, wealthy Silicon Valley tech workers searching for an affordable and "up and coming" place to live, as the property values in nearby San Francisco skyrocketed. Both the city government *and* SAIC were mutually profiting from the DAC. The theory of surveillance capitalism, then, is not adequate to explain what is happening in Oakland.

Oakland, then, is more than a continuity with past forms of discrimination systems: it is also something entirely new. The DAC is an exemplar of what I term *racialized surveillance capitalism*, an economic order that claims human experience as free raw material to be collected, extracted, and utilized differentially depending on the social, political, and economic status of the individual being surveilled.[40] Unlike surveillance capitalism, which is premised on a radical indifference toward data, racialized surveillance capitalism recognizes that data becomes more profitable when the capital it collects is not only economic but also political. Thus, it is activated by the state and operationalized by corporate actors, often for political ends. Crime data used to make policing more efficient, for example, carries not only the economic capital inherent to data collection but also the political capital inherent to tough-on-crime rhetoric. Those politicians who encourage the collection of such data will gain political support from both the companies they employ and the citizens to whom this rhetoric appeals. Thus, as critical race theorist David Theo Goldberg reminds us, the line between the political and the economic "likely collapses in the face of calculation, just as it is manufactured by and in the interests of those whose power is identified artificially on one or another side of the dividing line."[41]

When viewed in light of racialized surveillance capitalism, the story of Oakland becomes far more pernicious. It suggests that Oakland's surveillance center is not only the next step in a long history of policing and racism but also the genesis of rogue collaboration between the state and private industry.

Beyond Oakland

Margaret Atwood, author of the dystopic novel *The Handmaid's Tale*, once said: "Within every dystopia, there is a little utopia."[42] And indeed, utopias are dependent on dystopias, just as suburbia would not exist without the ghetto, the citizen would not exist without the illegal alien, and the rich would not exist without the poor. In reinforcing the continuity between historical practices of redlining and policing, the DAC in Oakland is a clear reflection of past discrimination systems. Yet, in its use of racialized surveillance capitalism, the DAC makes these systems profitable. The simultaneous reflection and refraction of California's history combine to produce conditions under which the DAC can be both a utopic security state for the City Council members *and* the dystopic "police state with surveillance" for the Oakland protesters.

These views are not contradictory. Rather, they are mutually reinforcing. As Jasanoff reminds us, "Multiple imaginaries can coexist within a society in tension or productive dialectical relationship."[43] In Oakland, the state's imaginary of security exists in combination with the resident's imaginary for a surveillance-free home. Racialized surveillance capitalism elevates the state's imaginary to reality. Surveillance of those at the margins of society for political or economic gain is a global problem, and an old one. The question then becomes: *what might we do to thwart it?*

Notes

1. Paul Beatty, *The Sellout,* 1st ed. (New York: Farrar, Straus and Giroux, 2015), http://www.netread.com/jcusers2/bk1388/507/9780374260507/image/lgcover.9780374260507.jpg.
2. "Domain Awareness Center (DAC) Draft Privacy Policy and Public Comments," City of Oakland, accessed January 12, 2019, https://www.oaklandca.gov/resources/dac-draft-privacy-policy-public-comments.
3. This preliminary meeting coincided with the fifth anniversary of Oakland's federal oversight, deemed necessary by a federal judge after four rogue police officers were found guilty of beating and framing drug suspects in Oakland and the department was found to seriously underestimate its use of force in reports.
4. Torin Monahan and Neal A. Palmer, "The Emerging Politics of DHS Fusion Centers," *Security Dialogue* 40, no. 6 (2009): 617–36.
5. Monahan and Palmer.
6. SAIC, "About SAIC," accessed December 15, 2018, http://www.saic.com/who-we-are/about-saic.
7. "DAC Videos," Oakland Privacy, accessed November 16, 2018, https://oaklandprivacy.org/videos-community-members-speaking-oakland/.
8. Michel Foucault, *Discipline and Punish: The Birth of the Prison/Michel Foucault;* translated from the French by Alan Sheridan, 2nd Vintage Books ed. (New York: Vintage Books, 1995). See also Jon Penney, "Understanding Chilling Effects," SSRN Scholarly Paper (Rochester, NY: Social Science Research Network, May 28, 2021), https://papers.ssrn.com/abstract=3855619.
9. Frantz Fanon, *Black Skin, White Masks/Frantz Fanon;* translated from the French by Richard Philcox, 1st ed., new ed. (Berkeley, CA: Distributed by Publishers Group West, 2008), http://www.loc.gov/catdir/enhancements/fy0712/2006049607-d.html.
10. See Khalil Muhammad's *The Condemnation of Blackness* for a thorough discussion of the creative ways in which data has been used to criminalize Black and Brown populations. Khalil Gibran Muhammad, *The Condemnation of Blackness: Race,*

Crime, and the Making of Modern Urban America, with a New Preface (Cambridge, MA: Harvard University Press, 2019).

11. For more examples of power embedded in maps and architecture, see the *Funambulist* magazine: https://thefunambulist.net/.

12. Barbara Reskin, "The Race Discrimination System," *Annual Review of Sociology* 38, no. 1 (2012): 17–35, https://doi.org/10.1146/annurev-soc-071811-145508.

13. Reskin, "The Race Discrimination System."

14. Elizabeth Korver-Glenn, "Compounding Inequalities: How Racial Stereotypes and Discrimination Accumulate across the Stages of Housing Exchange," *American Sociological Review* 83, no. 4 (2018): 627–56.

15. Safiya Umoja Noble, *Algorithms of Oppression: How Search Engines Reinforce Racism* (New York: New York University Press, 2018).

16. It is important to note here that even if the threat were real, we must question why that is so. We must understand the many social, political, and economic forces that produce crime and aim to eradicate those forces through social policy and collective action rather than band-aid solutions of policing.

17. Ruth Wilson Gilmore, *Golden Gulag: Prisons, Surplus, Crisis, and Opposition in Globalizing California*, American Crossroads, 21 (Berkeley: University of California Press, 2007).

18. Gilmore, *Golden Gulag.*

19. Gilmore.

20. "DAC Videos," Oakland Privacy, accessed November 16, 2018, https://oaklandprivacy.org/videos-community-members-speaking-oakland/.

21. "Oakland Privacy Advisory Commission Documents." Oakland Privacy, accessed November 16, 2018, https://oaklandprivacy.org/oak-advisory-commission/. Emphasis my own.

22. Will Kane, "Oakland OKs Money for Surveillance Center," *SFGate* (July 2013), https://www.sfgate.com/crime/article/Oakland-OKs-money-for-surveillance-center-4697146.php.

23. These maps often focus only on drug and violent crimes, while ignoring white collar crimes. A satirical play on these tools called the White Collar Crime Zone asks what these maps would look like if they relied on white collar crime data. To no surprise, the largest crime hotspot is Wall Street. For more on how neighborhood reputations are racialized, see Robert J. Sampson, *Great American City: Chicago and the Enduring Neighborhood Effect* (Chicago: University of Chicago Press, 2012).

24. Pierre Bourdieu, Loïc J. D. Wacquant, and Samar Farage, "Rethinking the State: Genesis and Structure of the Bureaucratic Field," *Sociological Theory* 12, no. 1 (March 1994): 1, https://doi.org/10.2307/202032; James C. Scott, *Seeing Like a State: How Certain Schemes to Improve the Human Condition Have Failed*, Yale Agrarian Studies (New Haven, CT: Yale University Press, 1998).

25. Bourdieu et al, "Rethinking the State."

26. J. B. Harley, "Deconstructing the Map," *Cartographica* 26, no. 2 (Summer 1989).

27. Europeans came into contact with those they would later subordinate through the very project of mapping. Maps have come to structure scientific, anthropological,

and religious movements instrumental to our continued understanding of racial difference. David N. Livingstone, "Cultural Politics and the Racial Cartographies of Human Origins," *Transactions of the Institute of British Geographers* 35, no. 2 (2010): 204–21.

28. I use the term "deconstructionist" as Harley does to describe the practices of "Read[ing] between the lines of the map" to "go beyond the stated purpose of cartography." In so doing, he recognizes that "much of the power of the map, as a representation of social geography, is that it operates behind a mask of a seemingly neutral science. It hides and denies its social dimensions at the same time as it legitimates them." Harley, "Deconstructing the Map."

29. Simone Browne, *Dark Matters: On the Surveillance of Blackness/Simone Browne* (Durham, NC: Duke University Press, 2015).

30. Browne, *Dark Matters.*

31. "DAC Videos."

32. "DAC Videos."

33. "Office of the New York State Comptroller—Debt," accessed April 17, 2019, https://www.osc.state.ny.us/finance/finreports/fcr/2018/debt.htm.

34. SAIC, "About SAIC." My emphasis.

35. Shoshana Zuboff, *The Age of Surveillance Capitalism: The Fight for a Human Future at the New Frontier of Power* (New York: PublicAffairs, 2019).

36. Though there are many fusion centers throughout the state of California (including centers in Sacramento, Los Angeles, San Francisco, and San Diego), the only video surveillance center exists in Oakland. While the DAC collects video surveillance of its residents, the Northern California Regional Intelligence Center located in San Francisco, for example, only collects "tips, leads, suspicious activity reporting, and criminal information." "Northern California Regional Intelligence Center—NCRIC," accessed April 1, 2019, https://ncric.org/default.aspx.

37. Compared to the 72% of California's population that is white, Oakland's population is only 32% white. "U.S. Census Bureau QuickFacts: Oakland City, California," accessed February 17, 2019, https://www.census.gov/quickfacts /oaklandcitycalifornia; "U.S. Census Bureau QuickFacts: California," accessed April 1, 2019, https://www.census.gov/quickfacts/ca.

38. Here, terrorism is defined as "the threatened or actual use of illegal force and violence by a non-state actor to attain a political, economic, religious, or social goal through fear, coercion, or intimidation." Max Roser, Mohamed Nagdy, and Hannah Ritchie, "Terrorism," Our World in Data, April 26, 2018, https://ourworld indata.org/terrorism.

39. "DAC Videos."

40. A combination of Shoshana Zuboff's surveillance capitalism and Cedric Robinson's racial capitalism, racialized surveillance capitalism recognizes that race and capital are closely entangled, and in new ways in the digital age. For a call to research at this intersection, see Tressie McMillan Cottom, "Where Platform Capitalism and Racial Capitalism Meet: The Sociology of Race and Racism in the Digital Society," *Sociology of Race and Ethnicity* 6, no. 9 (2020): 441–49.

41. David Theo Goldberg, *The Racial State* (Malden, MA: Blackwell Publishers, 2002).

42. Constance Grady, "Margaret Atwood on the Utopias Hiding inside Her Dystopias and Why There Is No 'the Future,'" Vox, June 9, 2017, https://www.vox.com/culture/2017/6/9/15758812/margaret-atwood-interview.

43. Sheila Jasanoff and Sang-Hyun Kim, eds. *Dreamscapes of Modernity: Sociotechnical Imaginaries and the Fabrication of Power* (Chicago: University of Chicago Press, 2015).

CHAPTER 18

Punk Culture and the Rise of the Hacker Ethic

Elyse Graham

The nineties saw a vogue in universities for calling things "postmodern," and the computer underground was no exception. But we can be more precise: this world wasn't just postmodern; it was punk.[1] However, we might more accurately describe it as *punk*. This chapter shows how the punk subculture profoundly shaped the computer underground in the 1980s and 1990s by instilling the young people who became internet insiders in those years with a do-it-yourself ethos, accompanied by a healthy distrust of authority, that happened to correspond well with the hacker ethic. It helped the hacker ethic to catch on outside of the privileged spheres of Stanford and MIT.[2]

As Michael Mahoney argues, because computing technologies are unusually "protean"—Turing described a computer as a machine that imitates other machines—"The computer . . . has little or no history of its own. Rather, it has histories derived from the histories of the groups of practitioners who saw in it, or in some yet to be envisioned form of it, the potential to realize their agendas and aspirations."[3] Few groups of practitioners have exerted more influence over our collective understanding of high tech than the "punks" who fraternized on underground bulletin boards in the 1980s and 1990s. Drawing on the work of scholars who have discussed the means by which social movements constitute themselves through shared rituals and symbols of belonging, this chapter explores how the community that put *punk* in cyberpunk adopted rituals, habits, and norms that set its members apart from the outside world, creating a framework of attitudes, tactics, and symbols that were separate from technological knowledge but, in practice, were intertwined with it.

Historians have often remarked upon the intersection of punk culture and computer culture, but rarely have they dwelled on its implications. Drawing upon scholarship that emphasizes the material cultures of computing, this chapter challenges the tendency to treat the *punk* in cyberpunk as incidental.[4] Ultimately, I seek to deepen our understanding of cyberpunk and the early computer underground—and to provide a critical assessment of cyberpunk's mixed political legacy. As Bryan Pfaffenberger and others have noted, though members of early computing communities, such as phone phreakers, often claimed to be participating in a countercultural revolution, in fact they often sought to win validation from, and secure a place within, the most reactionary of institutions: massive telecommunications companies, capitalist startup culture, a social world consisting exclusively of white men.[5] This chapter situates a history of punk and hacking within that critical tradition.

The Hacker Ethic

In general use, the term *hacking* means taking unauthorized access of computers. For those who practice hacking, the term often refers more broadly to the practice of mastering, exploring, and experimenting with computers. "The hacker ethic," as Steven Levy terms it, first emerged in the 1940s and 1950s among the members of a student group at MIT called the Tech Model Railroad Club, which, as its name suggests, was a club dedicated to building electronic model railroads. The club's members took pleasure in building a glossary of insider slang that formulated their values in terms of self-reliance, creativity, juvenile humor, and intense dedication—above schoolwork, above other activities—to work on the system of relays and switches and wires that made the railroad run. As far as this lingo was concerned, time was best spent exploring the system, revamping the system, disassembling the system and trying to improve it, and breaking things along the way:

> When a piece of equipment wasn't working, it was "losing"; when a piece of equipment was ruined, it was "munged" (**m**ashed **u**ntil **n**o **g**ood); the two desks in the corner of the room were not called the office, but the "orifice"; one who insisted on studying for courses was a "tool"; garbage was called "cruft"; and a project undertaken or a product built not solely to fulfill some constructive goal, but with some wild pleasure taken in mere involvement, was called a "hack."[6]

The group's members also worked up an unspoken yet firmly held "hacker ethic," an ethos that still pervades tech cultures to this day. Levy summarizes the tenets of this ethos as follows: "All information should be free; mistrust authority—promote decentralization; hackers should be judged by their hacking, not bogus criteria such as degrees, age, race, or position; you can create art and beauty on a computer; computers can change your life for the better."[7]

The hackers of the Railroad Club were among the first teenagers anywhere to play with electronic computers: they broke into locked rooms that contained IBM computers and taught themselves to program using Hollerith punch cards. (MIT's administration has always dealt mildly with transgressions of this kind.) Many of them became programmers and computer engineers in the world beyond MIT, often building new cultures as deep with lore as the one they had left. At MIT, engineers built a science fiction library and developed a computer game called *Spacewar!*; at Stanford, the preferred idiom was high fantasy, with a Tolkienesque computer game and a lab printer that could produce documents in Elven lettering.[8] The unapologetic nerdiness of these computer cultures was no accident. One purpose of them was just to be different, to occupy the margins rather than the mainstream, which is why computer culture could admit so many different idioms: model building, science fiction, fantasy, the crunchy counterculture, the psychedelic counterculture.

Computers and the Counterculture

Early computing inside and outside of the academy participated in the cultural divide—and the cultural interdependency—between the establishment, especially the establishment of the military-industrial-academic complex, and the counterculture. During these years, a great deal of support for research and development in computing came from the Department of Defense, as it had since the Second World War.[9] In universities, usage of the military-funded ARPANET (predecessor of the internet) grew dramatically though the 1970s and 1980s. The military didn't fund the student hackers of MIT's Tech Model Railroad Club, but it did give ample funds to MIT, and the CIA kept office space in Tech Square on MIT's campus. During the Vietnam War and the Cold War, many antiwar protestors viewed computers as instruments of the war machine, and protestors held antiwar rallies at MIT, Stanford, and Princeton, where they shouted, "Kill the computer!"[10]

Fred Turner's classic account documents the grassroots computer movement that ordinary citizens were building during these same years. Groups like the Homebrew Computer Club, which began meeting in 1975 in Gordon French's home garage in Menlo Park, represented a growing community of hobbyists who built personal computers and ran conferences (bulletin boards) on networks like the Whole Earth 'Lectronic Link (WELL).[11] Often, these hobbyists worked for free because the online world offered a place to satisfy niche passions. Howard Rheingold reports that the WELL's "single largest source of income" was, for some time, a conference dedicated to the Grateful Dead. "Because of the way the WELL's software allowed users to build their own boundaries," Rheingold writes, "many Deadheads would invest in the technology and the hours needed to learn the WELL's software, solely in order to trade audiotapes or to argue about the meaning of lyrics."[12] The first big fandom that fueled the internet was the fandom for the Grateful Dead.

For the hyperspace hippies that Turner describes, an important part of the Grateful Dead's appeal may well have been that it was one of the few bands holding on to the countercultural spirit of the 1960s after the rest of the music scene, dispirited after the shooting at Kent State and the deaths of '60s legends like Jimi Hendrix, Jim Morrison, and Janis Joplin, moved on. Indeed, the psychedelic aspect of hippie culture lingered for decades in the computer underground. The cyber magazine *MONDO 2000*, which originated as a drug culture magazine called *High Frontiers*—and included Timothy Leary among its contributors—regularly discussed drugs from a consumer perspective and joked about their relevance to tech: "Software packages always come with a book that says 'user's guide' on it, even though in the rest of the culture a 'user' is usually doing drugs. Are people who buy software the same as the people who buy drugs?" (The implied answer was yes.)[13] As one of the magazine's contributors argued, this aspect of computer culture was partly a metaphor for the dream of hacking reality itself: "To me the political point of being pro-psychedelic is that this means being AGAINST consensus reality, which I very strongly am."[14]

As decades passed, new styles of music, fashion, and rebellion took over the interest of the young and restless. In the 1980s, the subculture of punk, which exploded in New York in the 1970s and in London in the 1980s, became the new center of the counterculture. Dick Hebdige, Lauraine Leblanc, and others have charted the histories, philosophies, and semiotics of punk, which had its most active cultural scenes in the worlds of art, fashion, and music, but influenced youth culture at large for generations.[15] Punk rockers of the 1970s adopted the word for themselves—taking inspiration from a New York zine called *Punk* that began appearing in 1975—precisely because it was derogatory; they used it in the same spirit that they gave themselves band names like the Rejects, the Unwanted, and the Worst.[16] The rage of working-class youth in Thatcherian England became the driving narrative of the politics of punk, which came to emphasize rebellion against authority, egalitarian fellowship, and the exhortation to free oneself from mainstream music labels, clothes manufacturers, and even employers, per the famous slogan "Do It Yourself." Punk rockers favored a style that mixed these themes of self-reliance, dissent, and disgust with the status quo, featuring ripped material, spikes, skulls, disfiguring makeup, and vulgar language: "things to whiten mother's hair with," in a phrase that Hebdige borrowed from Claude Lévi-Strauss.

As Lauraine Leblanc notes, though punk drew on the reggae underground and other Black music subcultures, and though women like Vivienne Westwood, Patti Smith, and many others played major roles in the scene, narratives about punk music usually focus on the white working-class men in the scene. In her book about gender and punk, she remarks on how many histories of punk "document, in sometimes excruciating detail, the doings of major and minor players in the punk scenes—(mostly male) band members, (male) promoters, (male) club owners, and (male) record-company executives. In all these texts, women and girls appear only in glimpses, in the margins of the marginal." Leblanc interviews a range of young women in punk who, like her, feel "troubled about the male-dominated gender dynamics in the punk subculture, a subculture that portrays itself as being egalitarian, and even feminist, but is actually far from being either."[17]

Nonetheless, despite the contradictions between the claims of punk politics and the lived experiences of real punks, punk has held wide and enduring appeal for youth in revolt. How punk became the center of grassroots computer culture is a question to which we now turn.

From Cyberpunks to Cyberpunk

In the 1960s and '70s, the term *hacker* mostly referred to computer geeks. In the 1980s, new attention to grassroots computer communities changed the meaning of the word *hacker* in popular culture. Journalists began to cover the youngsters staying up late on underground bulletin boards in the plum and scarlet tones of crime writing. In 1984, for example, the *Washington Post* described hacker communities using these terms: "Using long-distance telephone lines that they break into . . . the hackers log onto underground 'bulletin boards' to trade surreptitiously obtained corporate telephone numbers and passwords, or post valid credit card numbers, or carry on silent computer screen conversations at hours when good high school students are supposed to be in bed." Hackers considered such phrasing—commonplace in news coverage of the computer underground—to be unnecessarily overwrought. After all, it wasn't illegal to piggyback on telephone lines—to "break into" telephone lines, in the *Post*'s alarmist phrase.

Hackers viewed the establishment with as much paranoia as the media viewed them. The computer underground circulated news, ideas, and opinions in newsletters and zines (*2600, A.T.I., Cheap Truth, Cybertek, Intertek, Mondo 2000, Phantasy Magazine, Phrack, Phrack Classic, The Syndicate Report,* W.O.R.M.) and on bulletin boards (the Computer Underground Digest, Hack-Tic, Phantasy). These publications described the workings of telephone systems, neat tricks to apply to faxes, phones, computers, and radios, and places where readers could find spare parts on the cheap. But this technological information came embedded in dystopian narratives that exhorted the reader to use this information against the enemies who lurked on all sides. *Phrack* offered a cautionary—and typical—parenthetical aside in a profile of a hacker with the handle Karl Marx: "The Secret Service had bugged [the] hotel room and surprised them (always remember, SECRET service and ROOM service are not *that* different.)"[18] *Cybertek*, which described itself as a technical journal for "dystonauts," published an essay that advised readers to get ahead of the government's impending collapse into an Orwellian nightmare: "The only solution is to join the underground economy, to travel as light as you can, and to make yourself as scarce as possible."[19]

The author Bruce Bethke coined the word *cyberpunk* in his 1983 short story "Cyberpunk." By 1993, the term was appearing in *Time* magazine: "With virtual sex, smart drugs and synthetic rock'n'roll, a new counterculture is surfing the dark edges of the computer age. . . . They call it cyberpunk, a late-20th century term derived from *cybernetics,* the science of communication and control theory, and *punk,* an antisocial rebel or hoodlum. Within this odd pairing lurks the essence of cyberpunk's international culture—a way of looking at the world that combines infatuation with high-tech tools and disdain for conventional ways of using them."[20] Cyberpunks portrayed themselves as alienated misfits who seized such power as they could from below. "Cyberpunks are very withdrawn," one wrote; "they do not flaunt their talents in the open, they prefer to be in the shadows, their message scrawled on the walls, pirated over the air, or posted on a BBS. . . . [W]e believe that people have the right to know what is going on. When you deprive us of that right, then we sink underground. A place where we are hidden, able to spring up and say what we have to say."[21]

The Mentor, a member of the hacker collective The Legion of Doom, wrote in *Phrack* that, although his parents and teachers did not give him enough respect, he was heir to the new world emerging online:

> This is our world now . . . the world of the electron and the switch, the beauty of the baud. We make use of a service already existing without paying for what could be dirt-cheap if it wasn't run by profiteering gluttons, and you call us criminals. We explore . . . and you call us criminals. We seek after knowledge . . . and you call us criminals. We exist without skin color, without nationality, without religious bias . . . and you call us criminals. You build atomic bombs, you wage wars, you murder, cheat, and lie to us and try to make us believe it's for our own good, yet we're the criminals. Yes, I am a criminal. My crime is that of curiosity.[22]

It was not inevitable that the punk subculture become the new carrier of the style and values of the computer underground. Hip-hop and glam rock coexisted with the rise of punk rock; stoner culture could have served to channel countercultural dissent; teenagers online could have emulated their elders, the earth children of the *Whole Earth Catalog*. One reason that the punk subculture and the computer underground became so quickly and thoroughly implicated in each other was the readiness with which journalists invoked the term *punks*, meaning criminals, in stories about the dangers of the new online world. Punk rockers sought to appall, to give a thrill of danger, and journalists made use of that thrill of danger. The criminal portrayal of hackers thus grew out of a narrative circuit: some hackers *did* commit crimes, for instance, stealing long-distance service from phone companies; journalists knew that crime stories sold, and they highlighted that aspect of hacker culture; and hackers played up their reputation as antiheroes, in part for the fun of being a dangerous character and in part as a protest, via mockery, against that reputation's unfairness.[23]

Studies of hacker zines and bulletin boards in the 1990s took note of the way these spaces were pervaded with the language of hostility, alienation, and apocalypse.[24] Users often chose handles from a cluster of connected categories: high fantasy (*Dragon Lord, Knight Lightning, Ultimate Warrior, Unknown Warrior*), postapocalyptic fantasy (*Black Avenger, Death Stalker, Necron 99, Storm Bringer, Terminus*), puns on references to drugs or tech (*Acid Phreak, Ellis Dea, Hitch Hacker, Phelix the Hack, Phiber Optik*). Hacker collectives gave themselves similar names (*The Hall of Justice, The Legion of Doom, The Masters of Deception*), as did hacker bulletin boards (*Dragonfire, Forbidden Zone, PHBI, Phreakenstein's Lair, Shadowland*).[25] Hebdige observes the same language of hostility in the names of punk bands: Black Flag, Bikini Kill, Bratmobile, The Clash, Conflict, the Dead Kennedys, Dicks, DOA, MIA, Minor Threat, the Misfits, the Rejects, Seven-Year Bitch, the Sex Pistols, the Unwanted, the Worst.[26]

Do It Yourself

Literature does not merely reflect historical events; literature can also cause historical events. Cyberpunk literature influenced the culture of the computer underground since before the genre had a name. As social movement theorists have discussed extensively, social movements develop characteristic *repertoires*—defined as tactics, symbols, knowhow—that, among other uses, help their members to live out the movement's ethos and negotiate the relationship between a personal identity and a political identity within the movement. ("Personal identity and political identity differ," writes Mabel Berezin. "Who am I becomes who are we? Who is one of us and who is not?")[27] Contributors to hacker zines and bulletin boards attributed the most distinctive discourses, identities, and narratives belonging to hacker communities, and even generational differences *within* hacker communities, in large part to the influence of decades of protocyberpunk and cyberpunk literature, from John Brunner's *Shockwave Rider* (1975) to Vernor Vinge's

True Names (1981) to William Gibson's *Neuromancer* (1984) and beyond. A contributor to *Phrack* wrote of Brunner's book, "It inspired a whole generation of hackers including, apparently, Robert Morris, Jr. of Cornell virus fame. The Los Angeles Times reported that Morris' mother identified 'Shockwave Rider' as 'her teen-age son's primer on computer viruses and one of the most tattered books in young Morris' room." The reporter added, "I am particularly struck by the 'generation gap' in the computer community when it comes to 'Neuromancer': Virtually every teenage hacker I spoke with has the book, but almost none of my friends over 30 have picked it up."[28]

But the subculture that grew up around punk music also helped to define and reinforce the values of the computer underground. Most notably, punk's credo "Do It Yourself" (DIY), which asked punks to make their own clothes, to make their own media, to make their own music, to build a whole world for themselves that the establishment didn't underwrite or control, reinforced the hacker mentality that the best system is the system you build or write yourself.[29] Dave Grohl, a member of the band Nirvana, which saw itself as a punk band first and foremost, said punk music resonated with him as a kid not because of the sound alone but because of its DIY politics:

> More than the noise and the rebellion and the danger, it was the blissful removal of these bands from any source of conventional, popular, corporate structure and the underground network that supported the music's independence that was totally inspiring to me. At thirteen years old, I realized I could start my own band, I could write my own song, I could record my own record, I could start my own label, I could release my own record, I could book my own shows, I could write and publish my own fanzine, I could silk-screen my own tee shirts—I could do this all by myself. There was no right or wrong, because it was all mine.

The very design of hacker zines, in their imitation of the DIY-driven design of punk zines, rendered the DIY ethos as embodied spectacle. Dick Hebdige writes of punk zines, which delivered their messages of protest and discontent in a distinctive aesthetic of cut-up magazine letters, hand-drawn pictures, and artifacts from photocopier machines,

"The overwhelming impression was one of urgency and immediacy, of a paper produced in indecent haste, of memos from the front line."[30] Visually, hacker zines placed themselves in the same genre: for example, *Cybertek* was printed (badly on purpose, it would appear) from a dot-matrix printer and annotated with notes written using a ballpoint pen. In the first issue, *Cybertek* argued that a hands-on, self-reliant approach to technology is at the heart of the term *cyberpunk*:

> A few of you out there are probably wondering what this enterprise is really about, and just what does "Cyberpunk" mean? For starters, Cyberpunk is more of an attitude than anything else at this point. It basically states that people should grab onto the practical, street aspects of technology with both hands and hang on! It's a middle ground where underground sources, symbolized by the punk movement of the '70s, join with the technological wizardry associated with the hacker movement.[31]

In time, punk and cyberpunk works of art, writing by outsiders about punk and cyberpunk subcultures, and the activities of members of punk and cyberpunk subcultures influenced each other in a reciprocal loop. An article in *Cybertek* about how to distinguish real cyberpunks from posers ("Real cyberpunks listen to Ministry, The Cure, Skinny Puppy, The Misfits, Rush, Pink Floyd, etc.") advised hackers to study a study of hackers in order to better understand the hacker ethic: "Hackers by Steve Levy—A good guide for the Ethics of a Hacker. . . ."[32] Conversely, Billy Idol, a god of punk music since the 1970s, gave his 1993 album the title *Cyberpunk* and opened it with a spoken-word manifesto that rehearsed major tenets of the hacker ethic: "All information should be free. It is not. Information is power and currency in the virtual world we inhabit. So mistrust authority. Cyberpunks are the true rebels. . . ."[33] Art imitated life imitated art.

"We Are as Gods"

Despite the narratives of government persecution that pervaded hacker bulletin boards, the reality was that authorities often treated these young men gently. Administrators at MIT winked at the burglaries of the Tech Model Railroad Club, whose members broke into locked rooms in order to access university computers. When the teenaged Sean Parker, who later created the piracy website Napster, hacked into a government network—a breach that prompted a raid on his house, the next day, by the FBI—his father, a government employee, managed to secure him a lenient sentence of community service. The venture capitalist Peter Thiel, in his book *Zero to One: Notes on Startups, or How to Build the Future*, makes the following boast about what he saw as his co-workers' creativity and daring—which most others would see as frankly disturbing: "Of the six people who started PayPal, four had built bombs in high school."[34] In many cases, the very breaches that early hackers and computerphiles used to substantiate their claims to be *outsiders, rebels, children of the counterculture*, showed the extent to which the power structures of the state were working on their behalf.

The English word *privilege* derives from an identical French word which means, literally, "private law." As Robert Darnton writes of the legal history of this concept, "Privilege was the organizing principle of the Ancien Régime not only in France but throughout most of Europe. Law did not fall equally on everyone, for everyone assumed that all men (and, even more, all women) were born unequal. . . . Law was a special dispensation accorded to particular individuals or groups by tradition and the grace of the king." Thus, for example, members of the publisher's guild during the ancien régime held a *privilege*, or legal right, to publish books, which was illegal for everyone else.[35] Today, the law no longer formally enshrines the idea that some classes of people are subject to the law and others are not, but in practice, the justice system often treats people differently on the basis of factors such as skin color—a fact that observers have resurrected the term *privilege* in order to describe. "We exist without skin color," The Mentor wrote; but the hacker community he spoke for was overwhelmingly white, and whether they knew it, their whiteness afforded them the protection they needed to venture as far, to break as many rules, as they did.[36]

In a 1988 article about the hacker community, the tech historian Bryan Pfaffenberger noted the pervasive desire among hackers to gain success within the same institutions they purported to rebel against.

Hackers told journalists that they hoped that Ma Bell or the FBI, whose systems they broke into, would hire them as technology experts. "This strange combination of motives—to seize the phone system for oneself, to appropriate its power without its permission . . . and yet to wish to improve the system and thereby gain its acknowledgment and approval—was common," Pfaffenberger writes. "For the phone phreakers, hackers, and (later) early personal computer users, the goal was not to overthrow the System, but rather the more conservative aim of gaining entry to the System."[37] In this respect, early hackers were far more conservative in their goals than they claimed to be. Their presumption that the system would ultimately embrace them was not incorrect—as we can see, for example, in the fact that so many budding teenage bombers went on to Silicon Valley fame as founders of PayPal.

This was not the only contradiction in the hacker community's identification with the punk ethos. As Pfaffenberger, Emily Chang, Cynthia Cockburn, Lori Kendall, Sherry Turkle, and others have discussed, in computing culture from the early hacker era to the present, "technical skill and high-tech artifacts are widely considered to be the exclusive possession (and constitutive symbol) of male maturity, potency, and prestige."[38] Hacker boards and zines presumed a male audience and rarely mentioned women at all: an observer wrote in *Computer Underground Digest*, "Computer Underground Digest, like the CU [computer underground] in general, is a male bastion. Sexist language, male metaphors, and if I'm counting correctly, not a single self-announced female contributor."[39] Early acts of mass trolling on Usenet often targeted women and female-dominated communities, sending a frightening message of unwelcome.[40]

This was not the case for the punk subculture, which, although it has often forgotten the women in its own scenes, in theory celebrated men and women alike. By contrast, many of the principal figures in recent Silicon Valley history have been explicitly misogynist. As an undergraduate at Stanford, for example, Peter Thiel started a campus publication with the aim of "fight[ing] feminism and political correctness on campus." Later, he said in a speech at the Cato Institute, "Since 1920, the vast increase in welfare beneficiaries and the extension of the franchise to women—two constituencies that are notoriously tough for libertarians—have rendered the notion of 'capitalist democracy' into an oxymoron."[41] His solution was not, it seems, to ask why these groups resisted libertarianism but rather to seek to disenfranchise them. The people whose power was worth fighting for turned out to be those who already had it.

Did these factors in the lives of early hackers—their sense of a promised place in the world, their middle-class ease with *the System* in its largest sense—inform their feeling of entitlement to the computers of the world and the information stored in them? Did hacker culture's disdain for rules, when combined with the wealth and power that the growing computer industry accumulated, bring about Silicon Valley's culture of destructive disruption? In a 1985 book published for the use of new hackers, which provides an overview of the hacker community, instructions on connecting to confidential networks, and definitions of terms such as *carding, crashing, hacking,* and *phreaking,* Michael Harry offers a picture of the demographics of hacker bulletin boards, based on a survey that he himself conducted.[42] "Most, to no one's surprise," he says, "come from upwardly-mobile, middle-class families, many of which are two-computer homes." His survey found that "we have a group of people, most of whom are in high school, whose primary activities are piracy and phreaking. . . . Most of the group have a year or less of actual computer experience, which may indicate that underground activities tend to be attractive primarily to computer novices. . . . Meanwhile, there are a handful of experts, the crackers and professional phreakers who provide the means for youngsters to pursue their computer banditry."[43]

This account of the hacker community as dominated by technological novices is notable coming from a sympathetic insider, which Harry certainly is. It coheres with Sherry Turkle's observation that early buyers of personal computers often had little experience with computing but loved the idea of possessing the coolness and power that computers seemed to offer. They saw the computer as "a machine that lets you see yourself differently, as in control, as 'smart enough to do science,' as more fully participant in the future."[44] This impetus—the computer as an aspirational totem; the hacker as someone who wants to be cool—is not nearly as cool as being a dystopian antihero, which may be why the motivations for hacking that Harry uncovered in his survey never generated much talk on hacker boards: "We . . . tried to learn what motivated respondents to participate in computer crimes. Two motives seemed to surface as the most important. One, to increase popularity among our peers, and the second, to relieve boredom. These two motives were equally important to the group as a whole."[45] Better to talk about fighting power from below: about boxing, pirating, trashing, wiretapping, ARPANET hopping, fun pranks, thrilling escapes, accidental calls to the president's bomb shelter, telephone systems, government surveillance, how to go off the grid, and what to do if your friend turns out to be an FBI agent.

In short, the punk ethos of rebellion helped to shape the values of hackers, but it also served as entertainment for them, and the counterculture's precepts were—in the end—all too easily hijacked by technologists who sought to consolidate power rather than dismantle it. The opening line of Stewart Brand's *Whole Earth Catalog* (1968), which meant to celebrate the power that tech gives us to shape our lives independently of big institutions, now reads quite differently—no longer possible as a statement from ordinary computer users, whose lives online are tracked, commodified, and manipulated by data brokers, though perhaps possible as a statement from the brokers at the very top: "We <u>are</u> as gods, and we might as well get used to it."[46]

Conclusion

Today, we can trace a multitude of values that define the high-tech industry at large to the power-from-below philosophy that united the punk subculture and the computer underground. The *punk* in cyberpunk helped to make the values of hacker culture legible to people who were new to computing or were outsiders to computing, and they helped to cement the idea, even among those who belonged to other subcultures or identity groups, that computer culture was the counterculture. A science fiction writer, talking about attending a literary convention in the 1990s after he had tiptoed into the genre of cyberpunk, described his surprise at finding that readers of cyberpunk had a fully developed style and ethos that he had to catch up to: "I'd expected them to be snobby arty/literary East Coast types, but they weren't like that at all. They were reality hackers, nuts, flakes, entrepreneurs, trippers, con-men, students, artists, mad engineers—Californians with the naïve belief that (a) There Is a Better Way, and (b) I Can Do It Myself."[47]

There is a better way, and I can do it myself: these beliefs have informed a broad spectrum of attitudes and practices in Silicon Valley: the love of reinvention; the contempt for regulation; the oil rush to exploit the world's data (your personal information, too, wants to be free); the uniform of hoodies, tee shirts, and jeans, which rejects, in Hebdige's phrase, "the adult world of work"; the "'me first' entrepreneurialism" of *Wired* magazine, which Langdon Winner memorably describes: "A *Wired* world is depicted as a realm of boundless creativity, self-indulgence, profit seeking, and free-floating ego. A perfect mascot for this colorful ideology would be Peter Pan, a little boy now seen flying through a Neverland of digital bits, a place where he can do what he pleases and never grow up."[48]

For those of us who both admire the defiance of punk and worry about the future envisioned by the tech industry, the past of cyberculture may be hard to reconcile with its current trajectory. Maybe cyberpunk was always a flawed vehicle for the ideals that cyberpunks purported to hold. For example, posters on hacker forums used the hostile posturing typical to those forums not just to signal rebelliousness but to drive away anyone who didn't belong to the in-group.[49] Or again, the paranoia that pervaded hacker forums may have been genuine for some, but for others may have been thrilling fantasies of persecution, which they indulged in the knowledge that the justice system tended to be gentle on people like themselves. Or again, the arrogation of punk style and rhetoric by the corporate mainstream—the sale of punk everything in stores—may have prepared us to accept the corporate arrogation of the style and rhetoric of the computer underground. Perhaps it's time to revisit this history not clinically but with anger over promises made but not kept. Doing so would certainly be in the spirit of punk. From a punk perspective, the past is corrupt, the present is complicit, but the future has possibilities. The future is science fiction.

Notes

1. See, for example, Gordon Meyer and Jim Thomas, "The Baudy World of the Byte Bandit: A Postmodernist Interpretation of the Computer Underground," *Computers in Criminal Justice*, ed. F. Schmalleger (Bristol, IN: Wyndham Hall, 1990); Elizabeth M. Reid, "Electropolis: Communication and Community on Internet Relay Chat" (honors thesis, University of Melbourne, 1991); and Paul Taylor, *Hackers: Crime in the Digital Sublime* (London: Routledge, 1999), 169–70.
2. The epigraph to this chapter comes from one of the movement's popular publications, *Mondo 2000: A User's Guide to the New Edge,* Rudy von Bitter, R. U. Sirius, and Queen Mu, eds. (New York: HarperCollins, 1992), 128.
3. Michael Mahoney, "The Histories of Computing(s)," *Interdisciplinary Science Reviews* 30, no. 2 (2005): 119.
4. Cyrus C. M. Mody and Andrew J. Nelson, "'A Towering Virtue of Necessity': Interdisciplinarity and the Rise of Computer Music at Vietnam-Era Stanford," *Osiris* 28 (2013): 254–77.
5. Pfaffenberger quotes a famous phone phreaker who said his goal was to be hired by the same phone company he was ostensibly fighting against. Bryan Pfaffenberger, "The Social Meaning of the Personal Computer: Or, Why the Personal Computer Revolution Was No Revolution," *Anthropological Quarterly* 61, no. 1 (1988): 39–47, 39. See also Lori Kendall, *Hanging Out in the Virtual Pub: Masculinities and Relationships Online* (Berkeley: University of California Press, 2002); Megan Condis, *Gaming Masculinity: Trolls, Fake Geeks, and the Gendered Battle for Online Culture* (Iowa City: University of Iowa Press, 2018); and E. J. White, *A Unified Theory of Cats on the Internet* (Stanford, CA: Stanford University Press, 2020).
6. The term *hack* predated the club, as Levy notes, having long been in use at MIT to describe "the elaborate college pranks that MIT students would regularly devise, such as covering the dome that overlooked the campus with reflecting foil." Steven Levy, *Hackers: Heroes of the Computer Revolution* (New York: Doubleday, [1984] 1994), 6–8.
7. Levy, *Hackers*, 23–26.
8. Levy, 139–42.
9. Hermann Goldstine, *The Computer from Pascal to Von Neumann* (Princeton, NJ: Princeton University Press, 1973).
10. Mody and Nelson, "'A Towering Virtue of Necessity,'" 259–60; William Barksdale Maynard, *Princeton: America's Campus* (University Park: Pennsylvania State University Press, 2012), 193.
11. Fred Turner, *From Counterculture to Cyberculture: Stewart Brand, the Whole Earth Network, and Digital Utopianism* (Chicago: University of Chicago Press, 2006).

12. Howard Rheingold, *The Virtual Community: Homesteading on the Electronic Frontier* (Cambridge, MA: MIT Press, 2000), 49.

13. Reprinted in von Bitter et al., *Mondo 2000: A User's Guide*, 9.

14. On the place of "reality hacking" in the history of artificial intelligence, see, for example, John Markoff, *What the Dormouse Said: How the Sixties Counterculture Shaped the Personal Computer Industry* (New York: Penguin Random House, 2005).

15. See, for example, Dick Hebdige, *Subculture: The Meaning of Style* (New York: Routledge, 1979), and Lauraine Leblanc, *Pretty in Punk: Gender Resistance in a Boy's Subculture* (New Brunswick, NJ: Rutgers University Press, [1999] 2002).

16. Elyse Graham, "The Long and Fascinating History of *Punk*," *OxfordWords* blog (October 24, 2018).

17. Leblanc, *Pretty in Punk*, 472–73, 102.

18. Taran King, "Phrack Pro-Phile XXII: Karl Marx," *Phrack* 2, no. 22 (1988).

19. John J. Williams, "Hiding Yourself," *Cybertek* vol. 5 (1991).

20. Quoted in R. U. Sirius, "Cyberpunk," reprinted in von Bitter et al., *Mondo 2000: A User's Guide*, 64.

21. Phantom Writer, "Why Cyberpunk?" *Cybertek* vol. 3 (1990).

22. The Mentor, "The Conscience of a Hacker" (January 8, 1986), republished in *Phrack* vol. 14 (1987) with the note, "The following file is being reprinted in honor and sympathy for the many phreaks and hackers that have been busted recently by the Secret Service.—KL" [Knight Lightning].

23. Gordon Meyer and Jim Thomas, "The Baudy World of the Byte Bandit: A Postmodernist Interpretation of the Computer Underground," *Gordon's Desktop Publications*, June 10, 1990, http://hacker.textfiles.com/papers/baudy.html.

24. See, for example, Meyer and Thomas, "The Baudy World of the Byte Bandit"; Thomas J. Holt, "Hacks, Cracks, and Crime: An Examination of the Subculture and Social Organization of Computer Hackers" (PhD diss., University of Missouri–St. Louis, 2005), 11–28.

25. Meyer and Thomas, "The Baudy World of the Byte Bandit."

26. Hebdige, *Subculture*, 109.

27. Mabel Berezin, "Emotions and Political Identity," in *Passionate Politics: Emotions and Social Movements*, Jeff Goodwin, James Jasper, and Francesca Polletta, eds. (Chicago: University of Chicago Press, 2009): 85.

28. Paul Saffo, "Consensual Realities in Cyberspace," *Phrack* 3, no. 30 (December 24, 1989).

29. Dave Grohl, keynote speech, South by Southwest festival, 2013.

30. Hebdige, *Subculture*, 111.

31. *Cybertek* vol. 1 (1990): 1.

32. *Cybertek* vol. 15 (1996): 2.

33. Karen Schoemer, "Seriously Wired: Billy Idol, the Bad Boy of Punk Pop, Hopes to Stay Relevant into Middle Age by Making Cyberpunk His True Rebel Yell," *New York Times*, August 8, 1993.

34. Quoted in Jonathan Taplin, *Move Fast and Break Things: How Facebook, Google, and Amazon Cornered Culture and Undermined Democracy* (New York: Little, Brown and Company [2017] 2018): 76, 90.

35. Robert Darnton, *Censors at Work: How States Shaped Literature* (New York: W. W. Norton, 2014), 29.
36. The centering of whiteness in cyberpunk operated in tandem with a simultaneous fascination with, and othering of, Asian culture that historians have described, aptly, as "techno-Orientalism." See, for example, David Roh, Betsy Huang, and Greta Niu, eds., *Techno-Orientalism: Imagining Asia in Speculative Fiction, History, and Media* (New Brunswick, NJ: Rutgers University Press, 2015).
37. Pfaffenberger, "The Social Meaning of the Personal Computer," 39–41.
38. Pfaffenberger, 44. See also Emily Chang, *Brotopia: Breaking Up the Boys' Club of Silicon Valley* (New York: Penguin Random House, 2019); Cynthia Cockburn, *Brothers: Male Dominance and Technological Change* (London: Pluto Press, 1983); Cynthia Cockburn, *Machinery of Dominance: Men, Women, and Technological Know-How* (London: Pluto Press, 1985); Kendall, *Hanging Out in the Virtual Pub;* and Sherry Turkle, *The Second Self: Computers and the Human Spirit* (New York: Simon and Schuster, [1984] 2005).
39. Liz E. Borden, "Sexism and the Computer Underground," *Computer Underground Digest* 3, no. 3.00 (January 6, 1991).
40. See Elyse Graham, "Boundary Maintenance and the Origins of Trolling," *New Media and Society* 21, no. 1 (2019).
41. Thiel quoted in Taplin, *Move Fast and Break Things,* 70.
42. Michael Harry, *The Computer Underground: Computer Hacking, Crashing, Pirating, and Phreaking* (Port Townsend, WA: Loompanics Unlimited, 1985).
43. Harry, *The Computer Underground,* 4–5.
44. Sherry Turkle, *The Second Self* (Cambridge, MA: MIT Press, 1984), 25.
45. Harry, *The Computer Underground,* 4.
46. Stewart Brand, "Purpose," *Whole Earth Catalog* (September 1968), 3. Quoted in Taplin, *Move Fast and Break Things,* 52. Emphasis in the original.
47. Reprinted in von Bitter et al., *Mondo 2000: A User's Guide,* 11.
48. Hebdige, *Subculture,* 61; Langdon Winner, "Peter Pan in Cyberspace," *Educause Review* 30, no. 3 (May/June 1995).
49. Graham, "Boundary Maintenance and the Origins of Trolling."

CHAPTER 19

The Computer as Prosthesis?

Embodiment, Augmentation, and Disability

Elizabeth Petrick

The computer is an intellectual prosthesis; an information prosthesis; a communication prosthesis:[1] variations on this metaphor show up throughout scholarship on the relationship between users and technology, from many different fields: history, philosophy, science and technology studies, media studies, education, and more. The metaphor perhaps began with Seymour Papert and Sylvia Weir's 1978 paper, "Information Prosthetics for the Handicapped," which suggested that "the computer can become an extension of the operator who can now do 'anything a computer can do' such as draw, compose music, gain access to information libraries, put text on permanent file and so on."[2] Notable figures in computer history have had the prosthesis metaphor applied to their understanding of computers and users. Lev Manovich describes J. C. R. Licklider's idea behind "Man-Computer Symbiosis" as: "an interactive digital computer can act as a kind of metaprosthesis that augments our memory, perception, decision making, and other cognitive operations."[3] Likewise, Thierry Bardini explains that Douglas Engelbart's computer interface with its mouse and chord keyset was "based on the premise that computers would be able to perform as powerful prostheses, coevolving with their users to enable new modes of creative thought, communication, and collaboration providing they could be made to manipulate the symbols that human beings manipulate."[4] With these and in other works, the metaphor usually goes unexplained, left to speak for itself in exactly what ways a computer can be a prosthesis for a human.[5]

What all these examples have in common is an attempt to describe the capability of the computer to augment human abilities. As Papert and Weir note—possibly referencing Marshall McLuhan—the computer can be an extension of a person, expanding their abilities and senses. The computer as extension explains why the prosthesis has been so widely latched onto as a metaphor, as it seems to provide a concrete, real-life example of what the computer is doing invisibly. We reach out via the computer, intellectually and with our senses, in a similar way as someone with a prosthetic arm reaches out to grasp an object. The prosthesis implies something like the opposite of an autonomous technology; it is integrated fully with the body, human and computer becoming one.[6] Going further with the possibilities implied by the metaphor, Lucy Suchman looks to near-future technologies to mesh computers with human bodies, with the work of Steve Mann and "the intersection of the wearable computer as environment and as prosthesis," where "the mirroring of environments and bodies in the projects of the disappearing and wearable computer suggests a desire always to be recognized, connected to familiar environments, while at the same time being fully autonomous and mobile."[7] Prosthetic technology provides a technological way of enacting human autonomy, extending the body and senses, while the user feels in control.

Flipping the metaphor, in 1990, John Perry Barlow wrote of cyberspace that "I don't know what to make of it, since, as things stand right now, nothing could be more disembodied or insensate than the experience of cyberspace. It's like having had your everything amputated."[8] The user is augmented by feeling like the body has been left behind entirely, leaving the mind floating untethered online, to do as it will. Employing these metaphors offers a way to get at what it feels like to use computers, to be able to do so many things our bodies, including our minds, could not without technology. Yet, what is missing from nearly all of these examples of the metaphor is disability and the reality of prosthetics. A prosthesis is assumed to be a perfect fit when the computer is talked about as one, a way for the user to reach beyond their body. The prosthesis here is a way of beating nature—technology conquering the incapabilities of the body.

With this fantastical image of the prosthesis, it is perhaps unsurprising that, when the metaphor is usually employed, it is not in consideration of actual computer users with disabilities: those who may use actual prostheses as they operate computers, along with those with other kinds of disabilities also having their abilities augmented by the computer. Or, as Katherine Ott puts it, "Cyborg theorists who use the term 'prosthesis' to describe cars and tennis rackets rarely consider the rehabilitative dimension of prosthetics, or the amputees who use them."[9] The full complexity of what a prosthesis means is absent from its part in the metaphor. Furthermore, to consider the subject literally, only a few recent works look at the relationship between computer technology and actual prosthetics, specifically attempts to impart a sense of touch to prosthetic devices using electronic technology.[10] Instead, the prosthesis—and, by extension, disability—often remain nothing more than a metaphor when technology is discussed. People with disabilities are rarely considered as computer users, and their place in computer history is not analyzed. The user is almost never one who wears a prosthesis in real life; they are instead able-bodied and their prosthesis only a play on words, not a technology replacing an actual missing limb. When a marginalized group is treated as merely a metaphor, they become further erased from the history they were a part of.

Disability studies scholars have published widely on the problems with using words and concepts related to disability as metaphors, particularly in literature and media. Carrie Sandahl argues, "Nondisabled artists in all media and genres have appropriated the disability experience to serve as a metaphor expressing their own outsider status, alienation, and alterity, not necessarily the social, economic, and political concerns of actual disabled people."[11] Disability metaphors also appear in politics, as Emily Russell has shown: "Despite representative democracy's consistent exclusion of disabled *individuals, figures* of anomalous bodies are often pressed into service as a metaphorical representation of the body politic."[12] Similar to the prosthesis metaphor, these other metaphors of disability rarely include actual people with disabilities but are instead appropriated by nondisabled people to mark some outsider status. The end result is, as Julie Avril Minich explains, that "pervasive disability representations like the overcoming narrative serve as metaphors for problems faced by able-bodied people or to reinforce the marginalization of disabled people."[13] These issues remain true for the prosthesis metaphor as well.

Vivian Sobchack suggests that an alternative approach to treating "technology as prosthesis" is to treat "prosthesis as technology."[14] This reversal opens the door to dig into the technologies that enable the computer to be a metaphorical prosthesis, those that allow access to it. All people require technologies of access to use the computer to its fullest: preferred input devices, ways of displaying information on the screen or through audio and other media, and various peripherals that allow us to enter something into the computer and receive something back. We all have different preferences and needs when it comes to the best ways to set up our computers for us to interact with them. This is all access—the technologies that connect us, body and mind, with computers. Thinking about human-computer interaction in this way leads to a shift in how we conceive of the interface, where the interface becomes the system of human and computer, along with all the various technologies the two need to communicate with each other. One way to understand these technologies is to center computer users with disabilities and the accessible technologies they use to operate the computer. This allows us to get at the diversity of bodies interacting with computers and what that might mean for the user-technology relationship. Frequently, when people with disabilities are mentioned as technology users, they are not talked about as normal users; instead they are a special case or a footnote to a standard history. Even scholarship focused on marginalized people and the relationship between different bodies and technology usually leaves out people with disabilities. Yet, there is a rich history of people with disabilities and their families tinkering with and adapting technology to fit their needs.[15] This is often done to make the built environment and consumer technology fit people who were not thought of as intended users. People with disabilities have always been technology users, and by understanding how technologies have been made to work for people with different bodies and needs we gain a more complete picture of the relationship between technology and users.

There is a well-known critique of the prosthesis metaphor from Sarah Jain, in which she argues that the metaphor itself falls apart as soon as you know anything about how prostheses work in reality.[16] She suggests some necessary questions to be answered that are left out of the metaphor: "Which bodies are enabled and which are disabled by specific technologies? How is the 'normative' configured? How does the use of the term prosthesis assume a disabled body in need of supplementation? How might the prosthesis produce the disability as a retroactive effect? Where and how is the disability located, and in whose interests are 'prostheses' adopted?"[17] Without such questions being addressed, the metaphor deflates, without any real substance. Jain also brings in an issue I will return to later, to not ignore the fact that technology—even metaphorical prostheses meant to augment human ability—also affect and change the body of the person using it: "Wearing glasses adjusts vision but also changes the comportment of the head and neck and over years changes the contour of the muscular-skeletal infrastructure, and the use of a thirty pound artificial leg strapped over the shoulder in the early century would have changed the weight distribution and physiology of the body."[18] This is as true for computers as for these literal prostheses, as Laine Nooney shows in this very edited volume, with all the forms of pain the computer has brought to the body.[19] Hunched shoulders, carpal tunnel and other repetitive stress injuries, and eye strain are all changes to our bodies that computer use has wrought in us.

I would like to take a different tack from Jain, however, by focusing on what we can learn about the relationship between people and computers when people with disabilities are foregrounded as technology users. Specifically, I argue that the prosthesis metaphor's focus on augmentation through embodied computer use can become a lens for critical analysis when we consider computer users with diverse bodies. By examining people with disabilities as normal computer users, we see not only the range of different bodies that must be made to fit with computer technology but also the creativity and frustrations behind trying to accomplish such. Once fit is achieved, then possibilities for augmentation open up, and yet, that in itself is called into question, in terms of what exactly about ourselves is being augmented by the computer. I find that we can keep something of the prosthesis metaphor in mind by considering questions of fit between the bodies of users and the technologies that allow them to access the computer. Explicit consideration of the role of the body in interaction with and part of the computer interface is still uncommon in the historiography of human-computer interaction, although more scholars have recently been centering the body in this fashion.[20] Doing so calls into question our understanding of exactly what the computer interface is and highlights the need to more fully understand the computer as an embodied technology.

I examine three case studies, of different computer technologies from the late 1970s to 2009 that could be understood as augmenting human abilities. This is a period of significant change in computer technology but also for the place of people with disabilities in society. This period begins at a low point for the disability rights movement, with a lack of progress in gaining further civil rights protections after earlier victories in the 1970s, but then surges forward with the passage of the Americans with Disabilities Act of 1990 and all the successes and challenges that have followed from it.[21] People with disabilities experienced far greater access to social participation over this timespan, which included new forms of participation through computer technology and the growing Internet. This increase in visibility and social activity also led to growing expectations among people with disabilities to no longer be just metaphorical outsiders.

Case 1: Unicorn Keyboard

The Unicorn Keyboard,[22] created in 1979 by Steve Gensler, was perhaps "the first widely used alternative keyboard."[23] Gensler invented the keyboard so that a friend of his with cerebral palsy could use a computer. The keyboard consisted of 128 programmable switches that could configure the keyboard however the user needed. This made the keyboard both simple and significant for people with disabilities. Because it was completely programmable, once the switches were covered with an overlay that was divided into keys, it could contain however many keys the user wanted, each doing whatever they needed. An overlay could be put on top replicating a typical computer keyboard and all of its keys, an overlay could just contain one key covering all of the switches that acted as a single-switch input, or the keyboard could be programmed to be anything in between. This total level of customizability allowed the keyboard to adapt to the users' abilities and needs.[24] The adaptability of the Unicorn Keyboard made it a technology of augmentation in a way that allowed it to fit the body of the user but also change with the user, making it particularly useful as an educational technology for children with multiple disabilities. A child could start with only one or two keys to learn cause and effect with how the computer responded to those key presses. From there, the overlays could be made more complex, to teach the user not only how to operate a computer but also help the user develop both cognitive and motor skills, making it an embodied technology in multiple ways.

Shoshana Brand was given one of the first Unicorn Keyboard prototypes when she was around five years old. Gensler met Shoshana's father at a computer class at which they were both trying to learn more about computer technology that might benefit people with disabilities in their lives.[25] Shoshana was born with cerebral palsy and had vision impairments; her parents, Jackie and Steve, hoped that the computer might help her learn to communicate. They got the Unicorn Keyboard working with their Apple II computer and started by having the keyboard contain only one key that had the computer play music whenever Shoshana pressed anywhere on the overlay.[26] This initial setup allowed her parents to determine Shoshana's abilities in terms of vision and motor control, while showing her that she could control basic cause and effect. She quickly advanced to pressing keys of different colors and animal pictures (the latter caused the computer to play the sound the animal made). As Jackie describes her daughter using the keyboard:

> Eventually the keys got smaller and smaller, there were more and more divisions on that board, until she had essentially a full keyboard to work with. Had we shown her that full keyboard right at the beginning, there was no way she could have done it. . . . It's like showing a very young child a standard keyboard and they go banging on it because they don't have the fine motor skills yet, and then they get bored and that's it. Instead, the computer became a real learning tool for her as it could develop and evolve on the keyboard as she developed and evolved both in a physical sense and gain the fine motor skills—and also in a cognitive sense as she went through the developmental stages.[27]

The Unicorn Keyboard was able to fit with Shoshana's body, to a striking degree, in both physical and cognitive ways. However, in spite of the promise of the keyboard, it still had to work alongside other technologies that may not fit as well; Steve Brand bemoaned the fact that so much software at the time was based on assumptions that the user could see the screen well, providing only visual output and not accommodating users with visual impairments.[28]

Another example of a Unicorn Keyboard user further illustrates how it could augment communication abilities. In 1992, a 15-year-old boy, Eric, with both hearing and vision impairments used the Unicorn Keyboard to learn Braille as his vision decreased, in order to complement his communication in sign language.[29] The keyboard was covered with an overlay of keys with raised dots on them, in order to get Eric used to feeling for the numbers of raised dots in Braille. He was then given tasks, such as to find and press a key with three dots, which would make the screen flash in response. Once he could correctly identify the numbers of dots on a key, an overlay with the partial Braille alphabet was introduced so that he could begin to learn the different letters in Braille. The keyboard could adapt to Eric's abilities and learning speed to effectively teach him a new form of communication that he would soon need to rely on.

While the Unicorn Keyboard held great potential to work with different people's bodies and abilities to allow them to access a computer, it also presented certain technical obstacles that needed to be over-come. In addition to the problems with inaccessible software, the key-board could not communicate directly with a computer until later in the 1990s, instead requiring an often expensive Adaptive Firmware Card (advertised for $520 in 1992)[30] that would translate between the keyboard and computer. While this workaround became easier to use over time, the Unicorn Keyboard also always required a significant level of expertise to set up and program the keys. As Jackie Brand described trying to make everything work together for her daughter: "We also re-alized that this was not easy stuff to do. It would have to be a lot easier to use before many people would benefit from it."[31] The keyboard may have fit the body well, but only after a significant amount of work was put in to make it do so.

Case 2: Macintosh GUI

This second case study[32] concerns a computer technology not intended specifically for disabled users but which significantly affected people with different kinds of disabilities: the graphical user interface (GUI). When the Apple Macintosh brought the GUI to a broad consumer au-dience in 1984, it carried with it new ways of interacting with personal computers: via a mouse and by clicking on icons, as opposed to enter-ing text commands. This change, for the most part, fit users better. It was more user-friendly and easier to learn than text-based interfaces. However, it did not work with all bodies, as the GUI carried assump-tions with it that users could see the screen in order to navigate with a mouse. The GUI demonstrates how a technology can be embodied in very different ways for different people, so that for some, there are new possibilities for augmentation and for others, a lack of access altogether.

An example of how the GUI was beneficial for people with certain kinds of disabilities can be seen with Mike Matvy, a psychologist with learning disabilities that affected his ability to read and write. He went to a disability and technology resource center after getting a job that required him to be able to use a personal computer to write reports and professional materials, and read patient records and office memos. In 1990, he detailed his experiences finding a computer that worked for him, after learning to use both an IBM PC and an Apple Macintosh:

> I could see why blind peaple would find IBM best for them. An IBM with voice out put does not require visusl skills. A person would need spelling skills, good memory for details, and ability to move through a system without visuat refrences. I could also see why I was able to move through the MAC system with such speed and eaze. It is built on a visual system, but it requires no spelling and verry little reading to oparate it. The fue writen words in the pull down minues and the dialog boxes are repeated identicly in all aplications. They are also kept with in a pictoral context which helps me know what the words are.[33]

A text-based operating system, like the IBM had with MS-DOS, could accommodate the needs of blind computer users through specialized screen reader software. There were not extensive graphics available on such computers, and a keyboard was used as input. Whereas the Macintosh GUI could accommodate the abilities of someone like Matvy much better, with its limited use of text in menus, consistency across software, and graphical representations in the form of icons. Matvy went so far as to say, "From what I have learned about IBM and MAC, it seams to me that, it is as if MAC were designed spicificly with my needs in mind."[34]

As more and more personal computers switched to GUIs in the 1990s, people like Matvy found an embodied relationship with computers that fit them well. However, blind computer users were increasingly denied access to new computers, as screen readers could not translate information on the GUI interface. There would eventually be a technological solution to this problem, but it would take until the late 1990s for screen readers to be able to work reliably with different GUI operating systems and commonly used software running on them. The GUI represents the way one technology can have very different kinds of embodied relationships with users. For many people, the GUI fit them, accommodating people's needs and generally being more user-friendly. But for others, there was a complete lack of fit, such that they were denied access to personal computers with GUIs altogether for a number of years. Who has access to a technology at all is based on assumptions of how people's bodies work in terms of how they control input and process output. When differences in bodies are not considered, some people are left out of a relationship with the technology and its potential to augment abilities.

Case 3: DiamondTouch

The final case study deals with the DiamondTouch tabletop device and children with autism as computer users. Researchers at the Mitsubishi Electric Research Laboratories developed the DiamondTouch in the early 2000s for collaborative work,[35] and it quickly attracted attention from autism researchers and therapists. It was a touch-sensitive surface that acted as a computer input device with information displayed on it via a projector from above. Unusual for an input device, it allowed multiple users to interact with it at the same time and could even distinguish between touches of different individuals. This allowed, for example, for interactions that would require more than one person to act in concert to accomplish some task, while also making it so that one user could not jump in and do a task that someone else was assigned to perform. This technology created a unique environment for embodied use, bringing multiple people into human-computer interaction at the same time.

In 2009, a group of researchers created software for the DiamondTouch to help children with autism practice socialization skills. Their software, StoryTable, used the tabletop as an interface for collaborative narrative construction in which two children took turns adding visual pieces to a larger story (e.g., background images, characters, or objects). The researchers found that this program encouraged children to initiate social interactions with their peers and even showed some generalization of improved socialization beyond these controlled activities. The researchers also noticed another effect of the technology: a reduction in what they refer to as "autistic behaviors" (such as repetitive movements or stimming). From the researchers' perspective: "Stereotyped movements have been suggested to serve as a coping response; the extent to which they are performed is influenced, in fact, by how much an individual is affected by sub-optimal stimuli in the environment, i.e., by an environmental setting that is perceived as either under-stimulating or over-stimulating."[36] StoryTable apparently struck the right balance in terms of stimulation for its users. There is an assumption here that such behaviors are inherently negative, a symptom of imbalance that the computer helps correct. As Meryl Alper has argued, stimming is not necessarily a negative thing for autistic people: "Many autistic people report that the repetition of physical movements or movement of objects helps them maintain emotional balance, regulates their senses, and provides pleasure."[37] Computer technology, then, is capable of disciplining or regulating the body, and perhaps, to read this critically, even helping to make more acceptable bodies out of users. This could be a form of positive augmentation in achieving greater balance or possibly as a more detrimental form of control, depending on what is actually best for the children to be doing with their own bodies.

The DiamondTouch can be seen as a technology of augmentation, in the way that it was a tool through which cooperative socialization could be experienced. The children, as technology users, could interact with each other through and with the tabletop, controlling its virtual objects and text by touch. The tabletop was not only a communication system between the children, enabling the children to learn new skills as they operated it, but also helped them develop ones that they could take with them when not using it. It fit them and their abilities by being a computer technology they wanted to interact with that had an intuitive interface. But its embodied relationship extended further than just augmentation, also affecting bodily activity seemingly external to its operation.

Conclusion

These three case studies of computer users with disabilities operating different kinds of computer technologies demonstrate a way to get at issues of embodiment and augmentation by centering people with diverse bodies and abilities as users. Embodiment takes place through hands, eyes, ears, and brains, enabling relationships between humans and computers where cognitive, sensory, and perhaps even social abilities can be augmented. To return to the beginning then, can a computer ultimately be thought of as something like a prosthesis? Does this metaphor offer anything in attempting to understand the relationship between humans and computers, and the possibilities for augmentation of human abilities? Don Idhe has criticized the prosthesis metaphor for missing out on the lived reality of prosthesis use: "But actual users of prosthetic devices know better—prostheses are better than going without (the tooth, the limb, the hand), but none have the degree of transparent, total 'withdrawal' of a tool totally embodied. All remain simply more permanently attached ready-to-hand tools. Yet when one's body fails or is irreparably injured, or parts of it are removed, the prosthesis becomes a viable and helpful compromise."[38]

A prosthesis replaces a missing or damaged body part, to get as close as possible to the original or the idea of normal. To say that something is missing or damaged about the computer users in the examples I have given implies a wrongness in disabled bodies that needs to be corrected and would imply the same for nondisabled users as well. The question is if this is the connotation we want to embrace to understand computer technology and human bodies.

Yet an alternative way to use the metaphor may be possible if the body itself is explicitly centered in the analysis. The computer, then, is not about perfectly transcending an inherently flawed body. Augmentation is possible, but it is never a straightforward relationship between such technologies and the bodies that use them. The case studies I have offered here exhibit the complexity of different bodies and the ways they are augmented or not by computer technologies. Vivian Sobchack, herself a prosthesis user, describes her own feelings toward her technological leg and other technologies that augment abilities: "I have not forgotten the limitations and finitude and naked capacities of my flesh—nor, more importantly, do I desire to escape them. They are, after all, what ground the concrete gravity and value of my life, and the very possibility of my partial transcendence of them through various perceptual technologies—be they my bifocals, my leg or my computer."[39] Bringing the body into focus offers a reminder of both its power and its flaws. Technologies of augmentation are not about trying to escape the body but to go beyond what it can do on its own in certain small, but significant, ways.

A way forward with the metaphor is suggested by Robert Rawdon Wilson, in remembering the messiness of the body and what an amputation and prosthesis actually mean for people, where "any consideration of prostheses has to take into account their potential failure and, even, the conditions under which they might go wrong or turn against their users."[40] No prosthesis is a perfect fit, as is no metaphorical prosthesis. Tobin Siebers reminds us that "when prostheses fit well, they still fit badly. They require the surface of the body to adjust—that is rarely easy—and impart their own special wounds."[41] A prosthesis is not about making its user superhuman; it grants abilities but also rubs up against the body in harmful ways. To push the body-technology relationship with the metaphor, Wilson argues that "an appended body part not only recalls the previous, now missing, organic part, but actively calls into question the body's integrity."[42] This forces us to broaden the human-computer relationship to consider what we are without computers in our lives, with that augmentation missing. Further, if the computer can be a prosthesis, its negative effects on the body must also be understood, alongside its possibilities for augmentation. There is no single, normal body for the computer user. For the diversity of bodies out there, computers are always an imperfect fit. As previously mentioned, they cause carpal tunnel, bad posture, headaches, and other problems in their users, as shown by Nooney, along with Jennifer Kaufmann-Buhler.[43] But they also allow us to do things our bodies are incapable of, just like all technologies do. They augment us and allow us to go beyond what our bodies can do on their own—without ever leaving such behind.

Notes

1. With thanks to Cathy Gere, John Alaniz, and my colleagues at Rice University, especially Lan Li, Rodrigo Ferreira, and Elizabeth Brake. This work was supported in part by the US National Science Foundation under grant 1928627.
2. Seymour A. Papert and Sylvia Weir, "Information Prosthetics for the Handicapped," September 1, 1978, 2, https://dspace.mit.edu/handle/1721.1/6308.
3. Lev Manovich, "Visual Technologies as Cognitive Prostheses: A Short History of the Externalization of the Mind," in Marquard Smith and Joanne Morra, eds., *The Prosthetic Impulse: From a Posthuman Present to a Biocultural Future* (Cambridge, MA: MIT Press, 2006), 206.
4. Thierry Bardini, *Bootstrapping: Douglas Engelbart, Coevolution, and the Origins of Personal Computing* (Stanford, CA: Stanford University Press, 2000), 143.
5. Other examples can be found in Elizabeth Grosz, "Bodies-Cities," in Heidi Nast and Steve Pile, eds., *Places Through the Body* (New York: Routledge, 1998), 42–51; Michael Hardt and Antonio Negri, *Empire* (Cambridge, MA: Harvard University Press, 2000); Lisa Nakamura, *Cybertypes: Race, Ethnicity, and Identity on the Internet* (New York: Routledge, 2002); Mark B. N. Hansen, *Bodies in Code: Interfaces with Digital Media* (New York: Routledge, 2006).
6. Joseph Weizenbaum contrasts prosthetic machines with autonomous machines—the former act as extensions of human bodies, while the latter run on their own based on some model of the external world. He considers the computer to be an autonomous machine, and humans make decisions using constrained computer output. However, as David Mindell has shown, technologies that are described as autonomous are replete with many kinds of human actions and decision-making. See Weizenbaum, *Computer Power and Human Reason: From Judgment to Calculation* (San Francisco, CA: W. H. Freeman, 1976), 20–24, 38, and Mindell, *Our Robots, Ourselves: Robotics and the Myths of Autonomy* (New York: Viking, 2015).
7. Lucy A. Suchman, *Human-Machine Reconfigurations: Plans and Situated Actions*, 2nd ed. (New York: Cambridge University Press, 2007), 223–24.
8. John Perry Barlow, "Being in Nothingness: Virtual Reality and the Pioneers of Cyberspace," *Mondo 2000* 2 (1990): 42.
9. Katherine Ott, "The Sum of Its Parts: An Introduction to Modern Histories of Prosthetics," in Katherine Ott, David Serlin, and Stephen Mihm, eds., *Artificial Parts, Practical Lives: Modern Histories of Prosthetics* (New York: New York University Press, 2002): 1–42.
10. Lynette Jones, *Haptics* (Cambridge, MA: MIT Press, 2018); David Parisi, *Archaeologies of Touch: Interfacing with Haptics from Electricity to Computing* (Minneapolis: University of Minnesota Press, 2018). For a history of prosthetics prior to the use of computer technology, see Ott, Serlin, and Mihm, eds., *Artificial Parts, Practical*

Lives. Yulia Frumer has explored the history of robotics and prosthetics, notably recontextualizing Mori Masahiro's concept of the "uncanny valley." Frumer, "Cognition and Emotions in Japanese Humanoid Robotics," *History and Technology* 34, no. 2 (April 3, 2018): 157–83.

11. Carrie Sandahl, "Black Man, Blind Man: Disability Identity Politics and Performance," *Theatre Journal* 56, no. 4 (2004): 583.

12. Emily Russell, *Reading Embodied Citizenship: Disability, Narrative, and the Body Politic* (New Brunswick, NJ: Rutgers University Press, 2011), 5–6.

13. Julie Avril Minich, *Accessible Citizenships: Disability, Nation, and the Cultural Politics of Greater Mexico* (Philadelphia, PA: Temple University Press, 2013), 156.

14. Vivian Sobchack, *Carnal Thoughts: Embodiment and Moving Image Culture* (Berkeley: University of California Press, 2004), 216.

15. Beth A. Robertson, "'Rehabilitation Aids for the Blind': Disability and Technological Knowledge in Canada, 1947–1985," *History and Technology,* May 20, 2020, 1–24. Bess Williamson, "Electric Moms and Quad Drivers: People with Disabilities Buying, Making, and Using Technology in Postwar America," *American Studies* 52, no. 1 (October 22, 2012): 5–29. Jaroslav Švelch explores a related kind of tinkering in "Power to the Clones: Hardware and Software Bricolage on the Periphery" in this volume: he applies the concept of bricolage to the creation of hardware and software clones in Soviet bloc countries. The bricolage of Soviet computer developers, another group of what Švelch calls "marginal and marginalized historical actors," was similar, in many ways, to the work of people who make computer technology fit diverse bodies.

16. Sarah S. Jain, "The Prosthetic Imagination: Enabling and Disabling the Prosthesis Trope," *Science, Technology, & Human Values* 24, no. 1 (Winter 1999): 47.

17. Jain, "The Prosthetic Imagination," 33.

18. Jain, 41.

19. Laine Nooney, "'Have Any Remedies for Tired Eyes?': Computer Pain as Computer History," in this volume.

20. See the authors discussed in Elizabeth R. Petrick, "A Historiography of Human-Computer Interaction," *IEEE Annals of the History of Computing* 42, no. 4 (2020): 8–23.

21. Joseph P. Shapiro, *No Pity: People with Disabilities Forging a New Civil Rights Movement* (New York: Times Books, 1993), 73.

22. Part of this case study is adapted from my book, Elizabeth R. Petrick, *Making Computers Accessible: Disability Rights and Digital Technology* (Baltimore, MD: Johns Hopkins University Press, 2015), 47–57.

23. Denis K. Anson, *Alternative Computer Access: A Guide to Selection* (Philadelphia: F. A. Davis Company, 1997), 127, http://archive.org/details/alternativecompuooanso.

24. Michael J Silva et al., Membrane computer keyboard and method, US Patent 5,450,078, filed Oct. 8, 1992, and issued Sept. 12, 1995.

25. Jacquelyn Brand, "Parent Advocate for Independent Living, Founder of the Disabled Children's Computer Group and the Alliance for Technology Access," an oral history conducted in 1998–99 by Denise Sherer Jacobson, in *Builders and Sustainers of the Independent Living Movement in Berkeley,* vol. 5, Regional Oral History Office, The Bancroft Library, University of California, Berkeley (2000): 52–53.

26. Jacquelyn Brand, "The Disabled Children's Computer Group: Families Working Together," *Exceptional Parent* 15, no. 6 (October 1985): 18.

27. Brand, "Parent Advocate for Independent Living," 54.

28. Harvey Pressman, "The National Special Education Alliance: Applying Microcomputer Technology to Benefit Disabled Children and Adults" (Cupertino, CA: National Special Education Alliance, Apple Computer, Inc., 1987), 9–10, box 1, folder 1, Coll. BANC MSS 99/248c, Bancroft Library, University of California, Berkeley.

29. Nancy Sall and Harvey H. Mar, "Technological Resources for Students with Deaf-Blindness and Severe Disabilities," 1992, 27, https://archive.org/details/ERIC _ED360794.

30. Sall and Mar, "Technological Resources," 73.

31. Brand, "Parent Advocate for Independent Living," 55.

32. This case study is adapted from my book, *Making Computers Accessible,* chapter 5.

33. "Impact!: Working Documents," Section 5.3.1.3, Spring 1991, box 2, folder 3, Coll. BANC MSS 99/248c, Bancroft Library, University of California, Berkeley. Matvy's original spelling is included, but spaces have been added between words for readability.

34. "Impact!: Working Documents," Section 5.3.1.3.

35. Paul Dietz and Darren Leigh, "DiamondTouch: A Multi-User Touch Technology," in *Proceedings of the 14th Annual ACM Symposium on User Interface Software and Technology,* UIST '01 (Orlando, Florida: Association for Computing Machinery, 2001), 219–26.

36. Eynat Gal et al., "Enhancing Social Communication of Children with High-Functioning Autism through a Co-Located Interface," *AI & Society* 24, no. 1 (August 2009): 82.

37. Meryl Alper, *Giving Voice: Mobile Communication, Disability, and Inequality* (Cambridge, MA: MIT Press, 2017), 6.

38. Don Ihde, *Bodies in Technology* (Minneapolis: University of Minnesota Press, 2002), 14.

39. Vivian Sobchack, "Beating the Meat/Surviving the Text, or How to Get Out of This Century Alive," in *Cyberspace/Cyberbodies/Cyberpunk: Cultures of Technological Embodiment,* Mike Featherstone and Roger Burrows, eds. (London: SAGE Publications Ltd., 1995), 210.

40. Robert Rawdon Wilson, "Cyber(Body)Parts: Prosthetic Consciousness," in *Cyberspace/Cyberbodies/Cyberpunk,* 242.

41. Tobin Siebers, "Disability in Theory: From Social Constructionism to the New Realism of the Body," *American Literary History* 13, no. 4 (2001): 753, n. 10.

42. Wilson, "Cyber(Body)Parts," 251.

43. Jennifer Kaufmann-Buhler, "The Politics and Logistics of Ergonomic Design," in Elizabeth Guffey and Bess Williamson, eds., *Making Disability Modern: Design Histories* (London: Bloomsbury Press, 2020), 177–92.

CHAPTER 20

"Have Any Remedies for Tired Eyes?"

Computer Pain as Computer History

Laine Nooney

In the ebbing days of 1980, Henry Getson of Cherry Hill, New Jersey, wrote a letter to his favorite computer hobbyist magazine, *Softalk*. Getson described himself as a computer user of "less than expert status" and expressed to the editors his appreciation for *Softalk*'s introductory tone and accessible articles, especially for someone like him, someone who had recently bought a personal computer and was just learning to program.[1] His letter closed with a short question, a stray thread dangled from the hem of heaping praise: "P.S. Have any remedies for tired eyes?"

It was a question legible to *Softalk*'s editors, who responded at length to this "problem that many computerists share."[2] "Some relief comes," they wrote, "from double folding a washcloth, saturating it with warm water, and holding it against your eyes for several minutes." In later issues, fellow readers volunteered their own tips for dealing with eye strain. A reader from Cypress, Texas, recommended Getson modify his screen with a piece of plexiglass covered in "the sun screen material found in auto stores."[3] Another reader, from Malibu, California, suggested buying light green theatrical gel sheets, the kind used to color stage lights, and taping one over the monitor.[4] We don't know how Getson resolved to treat his tired eyes, but certainly he had no lack of homespun options volunteered by computer users negotiating similar issues.

What Getson was discovering, like all the rest of the microcomputing early adopters of the 1980s, was just how much using computers *hurt*. Turns out, monitors caused eye strain. Or, to put it more accurately: the assemblage of computational life routinely strained eyes. Vision problems were the embodied human residue of natural interactions between light, glass, plastic, color, and other properties of the surrounding environment. When overhead lighting, strong task lighting, or daylight cast from behind a user hit the curve of a CRT monitor, the result was a glare, or reflection, over the display's specular surface.[5] The twentieth century's tradition of strong overhead lighting—optimal for paperwork, accounting, reading, all the traditional tasks of office labor—produced a variety of lighting issues that negatively affected human vision when that human sat down in front of the dark glass of a computer monitor.[6]

What the so-called computer revolution brought with it was a world of pain previously unknown to man; there was really no precedent, in our history of media interaction, for what the combination of sitting and looking at a computer monitor did to the human body. Unlike television viewing, which was done at greater distance and lacked interaction, monitor use requires a short depth of field and repetitive eye motions. And whereas television has long accommodated a variety of postures, seating types, and distances from the screen, personal computing typically requires less than 2–3 feet of proximity from both monitor and keyboard. The kind of pain Getson experienced was unique to a life lived on-screen and would become a more common complaint as desktop computers increasingly entered American homes over the course of the 1990s and into the early twenty-first century.

Of course, computer-related pain existed prior to the arrival of the first consumer-grade desktop microcomputers in the late 1970s. Mid-century mainframes and large-scale minicomputers, with their high energy consumption and cooling needs, whirling tape drives, and the co-presence of noisy teletypes and teleprinters, were known to cause stress on the auditory system.[7] Furthermore, prolonged use of teletypes—mechanical telecommunications devices for sending and receiving data in print, over phone lines or a computer network—specifically caused carpal tunnel syndrome and other musculoskeletal health issues common among the mostly female ranks of twentieth-century typists, telegraph operators, and other clerical occupations.[8] The locus of health concerns would shift from auditory to visual, however, once teletypes began converging with CRT monitors in the 1970s. Replacing crisp type with the fuzzy resolution of a screen prone to glare, the so-called "glass teletypes," "teletype terminals," or even just "computing terminals" compounded the occupational health deficits of repeated use.[9]

With the advent of microprocessors, early microcomputing designers began experimenting with computer designs that converged a central processing unit, monitor, and keyboard into a single consumer good. Steve Wozniak's 1976 Apple 1 circuit board was one of the first manufactured microcomputers to include a video display adapter as part of its design, as did Processor Technology's SOL-20, released that same year.[10] By 1977, the standardization of a keyboard and monitor as essential peripherals to a central computing unit was set in stone by the concurrent release of the first wave of truly mainstream consumer microcomputers—the Apple II, the TRS-80, and the Commodore PET. It was at this moment, at the tail end of the 1970s, that computer usage became identified with "desktop" computers, and took on the bodily postures we associate with it today: the constant bend of wrist over a keyboard, the staring at a monitor, and slightly later, the nudging of a mouse. As both desktop computers and networked terminals proliferated in offices, schools, and homes over the 1980s, chronic pain became their unanticipated remainder: wrist pain, vision problems, and back soreness grew exponentially. Desktop computing required dramatic affordances among the population at large, whether those be changes to household and office lighting, tolerating chronic discomfort, or the circulation of new domestic and occupational imaginaries.

Historicizing Computer Pain

While Getson's small query might be easily overlooked in the hundreds of letters and articles that cycled through 1980s computer magazines, the question of "tired eyes" offers an alternate terrain for mapping the dramas of computational life in the late twentieth-century Western world. To consider the history of computing through the lens of computer pain is to center bodies, users, and actions over and above hardware, software, and inventors. This perspective demands the computer history engage with a world beyond the charismatic object of computers themselves, with material culture, with design history, with workplace ethnography, with leisure studies. For all those computerists with "tired eyes," computer culture was not what happened on-screen or in-box, but rather what happened everywhere else: with, on, and around keyboards, televisions, joysticks, desks, offices, kitchens, tables, beds, hands, glasses, lightbulbs, windows, back supports, surge protectors, power supplies . . . and on and on.

By turning away from the computer, the assemblage of computer history changes. To do so, this chapter samples from a decentralized archive, because—how could it not? A history of computer pain is too diffuse to ever be centered in an archive, could never make itself apparent on its own. Rather, it requires reading *across* materials, and reading *for* moments where unusual but patterned juxtapositions happen. Getson's stray "P.S." became notable to me only because I was looking for moments when computers and bodies seemed inextricable, unresolved. Noticing it tuned me into something, made me sensitive to a set of historical concerns that had otherwise eluded me before: that the question of how to live with computers was *everywhere* in the historical record of personal computing, and yet never directly asked. As a historical pursuit, then, there is no grand narrative here, just fragments and scraps, but ones that might, through their assembly, elucidate something about how we learned to live with computers. This is not the history of killer apps, wild hacks, and the coding wizards who stayed up late, but something far quieter and harder to trace, histories as intimate as they are "unhistoric": histories of habit, use, and making do.

Such a historical repositioning shifts and expands what we imagine a "cultural" history of computing to be. Work on computing communities has largely focused on expert users, whether the programmers who defined programming's transition to white-collar work, the researchers who codified the functions of email, or the dedicated collectives who nurtured social communication on bulletin board systems, message boards, Usenet groups, and the like.[11] Yet most computer users from the late 1970s onward were *not* expert users—neither innovators nor maintainers. Rather, most computer users are simply *workers*, operating within technical systems that offer them little capacity for intervention and enforce a constrained set of environmental and postural demands. What becomes common to computing, then, is a way of feeling that has nothing to do with expertise, or even a desire to build culture through computer use. Rather, everyone whose job requires sitting down at a computer knows the accumulated effect of this history, feels it. That pain in your neck, the numbness in your fingers, has a history far more widespread and impactful than any individual computer, computing innovator, or sociotechnical community. No single computer changed the world, but computer pain has changed us all—though some of us more than others.

Thus, this is not just about getting away from the usual suspects of computer history. It is also about going toward something—in our case, an expanded knowledge of the relationship between the body and the many constructed environments it occupies, between who had the freedom to build their world and who was saddled with enduring it. And, as is so often the case, those who did the enduring were women, and in many cases, specifically, women of color. Despite a history of invention that has rendered the ascent of computing is a uniquely white male activity, women were there, everywhere—for it was their bodies that would be on the frontlines of the dramatic transformations in workplace automation wrought by computing terminals in the 1970s and personal computers in the 1980s. Unlike hobbyist and leisure users of home and personal computers like Henry Getson, both white women's and women of color's use of computing, at least in the United States, typically happened in a workplace context, as computing technology was pushed upon the clerical and administrative labor traditionally siloed to pink-collar workers (as Hicks's essay in this collection points out, the lower one is in occupational hierarchy, the less control one has over one's working conditions). As this chapter demonstrates, the history of computer pain reveals the subterranean power differentials that underpin the personal computer's emergence in the workplace.

Documenting Computer Pain

In 1981—just one year prior to *Time* magazine declaring the personal computer 1982's Machine of the Year—the journal *Human Factors* published an entire issue dedicated to the issue of computers in the workplace, noting that "the number of workers using display terminals [computer monitors] is large and is increasing rapidly."[12] Prior to the 1980s, computing terminals had never seen wide enough circulation within a worker population to generate such complaints; this research paper then, offers a window in time upon the workers who first negotiated the arrival of computers into their offices.

Included in this collection is the research paper "An Investigation of Health Complaints and Job Stress in Video Display Operations," which focused on the relationship between health complaints and the use of display terminals in clerical work.[13] To conduct their analysis, the researchers held interviews with and distributed questionnaires to both "professional" and "clerical" workers at several companies where video display terminals were used.[14] To produce a control group, the researchers also held interviews and distributed the same questionnaire to workers who were engaged in the same kind of work but did it manually, using typewriters and traditional indexing. Aside from gathering basic demographic data and asking a range of questions related to job stress, the questionnaires asked the participants to document an exhaustive range of visual, musculoskeletal, and emotional health complaints. Of those employees who reported their sex, 47% were women—though when considering clerical workers apart from professional, white-collar VDT workers, women composed 67% of the employment base. Furthermore, clerical VDT workers were disproportionately women of color (46% of all clerical workers who reported demographic data). The distinction between professional and clerical VDT workers is significant, as clerical workers had less control over the type of work they did or the management of their time on the computer.

In analyzing their data, the researchers found that "clerical VDT operators showed much higher levels of visual, musculoskeletal, and emotional health complaints, as well as higher job stress levels, than did control subjects and professionals using VDTs."[15] In every category of health complaint—from fainting to stomach pain to neck pressure to hand cramps—the percentage of complaints went up among clerical workers stationed at computer terminals, often doubling, tripling, or quadrupling in number. Blurred vision, blurring eyes, and eyestrain were reported by 70% to 90% of the sample, and some of the strong disparities between the clerical workers and the control subjects—such as with changes in color perception or stiff or sore wrists—were clear indicators of the impact of the soft repetitive strain of computer use. As the workers with the least degree of autonomy over their labor, the bodies of these women—predominantly women of color—found themselves most directly impacted by the physical toll of computer technology.

Yet there was another component to health complaints and stress that the social scientists documented in their research, but didn't quite know what to do with. In assessing levels of stress and job satisfaction between clerical workers placed at a computing terminal and those in the control group doing tasks by hand, the researchers determined that clerical employees using computer terminals reported higher degrees of monotony and fatigue and general job dissatisfaction versus those performing the same kind of work by hand. As they put it, "Stress problems reported that were by the clerical VDT operators are not solely related to the VDT viewing, but are related to the whole VDT work system."[16] Tasked with boring, repetitive labor, clerical VDT workers reported "low ratings of job involvement and job autonomy," and felt they had little control over their job requirements.[17] For the women pressed onto VDTs for clerical work, the problem was not simply the computer but the way the computer's so-called productivity diminished the satisfaction they took in their labor.

But what was it, precisely, about computers that caused work to *hurt* so much more? What these researchers were encountering in their data was the kind of psychic residue that the researchers' questionnaires and rote interviewing wasn't well suited to explain. Answers would have to wait until the publication of Shoshana Zuboff's landmark 1988 monograph *In the Age of the Smart Machine: The Future of Work and Power*, an ethnographic account of the impact of computer usage in work environments in the early 1980s.[18] Zuboff's book is a masterpiece of organizational workplace ethnography, in which she uses participant observation and interviews to assess how employees experienced changes to their work as the computer became an intermediary.

What Zuboff's investigations revealed were the psychophysical costs that shadowed the computer's entry into the workplace. In her grimly titled fourth chapter, "Office Technology as Exile and Integration," Zuboff documents the time she spent observing two administrative office sites where computers had just become integrated into clerical work practice (while Zuboff doesn't offer quantitative statistics on the employees she studied, she does note that they were predominantly women). Prior to the arrival of computer terminals, the women who handled these tasks described having a very material relationship to their work. They retrieved actual files and filled out physical pieces of paper; they moved folders back and forth between filing cabinets, they updated files by hand and left notes for themselves, and manipulated the intricacies of these files based on their personal knowledge of their clients, their accumulated know-how on the job, and through consultation with their fellow clerks and managers. The arrival of computers onto the desks of these workers was done with the intention of streamlining and speeding up the work these women engaged in by evaporating all of the small physical habits associated with their work—the walking and talking, the shuffling of paper, the flipping of pages, the personalized practices of self-annotation.

But in this effort to "simplify" these routines by making the office paperless, Zuboff found that the implementation of computers wound up eradicating the basis of the clerks' situated knowledge. Suddenly, making changes to a client's account meant simply inputting data in an order that was constrained by the computer itself. Work became a process of filling in blanks; there was no longer anywhere for the clerks to experience decision-making in their jobs. What Zuboff observed was that as intellectual engagement with the work went down, the necessity of concentration and attention went up. What the computer did was make the work so routine, so boring, so mindless, clerical workers had to physically exert themselves to be able to focus on what they were even doing. This transition, from work being about the *application of knowledge* to work being about the *application of attention,* turned out to have profound physical and psychological impact on the clerical workers themselves.

Zuboff was able to track the extent of this toll by asking the clerical workers to draw pictures of themselves at work before and after the computer.[19] These images reveal themselves, embodying a kind of juvenile terror in their simple lines and stark contrasts. The workers depicted themselves as happy in the times before the computer, and frequently in the company of others (fig. 20.1). What the computer brings to them is a kind of desolation: a worker who has become

Figure 4.1   Transfer assistant

Before

Figure 4.2   Transfer assistant

After
"The after picture is only the back of my head because it is a nonperson."

Figure 4.3   Transfer assistant

Before

Figure 4.4   Transfer assistant

After
"There's a lot of tension now, and that makes people get mean. We had more control before and less confusion. You could get things done. Every once and a while my head starts to throb. I can't take it."

Figure 20.1. Figures 4.1, 4.2, 4.3, and 4.4 from Shoshana Zuboff, *In the Age of the Smart Machine: The Future of Work and Power* (New York: Basic Books, 1988), 142–43.

nothing more than the back of her head; hair, ripped from the scalp; a deep sense of being alone. One of the most detailed drawings is accompanied by the caption: "No talking, no looking, no walking. I have a cork in my mouth, blinders for my eyes, chains on my arms. With the radiation I have lost my hair. The only way you can make your production goals is give up your freedom" (fig. 20.2).[20] The side of the desk is marked by the ascending arrow of a productivity chart. Another image depicts the worker in the striped uniform of a convict (fig. 20.3).[21] A phone ring ring rings on the desk and a flower in a vase droops beside the computer. The calendar is empty, and her supervisor watches from above. A sign, intended to be inspirational—"keep

Figure 4.8   Benefits analyst

After

"No talking, no looking, no walking. I have a cork in my mouth, blinders for my eyes, chains on my arms. With the radiation I have lost my hair. The only way you can make your production goals is give up your freedom."

Figure 20.2. Figure 4.8 from Shoshana Zuboff, *In the Age of the Smart Machine: The Future of Work and Power* (New York: Basic Books, 1988), 145.

Figure 4.12   Benefits analyst

After
"My supervisor is frowning because we shouldn't be talking. I have on the stripes of a convict. It's all true. It feels like a prison in here."

Figure 20.3. Figure 4.12 from Shoshana Zuboff, *In the Age of the Smart Machine: The Future of Work and Power* (New York: Basic Books, 1988), 147.

up MPH"—suggests the new emphasis on speed that the clerical workers have been asked to internalize.

What both Zuboff's research and the *Human Factors* article demonstrate is the extent to which the computer altered the environmental conditions of work. Physical pain and mental distress related to computer use was not simply a product of the tool itself but about a complex chain of interconnection between desks, chairs, keyboards, and lighting conditions, as well as shifting notions of efficiency, speed, attention, what "knowledge" even meant. This was not computing making good on latent promises of prosthesis or augmentation—a frequently conjured metaphor Elizabeth Petrick explores in this volume, in which computer technology is celebrated and esteemed for "conquering the incapabilities of the body."[22] As Zuboff's images make clear, the only thing being conquered is the body of the worker: if computers could change how much data a worker could process, then the human body no longer intervened on profitability with its pesky physiological limits.

Working against Computer Pain

Keeping computing profitable, however, meant finding ways to mitigate, negotiate, and address rising complaints of physical pain from its users. Beginning in the mid-1980s, the specialists in ergonomics, human factors, and physical health began turning their attention to desktop computer use. This is testified to by the publication of books like *Zap!: How Your Computer Can Hurt You and What You Can Do About It*, which presents the office or home office as an ecology, in which relations between monitors, keyboards, lighting, chairs, air quality, and work schedules had to be endlessly manipulated to acquire one's "perfect workstation" for safer computing.[23] Even physical fitness specialists could cash in on America's new attention to the ailing bodies of its workers—merely consider Denise Austin's late 1980s *Tone Up at the Terminals: An Exercise Guide for High-Tech Automated Office Workers* (fig. 20.4). Austin, a popular fitness personality with a workout show on ESPN, promoted an entire corporate fitness program, for which this free instructional booklet, published in a partnership between the New York State Library and Denise Austin Fitness Systems, served as both a government resource on ergonomics in the workplace as well as a marketing tease.

Austin's role, and her own booklet, is to serve as an enthusiastic guide for the reader, modeling how "high-tech automated office workers" can reduce tension and "nervous fatigue." Shoulders, arms, wrists, hands, waist, back, legs, ankles, feet, and posture are all addressed through a series of increasingly absurdist positions Austin manages to maintain while remaining seated in a knee-length tweed skirt (fig. 20.5). Austin never stands; surely employers did not want to see images of workers stretching their hamstrings on a walk to the water cooler. In Austin's feminine decorum, we are reminded of the women from Zuboff's study: the emphasis on being nondisruptive, on not taking up space, of maintaining the possibility of continuous work. And, of course, you end with a hug—after all, "YOU DESERVE IT!"

We know whose bodies such a document was designed to discipline. Just as Zuboff documented in her ethnographic work, the arrival of

Figure 20.4. Cover of *Tone Up at the Terminals: An Exercise Guide for High-Tech Automated Office Workers,* n.d.

computers into offices was often done as part of an initiative to auto-mate clerical, feminized labor like data entry and word processing. Fur-thermore, knowing how to type, which was a prerequisite skill for us-ing a computer, was the domain of clerical work; it was a skill taught to women in school, but not to men (anyone who has ever seen an older male programmer do "hunt-and-peck" with their index fingers has seen these histories in action). The uptake of the computer within execu-

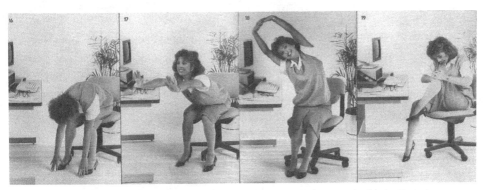

Figure 20.5. Close-up exercise spreads from *Tone Up at the Terminals: An Exercise Guide for High-Tech Automated Office Workers,* n.d., pp. 4–5, 8–9.

tive and managerial strata actually stalled for a period due to the cultural association of typing as fundamentally secretarial.[24] Advertising reflected these anxieties about gendered occupational roles throughout the 1980s: women were depicted typing on computers, while men pointed at screens, looked over a woman's shoulder, or merely posed with a computer on their desk.[25] It was not until the mainstreaming of the mouse in the late 1980s that these tensions began to ease.[26] With a mouse, a male executive could operate the computer without adopting the presumably demeaning posture of his secretary.

What all of this adds up to is a decades-long drama between body and machine, a uniquely gendered one at that. Probably not since the automobile has there been a technology that has so insistently reorganized how we use our bodies in day-to-day practice—and the long arc of these transformations is still being played out. As those reading this chapter are among the first generation of humans to come of age on the computer, the toll of this is persistently being felt in the now commonplace reality of carpal tunnel syndrome, chronic back pain, and vision problems among a workforce just beginning to hit middle age.

The Multitasking of Pain Management Today

Like Denise Austin's office workout routine, many of the interventions we're asked to adopt demand we internalize responsibility for our physical well-being, while never becoming a burden on the workplace or lowering our productivity. Our pain feeds whole new industries, blossoming in the form of standing desks, walking desks, adjustable keyboards, and ergonomic mice of every stripe; our aggrieved bodies have been a boon for voice recognition software (this entire chapter was written with voice recognition software). And we've sought help beyond our desks too. One of the most popular yoga YouTube personalities, Adriene Mishler, offers multiple videos that conjure Austin's legacy, including *Yoga at Your Desk, Office Break Yoga,* and *Yoga for Text Neck.*[27] Similar topics are popular among many YouTube channels, ranging from midtier health and wellness personalities to established institutions such as the Mayo Clinic. These practices have become critical parts of the way we have long been expected to take work with us—spending our offline, off-work hours repairing the damage done by our jobs.

The smartphone's insatiable demand on our attention is just the latest in a long dance between our psychic and emotional health and the computer. The posture of the head tilt is an index to the encumbrance of multitasking, a term now synonymous with what it even means to use a computer device—to slide between applications, to flick attention from one priority to the next with no delay for contextual readjustment, the seemingly seamless movement we now engage between our personal and our occupational lives. Multitasking was once something that belonged solely to the realm of the computer; it was a technical term, referring to the capacity for time-sharing systems to concurrently process the operations of multiple users by switching back and forth rapidly between jobs. It was over the course of the late 1980s and '90s, with the rise of the graphical user interface and the increasing gig-ification of the US workforce, that the term "multitasking" came to be applied to human labor, to the idealized state of being able to work on multiple tasks more or less simultaneously. The doldrum of Zuboff's clerical workers has become the endless noise of habituated computer use.

So the next time you experience "tired eyes," wrists tingling, neck cramps, or even the twinge of text neck, let it serve as a denaturalizing reminder that the function of technology has never been to make our lives easier, but only to complicate us in new ways. Computer-related pain, and the astounding efforts humans went to (and continue to go to), to alleviate it, manage it, and negotiate it, provide one thread through the question of how the computer became personal. The introduction of computers into everyday routines, both at work and at home, was a historic site of vast cultural anxiety around the body. To locate a history of computing that might be otherwise—one embodied, habituated, and distinctly spatial—we would do well to think about Getson's letter and consider what kind of histories of computing might be lying *around* the computer, rather than inside of it.

Notes

1. Henry Getson, letter to Open Discussion, *Softalk,* January 1981.
2. Getson, letter. For further reading on *Softalk* and the role it played among early Apple II users, see Laine Nooney, Kevin Driscoll, and Kera Allen, "From Programming to Products: Softalk Magazine and the Rise of the Personal Computer User," *Information & Culture* 55, no. 2 (2020): 105–29.
3. Randy Reeves, letter to Open Discussion, *Softalk,* March 1981.
4. Stephen R. Bosustow, letter to Open Discussion, *Softalk,* March 1981.
5. Don Sellers, *Zap!: How Your Computer Can Hurt You and What You Can Do About It* (Berkeley, CA: Peachpit Press, 1994): chapters 2, 5, 6. Modding a monitor with a filter or gel could resolve glare because it created two additional surfaces that ambient light had to pass through.
6. Relationships between monitors and workplace lighting come up in several office architectural books from the period. The blurriness of the letterforms also caused vision problems due to the low resolution of the monitors. See Marvin J. Dainoff, Alan Happ, and Peter Crane, "Visual Fatigue and Occupational Stress in VDT Operators," *Human Factors* 23, no. 4 (1981): 422.
7. Given the proportionally small number of people who worked directly with computing installations prior to the 1970s, such information is largely anecdotal. However, trace evidence can be found. A *New York Times* article from November 23, 1969, listed "computers and typewriters and tabulators" as just a few of the myriad machines polluting the noisescape of New York City. Anthony Bailey, "'Noise Is a Slow Agent of Death," *New York Times,* November 23, 1969. In the summer of 1970, *Datamation* reported that the National Bureau of Standards released a report on the dangers of hearing loss and computer centers. "What's All the Noise About," *Datamation* (August 15, 1970): 73. In 1997, US Department of Veterans Affairs denied a disability appeal from a veteran who filed for disability on the basis of hearing loss he claimed came from exposure to both mainframe computers and aircraft: https://www.va.gov/vetapp97/files5/9742889.txt. Many thanks to Cormac Deane, for

pointing me toward the sources on Twitter, as well as a piece from *Sounding Out*, about the role of computer noise in the IBM 360 episodes of *Mad Men*. See Andrew J. Salvati, "That Infernal Racket: Sound, Anxiety, and the IBM Computer and AMCs *Mad Men*," *Sounding Out!*, May 26, 2014, https://soundstudiesblog.com/2014/05/26/that-infernal-racket-sound-anxiety-and-the-computer-in-amcs-mad-men/.

8. On the overlap between telegraph and teletype, see Thomas C. Jepsen, *My Sisters Telegraphic: Women in the Telegraph Office, 1846–1950* (Athens: Ohio University Press, 2000). On carpal tunnel syndrome and occupational factors, see Barbara A. Silverstein, Lawrence J. Fine, and Thomas J. Armstrong, "Occupational Factors and Carpal Tunnel Syndrome," *American Journal of Industrial Medicine* 11, no. 3 (1987): 343–58.

9. On terminals and teletypes, see Matthew G. Kirschenbaum, *Track Changes: A Literary History of Word Processing* (Cambridge, MA: Belknap Press, 2016), 124.

10. While the Apple 1 did not come with a monitor or keyboard as part of its purchase price, the fact that adapters for such peripherals were built into the board was a technological innovation compared to prior hobbyist computing systems like the Altair 8800. The SOL-20, however, did include a keyboard as part of the computer, all manufactured in a single metal and wood case. This form factor is indebted to glass teletype terminals as evidenced by the fact that these microcomputers were sometimes known as "terminal computers" or "intelligent terminals." On the influence of glass teletypes on the design of early consumer grade microcomputers, see Laine Nooney, *The Apple II Age: How the Computer Became Personal* (Chicago: University of Chicago Press, forthcoming, 2023), chapter 1.

11. On expert users and communities of practice, see, from a vast literature, Finn Brunton, *Spam: A Shadow History of the Internet* (Cambridge, MA: MIT Press, 2013); Kevin Driscoll, "Social Media's Dial-Up Roots," *IEEE Spectrum* 53, no. 11 (2016): 54–60; Nathan Ensmenger, *The Computer Boys Take Over: Programmers, Programmers, and the Politics of Technical Expertise* (Cambridge, MA: MIT Press, 2012).

12. Throughout the issue, the term "video display terminal" (VDT) is used as a synonym for computer monitor. No particular differentiation seemed to be made between VDTs that were part of time-sharing systems versus VDTs that were part of a desktop personal computing system. Vivek D. Bhise and Edward J. Rinalducci, "Special Issue Preface," *Human Factors* 23, no. 4 (1981): 385–86.

13. Michael J. Smith, Barbara G. F. Cohen, Lambert W. Stammerjohn, Jr., and Alan Happ, "An Investigation of Health Complaints and Job Stress in Video Display Operations," *Human Factors* 23, no. 4 (1981): 387–400.

14. Smith et al., "An Investigation," 388–90.

15. Smith et al., 396.

16. Smith et al., 398.

17. Smith et al.

18. Shoshana Zuboff, *In the Age of the Smart Machine: The Future of Work and Power* (New York: Basic Books, 1988).

19. Zuboff, *In the Age of the Smart Machine*, 142–49.

20. Zuboff, 145.

21. Zuboff, 147.

22. Elizabeth Petrick, "The Computer as Prosthesis? Embodiment, Augmentation, and Disability," in this volume.

23. Sellers, *Zap!*

24. Paul Atkinson, "The Best Laid Plans of Mice and Men: The Computer Mouse in the History of Computing," *Design Issues* 23, no. 3 (2007): 59–61. Historian Jesse Adams Stein's master's thesis also explores the gendered perceptions surrounding keyboards and mice during the 1980s. See Jesse Adams Stein, "Domesticity and Gender in the Industrial Design of Apple Computer, 1977–1984" (master's thesis, School of the Art Institute of Chicago, 2009), chapter 2, 50–90.

25. Stein, "Domesticity and Gender," 64–71.

26. Atkinson, "Best Laid Plans," 61; Stein, "Domesticity and Gender," 84–86.

27. "Text neck" is the latest variation on computer-related body pain. A quick Google search for the term "text neck" generates an array of links to quasi-medical advice websites, including physio-pedia.org, healthline.com, and spine-health.com. The Text Neck Institute (which appears to be a doctor's office in Plantation, Florida) identified text neck as a "global epidemic" as early as 2015 (www.text-neck.com). At www.textneck.com, you are redirected to www.teknekk.com, the "ultimate parental remote-control app" that allows parents to manage screen time while also enforcing behavioral changes around smart phone posture. For just a sampling of published research on the subject, see Sunil Neupane, U. I. Ali, and A. Mathew, "Text Neck Syndrome—Systematic Review," *Imperial Journal of Interdisciplinary Research* 3, no. 7 (2017): 141–48; Woojin Yoon et al., "Neck Muscular Load When Using a Smartphone While Sitting, Standing, and Walking," *Human Factors* (2020).

Beyond Abstractions and Embodiments

Stephanie Dick and Janet Abbate

Who has a mind and who has a body? This question lies unspoken behind histories that emphasize some people's intellectual contributions and other people's labor. As historians increasingly foreground the *practices* of computing, this question might seem out of date. Bodies are, of course, everywhere; even the "lone geniuses" beloved of computer mythology have them. However, emphasizing the importance of embodiment does not negate or dissolve mind-body dualism, it rather underscores its historical importance: the dichotomy is reconstructed and reconstituted in different historical contexts *and in historical scholarship*, as a way of delineating social hierarchies and theories of historical change as propelled by heads or hands. This volume has advanced the argument that the dichotomy between abstractions and materialities, between mathematics and engineering, between concepts and labor, between ideas and people, between theory and practice, between mind and body, *is itself a social relation*. Individually, the chapters make clear where specific historical actors drew and navigated the line between abstractions and embodiments, and the many reasons they had for doing so. Collectively, the authors make a compelling case for the *centrality, mutability,* and *stakes* of that boundary for the social history of computing.

The artificiality of the mind-body dichotomy has long been an open secret in the history of computing, like the barely concealed homosexuality in Bletchley Park where Alan Turing became the "father" of modern computer science.[1] There, Turing—a man whose own body would soon come under hostile legal scrutiny—prescribed ways of thinking that hid the body from view. In 1950, Turing advanced his famous behaviorist test for machine thinking, which features a computer, a human interlocutor, and a human judge who puts written questions to both.[2] If the computer's responses can convince the judge that *it* is the human sufficiently often, then, Turing proposes, we should conclude that the computer is thinking. The test strips away all bodies from the question of intelligence—all that matters is the textual and conversational performance of thought. Turing notes that it would be unfair to allow the judge to see or hear the human interlocutor and the machine, which would give away their identities and tie intelligence to materiality in a way Turing believed was unnecessary. For him, the test "has the advantage of drawing a fairly sharp line between the physical and the intellectual capacities of a man." Turing's exclusion of bodies and materiality from the question of intelligence is often read as the foundational invocation of mind-body dualism from the early decades of modern digital computing. As Katherine Hayles put it in the opening to *How We Became Posthuman*, "Here, at the inaugural moment of the computer age, the erasure of embodiment is performed so that 'intelligence' becomes a property of the formal manipulation of symbols rather than enaction in the human lifeworld."[3] But as Hayles also notes, bodies in fact lurk throughout Turing's test.

The test is historically instructive not just because it posits a foundational mind-body dualism of the computer age but also because it signals how that dualism is reconstructed, reinscribed, blurred, deconstructed, and reproduced time and again. The "imitation game" for machine intelligence that Turing introduced in 1950 was based, in fact, on a parlor game for gender switching and performance. As Turing describes the game, a man and a woman leave the room, and a judge who remains asks them questions written on paper. The woman's job is to convince the judge that she is indeed a woman, while the man's job is to *fool* the judge into thinking rather that *he* is the woman. But what should you ask if you want to know who is the man and who is the woman? Likely, you would ask about their bodies! Students with whom we have played the "imitation game" for gender often ask about menstruation, the location of the nearest women's bathroom, or what brand of makeup contestants prefer; Turing himself suggests asking about hair styles. He reinvents this parlor game in "Computing Machinery and Intelligence" by replacing the man with the computer and assigning it the job of performing not gender, but rather "humanness." Turing then wonders, "Will the interrogator decide wrongly as often when the game is played like this as he does when the game is played between a man and a woman?" The question marks a continuity, rather than a break, with corporeal distinctions. Moreover, underneath its attempted objectivity the Turing test is irreducibly social, since the proof of success is not anything produced by the computer itself but rather the impression it makes in the mind of the human judge.[4] It is testament to Turing's mystique that few people have questioned why telling a convincing lie should be the measure of intelligence.

The centrality of bodies, gender, and social maneuvering in Turing's test becomes even more evident when one considers that Turing was gay at a time when homosexual acts were illegal in England. In 1952 Turing's home was burgled, and when he reported the crime the ensuing investigation led to *him* being charged with crimes of "gross indecency," because it revealed his sexual relationship with another man.[5] Turing was forced to undergo a hormonal treatment erroneously thought to "cure" homosexuality in men; he died two years later, an apparent suicide. Hayles notes that Turing's "conviction and the court-ordered hormone treatments for his homosexuality tragically demonstrated the importance of *doing* over *saying* in the coercive order of homophobic society with the power to enforce its will upon the bodies of its citizens."[6] The Turing test can be seen as a parable of a gay man who wanted to theorize "intelligence" as something independent of gender, sexuality, and embodiment, in part because of his own struggle with them.[7] But that sells the theorist short. Though his name and the eponymous test may now signal the erasure of bodies, Turing himself had a more nuanced understanding of the interdependence of mind and body.

Two years prior to publishing "Computing Machinery and Intelligence," Turing wrote a never-published piece called "Intelligent Machinery." Here he posits that the "surest way" to make an intelligent machine would be to give the machine a body, to "take a man as a whole and to try and replace all the parts of him by machinery."[8] He imagined a machine equipped with "television cameras, microphones, loudspeakers, wheels and 'handling servomechanisms' as well as some sort of 'electronic brain.'" Turing's machine-man would become intelligent in the same way that its flesh-and-bone counterparts do, namely, by engaging with the world: "In order that the machine should have a chance of finding things out for itself it should be allowed to roam the countryside." Yet machine embodiment posed many obstacles, not least of which was the inevitability of social encounters. The young Turing, clearly having read Mary Shelley's *Frankenstein*, imagined that if composed of late 1940s technology this machine-man would be of "immense size" and if allowed to roam the countryside, "the danger to the ordinary citizen would be serious." Turing also acknowledged that the body is not merely a conduit for sensory information but also a source of pleasure and meaning. He observed that because "the creature would still have no contact with food, sex, sport, and many other things of interest to the human being," its ability to develop like a person would be limited. Ultimately, Turing concluded that "although this method is probably the 'sure' way of producing a thinking machine," it would be too slow and impractical. Instead, he proposed to "try and see what can be done with a 'brain' which is more or less without a body."

What can a brain do without a body? Turing suggested mathematics, chess, and natural language processing—each of which, in turn, was a central area of focus in early artificial intelligence research. The scope of answers has only expanded from there. The belief that bodies can be dispensed with, and that brains and minds and the machines that simulate them can do a great deal in their absence, persists and is celebrated or assumed within much of computing, including by many who write its history.

Western culture, since the enlightenment, has prioritized mind over body, theory over practice, mathematics over materiality as *more fundamental.* These were epistemological priorities, but they were also social priorities. Race and technology scholars like Ruha Benjamin and Yarden Katz have explored how white, Western, and colonial systems of political and epistemic order repeatedly and predictably hold space at the top for the perspectives, designs, goals, and knowledge of those who already hold significant power—those whose *ideas* are often held up as the mechanisms of historical change.[9] Much of the rest of the world is then relegated to the "secondary," the "incidental," and often hidden roles of laborer, manufacturer, implementer, consumer, maintainer, miner, user. As such, "the computer" does not just bring together mathematical histories with technological ones, combining the logic of Turing and Boole with the technologies of twentieth-century information machines. The computer is also a technological manifestation of social, economic, and political order: a specific configuration of some peoples' minds and other peoples' bodies.

As Turing himself makes clear, however, abstractions and embodiments are often co-constituting. The essays in this volume have shown that there are no abstractions that are not entangled with bodies—classed, gendered, racialized, laboring—and the authors have traced when and how those bodies are foregrounded or erased, and the myths of abstraction and origin that their erasure supports. Abstractions are created through the erasure of context, of specificity, of people, of labor, of experiments on paper, of company bottom lines, of origins, of failed alternatives. Each of those erasures signals its own political economy. Whom did it serve to hide these people or alternatives from view? Whose life was made easier or whose product or theory was more saleable, more "objective," once that messiness was "gone"? Or, as Laine Nooney writes in this volume, "Who had the freedom to build their world and who was saddled with enduring it"? Conversely, as we recover labor, people, messiness, contingency, and context—as we recover embodiment—new abstractions also become visible. We see how material constraints can inspire new ideas, as Jaroslav Švelch explores in Soviet-era "bricolage"; observable divisions of labor alert us to the power of social symbols, as Kelcey Gibbons describes. The frictions we experience in our physical interactions with machines lead us back to the conceptual orders of gender, race, and (dis)ability that have shaped the interfaces we confront.

The inequalities foregrounded in these accounts have also made some histories harder to tell, through the uneven availability of sources, the dominance of English, and the fact that underrepresented groups may lack the resources or encouragement to document their experiences in computing. The recent surge of scholarship linking computing to Black and disabled perspectives has reframed the history of computing, both by including actors who were formerly invisible and, more radically, by challenging the universality and neutrality of computer technologies.[10] They ask who the implied user of a technology is and which users are considered normative; they look for the invisible laborers behind the products; and they ask how the opportunity to innovate—and even what counts as innovation—has been constrained by power relations in the larger society.

By foregrounding the repeated reestablishment of distinctions between abstractions and embodiments, we hope not only to give historical specificity to those demarcations but also to underscore the political work of separating mind and body (and of reconnecting them) in the history of computing. We offer this alongside other historiographic reframings in recent and forthcoming volumes in the history of computing, including *Captivating Technology, Your Computer Is on Fire*, and *Just Code*, in hopes of charting essential paths forward and backward.[11] These studies point to a methodology that resituates the computer, taking it out of its assumed place within a lineage of computing developments and placing it within larger histories of human achievement and struggle.

Notes

1. Jacob Gaboury, "A Queer History of Computing," in *Rhizome*, February 19, 2013, https://rhizome.org/editorial/2013/feb/19/queer-computing-1/.
2. Alan Turing, "Computing Machinery and Intelligence," in *Mind* 59, no. 236 (1950): 433–60.
3. Katherine Hayles, *How We Became Posthuman: Virtual Bodies in Cybernetics, Literature, and Informatics* (Chicago, IL: University of Chicago Press, 1999), xi.

4. Simone Natale has similarly pointed out how AI depends on anticipating and manipulating the user's reaction to it. Simone Natale, *Deceitful Media: Artificial Intelligence and Social Life after the Turing Test* (New York: Oxford University Press, 2021).

5. Alan Hodges, *Alan Turing: The Enigma* (Princeton, NJ: Princeton University Press, 2014).

6. Hayles, *How We Became Posthuman*, xii.

7. See Hayles, *How We Became Posthuman*, and Gaboury, "A Queer History of Computing."

8. Alan Turing, "Intelligent Machinery," AMT/C, Unpublished Manuscripts and Notes, 11, http://www.turingarchive.org/viewer/?id=127&title=1, image 20.

9. Benjamin, *Captivating Technology* and *Race after Technology* (Polity, 2019); Yarden Katz, *Artificial Whiteness: Politics and Ideology in Artificial Intelligence* (New York: Columbia University Press, 2020).

10. Elizabeth R. Petrick. *Making Computers Accessible: Disability Rights and Digital Technology* (Baltimore, MD: John Hopkins University Press, 2015).

11. Ruha Benjamin, *Captivating Technology: Race, Carceral Technoscience, and Liberatory Imagination in Everyday Life* (Durham, NC: Duke University Press, 2019); Thomas Mullaney, Benjamin Peters, Mar Hicks, Kavita Philip, *Your Computer Is on Fire* (Cambridge, MA: MIT Press, 2021); Jeffrey Yost and Gerardo Con Diaz, *Just Code* (forthcoming with Johns Hopkins University Press).

Editors

Janet Abbate is a Professor of Science, Technology and Society at Virginia Tech. Her books include *Inventing the Internet* (MIT Press, 1999) and *Recoding Gender: Women's Changing Participation in Computing* (MIT Press, 2012).

Stephanie Dick is an Assistant Professor in the School of Communication at Simon Fraser University. She is a historian of mathematics, computing, and artificial intelligence in the postwar United States.

Authors

Marc Aidinoff is a doctoral candidate in History and STS at the Massachusetts Institute of Technology. His dissertation, "A More Updated Union: New Liberals and Their New Computers in the New New South, 1984–2004," examines the computerization of the US welfare state. Marc is a fellow at Data for Progress, a senior strategist for OpenLabs, and an advisor to the Rosedale Mississippi Freedom Project.

Troy Kaighin Astarte is a Lecturer in Computer Science at Swansea University, affiliated with the Newcastle University Historic Computing Committee, and a Council member for the British Society for the History of Mathematics. They research the history of computing and mathematics, with an emphasis on the formal and theoretical elements of computer science, as explored in their dissertation, "Formalising Meaning: A History of Programming Language Semantics."

Ekaterina Babintseva is the Hixon-Riggs Early Career Postdoctoral Fellow in Science and Technology Studies at Harvey Mudd College. She received her PhD in History and Sociology of Science from the University of Pennsylvania. Her research exploring the history of computing and psychology in the twentieth century has been supported by the Charles Babbage Institute, Consortium for the History of Science, Technology, and Medicine, and Association for Computing Machinery.

André Brock is an Associate Professor at the School of Literature, Media, and Communication at Georgia Tech and author of *Distributed Blackness: African American Cybercultures* (NYU Press, 2020). His work explores Black technocultures, with an emphasis on community formation and everyday uses of networked technology, and it pushes against the default Whiteness at work within ideals of computing and within media archeology.

Maarten Bullynck is Associate Professor of History and Epistemology of Science in the Department of Mathematics and History of Science and member of the UMR 8533 IDHE.S at the University of Paris 8. His research examines the history of science and communication in the eighteenth century, the history of number theory, and the history of computing and its uses.

Jiahui Chan is a 2016 graduate of the history program at Nanyang Technological University in Singapore.

Gerardo Con Diaz is Associate Professor of Science and Technology Studies at the University of California, Davis. He is a historian of digital law, exploring the interplay among intellectual property, industrial change, and information technology. He is the author of *Software Rights: How Patent Law Transformed Software Development in America* (Yale University Press, 2019). He is Editor-in-Chief of the *IEEE Annals of the History of Computing.*

Liesbeth De Mol is a permanent researcher with the French National Center for Scientific Research (CNRS), affiliated to UMR 8163 Savoirs, Textes, Langage. She is a historian and philosopher of computer science and programming and was the founding President of the DHST/DLMPST Commission for the History and Philosophy of Computing (HaPoC). Currently, she is leading the international ANR project PROGRAMme focusing on the question, *What is a computer program?*

Kelcey Gibbons is a doctoral student in the History, Anthropology, Science, Technology and Society program at the Massachusetts Institute of Technology. She explores the history of computing in African American communities and relationships between the Civil Rights movement and the so-called personal computer revolution of the 1960s. Kelcey is the first recipient of MIT's L. Dennis Shapiro Graduate Fellowship in the History of African American Experience of Technology.

Elyse Graham is Associate Professor of Digital Humanities at Stony Brook University and the author of three books. Her book *The Republic of Games: Textual Culture between Old Books and New Media* (McGill–Queens University Press, 2018) explores the "gamification" of content-production in digital platforms and its consequences for the future of text.

Michael J. Halvorson is Benson Family Chair in Business and Economic History and Professor of History at Pacific Lutheran University in Tacoma, Washington. He worked at Microsoft Corporation from 1985 to 1993 and is the author of 40 books related to computer programming, software, and history, including *Code Nation: Personal Computing and the Learn to Program Movement in America* (ACM Books/Morgan & Claypool, 2020); and *Defining Community in Early Modern Europe* (Routledge, 2008), with Karen E. Spierling.

Mar Hicks is Associate Professor of History at Illinois Institute of Technology. Their award-winning book, *Programmed Inequality: How Britain Discarded Women Technologists and Lost Its Edge in Computing* (MIT, 2017), examines structural sexism's effects in the history of computing. They are co-editor of *Your Computer Is on Fire* (MIT, 2021), a collection of essays on urgent problems in high tech.

Scott Kushner is Assistant Professor of Communication Studies at the University of Rhode Island. He studies cultural infrastructures, the practices and machinery that shape everyday life. His current book project, *Enclosing Performance: Crowds, Control, and the Commodification of Culture,* shows how event venues convert chaotic crowds into calm, cooperative consumers. The Smithsonian Institution, New York Public Library, Hagley Museum and Library, Charles Babbage Institute, and Association for Computing Machinery have all supported his work.

Xiaochang Li is Assistant Professor in the Department of Communication at Stanford University. Her current book project explores the history of automatic speech recognition and natural language processing and examines in particular how the problem of mapping communication to computation shaped the rise of big data, machine learning, and related forms of algorithmic practice.

Zachary Loeb is a PhD candidate in History and Sociology of Science at the University of Pennsylvania. He works at the intersection of disaster studies and history of computing, in particular exploring technical, political, and social understandings of risk in the context of Y2K. He received the 2021–2022 Adelle and Erwin Tomash Fellowship in the History of Information Technology.

Lisa Nakamura is the Gwendolyn Calvert Baker Collegiate Professor and the Founding Director of the Digital Studies Institute at the University of Michigan. She is the author and editor of several books examining race and digital technologies, including *Digitizing Race: Visual Cultures of the Internet* (University of Minnesota Press, 2007); *Cybertypes: Race, Ethnicity, and Identity on the Internet* (Routledge, 2002); *Race after the Internet,* co-edited with Peter Chow-White (Routledge, 2012), and *Technoprecarious,* with the Precarity Lab Collective (MIT and Goldsmiths Press, 2020).

Tiffany Nichols is a licensed attorney in California and Washington, DC, and a registered patent attorney before the US Patent and Trademark Office. She is a doctoral candidate in the Department of the History of Science at Harvard University and graduate student fellow at the Black Hole Initiative. Her NSF-funded research focuses on topics at the intersection of gravitational wave physics, astrophysics, and multi-messenger astronomy.

Laine Nooney is Assistant Professor of Media and Information Industries in the Department of Media, Culture, and Communication at New York University. They are a computing and game historian and a Founding Editor and Managing Editor of *ROMchip: A Journal of Game Histories*. Their book *The Apple II: How the Computer Became Personal* will appear in 2023 with University of Chicago Press.

Elizabeth Petrick is Associate Professor of History at Rice University. She works at the intersection of computing history and disability studies, examining how technology relates to civil rights, use and users (especially those with disabilities) and notions of "access" and "ability." Her book *Making Computers Accessible: Disability Rights and Digital Technology* (Johns Hopkins University Press, 2015) received the 2017 Computer History Museum Book Prize.

Cierra Robson is a doctoral student of Sociology and Social Policy at Harvard University. Her work examines racial inequity, crime and punishment, and history of technology in the United States. She is the Associate Director of the Ida B. Wells JUST Data Lab at Princeton University and a Malcolm Hewitt Wiener PhD Research Fellow in Poverty and Justice. She has received awards from the National Science Foundation and the Mellon Sawyer Seminar on *Histories of AI: A Genealogy of Power* at Cambridge University.

Hallam Stevens is Associate Professor in the History Programme and in the School of Biological Sciences at Nanyang Technological University in Singapore, as well as Associate Director of the NTU Institute of Science and Technology for Humanity. His work explores the history of genomics, life sciences, big data, and computing. He is author of *Life Out of Sequence* (University of Chicago Press, 2013) and *Biotechnology and Society: An Introduction* (University of Chicago Press, 2016) and co-editor of *Postgenomics: Perspectives on Biology after the Genome* (Duke University Press, 2015).

Jaroslav Švelch is an Assistant Professor at Charles University, Prague. His work explores the history and theory of computer games, humor in games and social media, and the history, theory, and reception of monsters in video games. His book *Gaming the Iron Curtain: How Teenagers and Amateurs in Communist Czechoslovakia Claimed the Medium of Computer Games* (MIT Press, 2018) received the 2019 Computer History Museum Prize.

INDEX

Page numbers followed by f indicate a figure.

Abbate, Janet, 1, 6, 9, 34–35, 45, 435
Abrahams, Paul W., 349
Academy of Pedagogical Sciences (Soviet Union), 320, 322, 324, 325
Academy of Sciences (Soviet Union), 322
Adaptive Firmware Card, 407
The Admin Zone, 169
Advanced Research Projects Agency, 45, 49
advertising: algorithms for targeting of, 108–13, 113f; Black computer professionals and, 268; lurking and, 178; origin of technology terminology and, 162; of pain mitigation for computer-related pain, 429; web browsers and, 65
African Americans: Blackbird web browser and, 60–82; as computer professionals, 257–73; discrimination against, 117, 262–63, 364–68, 367f, 374–75; redlining and, 365–68, 367f; Shockley's views on, 231–32; technological redlining and, 102, 111, 113, 117. *See also* race and ethnicity
Agar, Jon, 6
agency. *See* human agency
Agyemang, Brianna, 181–82
AI. *See* artificial intelligence
Aidinoff, Marc, 12, 28, 40, 365
Aiken, Howard, 152
Air Force (US), 6, 264
Air France, 277, 292
Alberts, Gerard, 6
Alcatraz Occupation (1969–71), 250
Alexander, Michelle, 117
Algo-Heuristic Theory (AHT), 319–36
ALGOL (programming language), 126, 128, 130–32, 136, 138, 159–61
algorithms: conceptual history of, 2–3; data sets and, 107–11, 114–15; Landa's

Algo-Heuristic Theory, 319–36; origin and use of term, 146–64, 160–62f; patent on image search algorithms, 114–15; for race and ethnicity classifications, 102–20
Alper, Meryl, 410
Altair 8800, 191–92, 432n10
Amazon, 179
AMD, 232
American Federation of Information Processing Societies (AFIPS), 164
American Indian Movement (AIM), 232–33, 250–51
American International Assurance, 302
Americans with Disabilities Act of 1990, 404
Ames, Morgan, 9
The Angry Black Woman (blog), 71, 76
Annals of the History of Computing (journal), 210, 225n11
AOL (America Online), 90, 178
Apple, 190, 192–93, 210–11, 255n42, 407–9
AroundHarlem.com, 71, 77–78
ARPA: ARPANET, 7, 45, 55, 210, 383; Speech Understanding Research (SUR) program, 351, 357n16
Ars Technica, 71, 73–74, 79
artificial intelligence: algorithms and, 164; cybernetic psychology and, 322–24; instructional design and, 333–36; knowledge engineering and, 335–36; Landa's Algo-Heuristic Theory and, 15, 319–36; military and industrial logics in, 6; natural language processing and, 359n43, 438; nodes and, 326; operators and, 326; problem space and, 326; speech recognition and, 15, 343; template-matching systems, 346–47, 358n22
ASCC/Mark I, 152–53, 166n21, 166n23
Aspray, William, 3, 7

cybernetics, 7
cyberpunk culture: DIY ethic and, 389–90; rise of, 385–88
Cybertek (zine), 386, 390
Czechoslovakia: hardware systems cloning in, 213–15; microcomputer cloning in, 215–19, 218*f*; software cloning in, 219–23, 221*f*

da Costa Marques, Ivan, 209, 211
Daisey, Mike, 255n42
Darnton, Robert, 391
Datamation, 159, 161–63, 162*f*, 431n7
Davies, Donald, 45
Davis, Anne, 282, 283*f*
Davis, Randall, 334
Deane, Cormac, 431n7
DEC, 301
decentralized networking: Cold War development of, 43–46; hacker ethic and, 382; Jeffersonian politics of, 49–55; models of, 43–44, 44*f*; social rules for, 55
deconstructionist perspective, 371–72, 378n28
Defense Department (US): ARPANET and, 7, 45, 55, 210, 383; research and development funding from, 302, 383; surveillance technologies and, 362
de Gaulle, Charles, 280, 292
De Jager, Peter, 25–26, 27, 36
Deloria, Philip, 239, 251
Delta Sigma Theta Sorority, 270, 275n35
delurking, 176–77
DeMillo, Richard, 134
Democratic Party, 46–49, 58n30
De Mol, Liesbeth, 13, 122n17, 146, 321
Dick, Stephanie, 1, 7–8, 320, 332, 435
Didaktik Gama, 217–19, 218*f*
Digg.com, 69
Digital Millennium Copyright Act of 1998, 96
Dijkstra, Edsger, 128, 132, 138
disabilities: DiamondTouch device, 409–10; Macintosh GUI and, 407–9; prosthesis-technology metaphor and, 399–413; Unicorn Keyboard and, 404–7
discrimination: Black computer professionals and, 262–63; redlining and, 365–68, 367*f*; surveillance technologies

and, 117, 364–65, 374–75; women in technology fields and, 280, 293
distributed networking model, 43–44, 44*f*
DIY ethic of hackers, 388–90, 395
Dollis Hill Research Station, 281
Domain Awareness Center, 361–76
Domingo, Renee, 362
Douglas, Mary, 24
Downey, Greg, 205–6
Downing, Lewis, 261
DRAGON (data-related abend/garbage/or nothing), 26–27
Dreyfus, Hubert, 349
Dreyfus-Graf, Jean, 357n22
Driscoll, Kevin, 7
Driscoll, Paul, 241
Dryer, Theodora, 7
Duarte, Marisa, 233
Du Bois, W. E. B., 258
Dyer, Richard, 63, 65

Eastland, Jim, 47
Ebony, 266–67, 267*f*, 271, 272*f*
Eckert, John Presper, 154
EDSAC, 158
Education Technology (journal), 335
Edwards, Paul, 6, 24
Electronic Frontier Foundation (EFF), 43, 52, 53–54, 93–94
electronics manufacturing: Foxconn and, 255n42; Navajo women in, 231–53
Elwro 800 Junior computer, 217
embodiments, 2, 14–16
Engelbart, Douglas, 399
ENIAC, 153–54
Ensmenger, Nathan, 5, 34–35, 212
Erickson, Paul, 7
Erlich, Dennis, 84–98
Ervin, Lawrence, 266
Ethereal, Taja, 177
ethnicity. *See* race and ethnicity
Eubanks, Virginia, 9
European Economic Community (EEC), 279, 291
eye strain and fatigue, 416–17, 419, 422, 431n6

Facebook: algorithms used by, 119, 164; Blackbird web browser and, 66, 67–68; content generation and, 180; lurking

and, 178; patent linking racial and ethnic classifications to surveillance technologies, 115–18; Terms of Service, 117
facial recognition, 365
Fairchild Semiconductor, 231–53, 301
Fair Housing Act of 1968, 368
fair use doctrine, 85
FBI (Federal Bureau of Investigation), 391, 392
FCC (Federal Communications Commission), 111
Feigenbaum, Edward, 334, 335
Ferranti (computer company), 287
F International, 282
Firefox, 66–67, 73
First Amendment, 96
First Star Software, 220
FLDL (Formal Language Description Languages), 130–31, 132, 134, 139
Florida, Richard, 246
Fonovisa v. Cherry Auction (E.D. Cal. 1994), 95, 98
Ford Foundation, 305
FORTRAN, 131, 158, 304, 307, 309
40A (company), 60, 69, 70
Foucault, Michel, 363, 370
Foxconn, 255n42
France: Concorde and, 277, 279–81, 284, 295–96n4; European Economic Community and, 279, 292
Francis, Rick, 90–91
Freelance Programmers Ltd., 282, 287–91, 293–94, 297n30, 298n45
French, Gordon, 383
Friedman, Ted, 252
Frumer, Yulia, 414n10

Galison, Peter, 7–8, 45
Galloway, Alex, 251
game theory, 7
Garcia-Swartz, Daniel, 7
Garda, Maria B., 209
Garza, Alicia, 182
Gates, Bill, 204
Gavar, Ladislav, 222
Geismer, Lily, 58n30
gender: decentering of computing history and, 8–9; as digital resource, 240–51; wage gap, 292–93. *See also* women

General Dynamics, 240, 247
Gensler, Steve, 404–5
Geoghegan, Bernard, 9
Getson, Henry, 416–17
Gibbons, Kelcey, 14, 41, 121n11, 257, 439
Gibson, William, 389
gig economy, 294, 297n41
Gilmore, Ruth Wilson, 371, 372
Gingrich, Newt, 54–55
Giroux, H., 63
Glennie, Aleck, 158
Gödel, Kurt, 129
Goh Chok Tong, 299
Goldberg, David Theo, 375
Goldstine, Hermann, 155
Google, 68–69, 68t, 71, 114–15, 164, 359n43
Gore, Al, 43, 48–54
Graham, Elyse, 16, 32, 380
Grateful Dead, 383–84
Gray, Chris, 220
Great Eastern General Insurance, 302
Greene, Daniel, 9
Grohl, Dave, 389
GUI (graphical user interface), 64–65, 407–9

hackers, 380–96; computing history and, 6; criminalization of, 387–88; cyberpunk and, 385–88; decentralized networking and, 49; demographics of, 392–93; DIY ethic of, 388–90, 395; hacker ethic, defined, 381–83; punk culture and, 383–85; values system of, 390–94
Hack-Tic, 386
Haigh, Thomas, 9, 10, 168n61, 191–92
Halvorson, Michael J., 13–14, 121n11, 189
Hamilton, Alexander, 42
Haraway, Donna, 233–34
Harley, J. B., 378n28
Harriot, Michael, 182
Harrison, Greg, 254n20
Harry, Michael, 393–94
Hartree, Douglas, 155, 157, 158
Hauff, J. C., 148
Hayes, Frank, 33
Hayles, Katherine, 251, 436–37
HBCUs, 264–65
Heath, Edward, 292
Hebdige, Dick, 384, 385, 388, 389, 395

Masahiro, Mori, 414n10
materiality: cloning and, 212; of hardware, 5, 234; intelligence and, 436, 438–39
Matvy, Mike, 407–8, 415n33
Mauchly, John W., 153–54
May, Kenneth O., 150
Mayo Clinic, 430
McCarthy, John, 129–30, 131, 133, 134, 137, 139, 349
McElhaney, Lynette Gibson, 369
McGill, Meredith, 186n25
McIlroy, Doug, 139
McIlwain, Charlton, 9, 259
McLuhan, Marshall, 400
McPherson, J. L., 156
meaning-measurement relation, 342–43, 355
Means, Russell, 250
Mears, Margaret, 298n43
Medina, Eden, 6, 9–10, 209
Medina, Heidi, 177
Mercer, Robert, 353
Microsoft, 190–91, 195–206, 372
Microsoft Developer Network (MSDN), 196
microtargeting. See targeting advertisements
millennium bug, 23–36; computing world's response to, 25–27; government's response to, 28–30; public's response to, 30–33
Miller, Harris, 29
Mills, Mara, 357n14
Milne, Robert, 136
mind-body dualism, 435–37
Mindell, David, 413n6
Minich, Julie Avril, 402
Minsky, Marvin, 349
Mirzoeff, Nicholas, 252
Misa, Thomas J., 7
Mishler, Adriene, 430
MIT: hacker culture and, 381–82, 383, 390–91; "Management and the Computer of the Future" lecture series, 349–50; Tech Model Railroad Club, 381–82, 383, 390–91, 396n6; Whirlwind system at, 158
mitigation of, 427–31, 428–29f
MITS, 190, 191–92
Mitsubishi Electric Research Laboratories, 409
mobile phones, 430, 433n27

Moffatt, Ann, 285f, 286–93, 289f, 295, 296n29, 298n45
MONDO 2000 (zine), 384, 386
Montfort, Nick, 212, 251
Moreau, René, 3
Morella, Constance, 29, 33
Morris, Robert, Jr., 389
Morrison, Jim, 384
Morse code, 149
MOS 6502, 216
Mosaic, 64
Moynihan, Patrick, 28, 30
Mozilla, 66–67
Muhammad, Khalil, 376n10
Mullaney, Thomas S., 8, 301
multitasking, 429–31
Mumford, Lewis, 34, 35
MySpace, 66, 67

Nakai, Raymond, 239, 254n22
Nakamura, Lisa, 9, 14, 62, 231
Nantah Lee Kong Chian Computer Centre (Singapore): computer science studies and, 311–14; computer use at, 305–8, 306f; creation of, 299, 303–5; goals of, 306–7; legacy of, 314–16; training classes provided by, 308–11, 310f
Nanyang Computer Society, 309–11, 310f
Nanyang Technological University (NTU), 314
Nanyang University, 299–308, 312–14. See also Nantah Lee Kong Chian Computer Centre
Napster, 391
Natale, Simone, 441n4
National Academy of Sciences, 30
National Research and Education Network (NREN), 49
National Science Foundation, 49, 53
National Skills Bank (National Urban League), 268–70
National Technical Association, 261
National University of Singapore (NUS), 314
Naur, P., 132
Navajo women in technology industry, 231–53
NBS (National Bureau of Standards), 265, 431n7
neoliberalism, 47

problem space, 326
Processing Service Organizations, 163
program, origin and use of term, 146–64, 160–62f
programming languages: ALGOL, 126, 128, 130–32, 136, 138, 159–61; COBOL, 26, 131, 162–63, 312; CPL, 130; formal semantics and, 126–40; FORTRAN, 131, 158, 304, 307, 309; PL/I, 131, 135, 136, 137; Sal, 133
prosthesis-technology metaphor, 399–413; Macintosh GUI and, 407–9; Unicorn Keyboard and, 404–7
Public Life Insurance, 302
punk culture, 383–85; cyberpunk and, 385–88; DIY ethic and, 389
Puryear, Mahlon, 264
Pushkin, Veniamin, 324–26, 328

race and ethnicity: Blackbird web browser and, 60–82; Critical Race Intellectual Property and, 103; Democratic Party and, 47, 58n27; digital platforms and, 251–52; as digital resource, 240–51; electronic manufacturing and, 231–53; Facebook's patent linking racial and ethnic classifications to surveillance technologies, 115–18; Google's patent on image search algorithms and, 114–15; homoethnicity, 110; inventing Black computer professionals, 257–73; NetSuite's patent assigning racial and ethnic categories to users, 112–13, 113f; redlining and, 365–68, 367f; surveillance capitalism and, 374–75, 376, 378n40; Verizon's patent for automation of race and ethnicity classifications, 102–20
racialized surveillance capitalism, 374–75, 376, 378n40
Radin, Joanna, 7
Ramo-Woolridge, 162
RAND Corporation, 43, 44f, 168n56, 320, 321
Rankin, Joy, 7, 208, 300, 316, 337n6
RCA, 264–65, 269
Reagan, Ronald, 48–49
Reagle, Joseph, 180
Reddy, Raj, 358n24
redlining: residential, 365–68, 367f; technological, 102, 111, 113, 117

Remington Rand, 305
repetitive stress injuries, 422
Research Institute for Mathematical Machines (Prague), 213–14, 215
Reskin, Barbara, 364
Rheingold, Howard, 383
Rice, Randolf, 91, 92
risk perception: of computing world to Y2K, 25–27; government response to Y2K, 28–30; millennium bug and, 23–36; public response to Y2K, 30–33
Roberts, Dorothy, 109
Roberts, Ed, 192
Roberts, Lawrence, 45
Robinson, H. W., 163
Robinson, Markus, 75
Robson, Cierra, 15–16, 41, 117, 360
Rogers, Avian, 30
Rohde, Joy, 7
Roosevelt, Franklin D., 42
Rosenblith, Walter, 349
RTC v. Netcom (N.D. Cal. 1995), 85–98
Rubin, Howard, 27, 34
Rummelhart, David, 164
Russell, Andrew, 35, 45
Russell, Emily, 401
Rutishauser, Heinz, 157, 159–60

Sadovskiĭ, Vadim, 324–25
SAIC (Science Application International Corporation), 362, 368, 373
Sal (programming language), 133
Sandahl, Carrie, 401
Sandberg, Sheryl, 298n42
Sanders, James, 26, 27
Schafer, Valérie, 45
Schultz, James, 27
Schwarzenegger, Arnold, 368
Scientists and Technicians of Tomorrow program, 271
Scientology copyright infringement suits, 84–98
Scott, Dana, 132, 133, 135
Seagate Technologies, 301
segregation: redlining and, 365–68, 367f; surveillance technologies and, 364; technological redlining, 111, 113, 117
Selfridge, Oliver Gordon, 350
Sellers, Don, 427

semantics, 126–40

Senate (US): Small Business Committee, 261; Special Committee on the Year 2000 Problem, 29–30, 34, 36

Senese, Guy, 241

Shannon, Claude, 349, 350

shareware, 191

Sharp MZ-800, 220

Shaw, J. C., 338n24

Shell Eastern Petroleum, 303

Shelley, Mary, 438

Shermer, Dee, 285f

Shiprock, New Mexico, 237–40, 242, 249, 253, 255n29

Shirley, Stephanie "Steve," 281–85, 285f, 287–88, 290–91, 293, 295

Shockley, William, 231–32, 259

Siebers, Tobin, 412

Sikora, Mirosław, 214

Simon, Herbert, 128, 129, 319, 320, 325–26, 328, 332, 338n24

Singapore: Committee on National Computerisation, 299; computer science studies in, 311–14; first computers in, 302–3; Housing and Development Board, 302; IT industry development in, 299–316; National Computer Board (NCB), 299–300

Slayton, Rebecca, 6

Small, Thomas, 84

smartphones, 430, 433n27

Smith, Al, 42

Smith, Eleazar, 92

Smith, Kael, 175

Smith, Patti, 385

Smith, Roney, 71, 75

Smith v. California (1959), 92

Smutný, Eduard, 216–17

Snapchat, 180

Sobchack, Vivian, 402, 411–12

social media: lurking and, 177–78; as surveillance technology, 170. See also specific platforms

social relations, computing as, 3, 11–16, 435

Softalk (magazine), 416

software: cloning of, 219–23, 221f; Concorde black box software, 283–91; human agency obscured by patents for, 105–7, 122n17; origin and use of term, 160–62f,

161–63, 168n56; porting of, 220–21, 227n54. See also specific software programs

sound spectrograph, 345–48, 347f

Southern Industry Project (National Urban League), 264–68, 275n19

South Korea, hardware clones in, 211

Soviet Union: cloning in, 213–15; hardware clones in, 211; Landa's Algo-Heuristic Theory and, 320–21; microcomputer cloning in, 216–19, 218f

Spacewar! (computer game), 382

speech recognition, 341–56; artificial ears and, 344–48, 347f; computational symbiosis and, 348–52; engineered knowledge and, 348–52; noisy channel problem, 352–56, 354f; template-matching systems, 346–47, 358n22; vocal gestures and, 344–48

Sporck, Charlie, 237, 255n35, 255n39

Spurný, Antonín, 220–21, 221f

Stanford University, 232

Steele, Shari, 93

Stein, Jesse Adams, 433n24

Steinberg, J. C., 346, 358n23

Stennis, John, 47

Stephens, Robert P., 266

Sterne, Jonathan, 357n14

Stevens, Hallam, 8, 15, 209, 299

Stonesifer, Patty, 204–5

StoryTable software, 409–10

Strachey, Christopher, 130, 131–36

Streeter, Thomas, 46

Suchman, Lucy, 400

Suominen, Jaakko, 208

surveillance technologies: carceral state and, 360–76; economic and political incentives for, 372–75; patent linking racial and ethnic classifications to, 115–18; racialized surveillance capitalism, 374–75, 376, 378n40; social media as, 170

Švelch, Jaroslov, 8, 14, 208, 209, 300, 321, 414n15, 439

Swanson, Kara, 104

The Syndicate Report (zine), 386

System of Small Electronic Computers (Soviet Union), 215

Taiwan, hardware clones in, 211

Tandy TRS-80, 193, 418

Wildavsky, Aaron, 24
Wilkes, Maurice, 130, 157
Williams, Zoe, 298n42
Wilson, Robert Rawdon, 412
Winant, H., 63
Winner, Langdon, 395
Wolverton, Van, 195
women: decentering of computing history and, 8–9; hacker culture and, 392–93; Navajo women in technology industry, 231–53; pay gap and, 292–93; technochauvinism and, 293–94, 297n39; in technology industry, 281–91, 283*f*, 285*f*, 289*f*
Woodger, Mike, 138
Woodman, Pamela, 298n43
WordStar, 210
W.O.R.M. (zine), 386

Wounded Knee incident (1973), 250
Wozniak, Steve, 192, 418
Wu Wen-Tsun, 300–301
Wyatt, Willa, 154

Y2K. *See* millennium bug
Yahoo! Search, 71
Yost, Jeffrey, 5
YouTube, 179–80, 430

Zemanek, Heinz, 130, 131, 134
Zemmix, 211
Zilog Z80, 216
Zuboff, Shoshana, 170, 378n40, 423–26, 424–26*f*
Zuse, Konrad, 157
ZX Spectrum, 211, 217, 219